Contemporary Electronics

Contemporary Electronics

Edited by Norman Schultz

CLANRYE
INTERNATIONAL
www.clanryeinternational.com

Clanrye International,
750 Third Avenue, 9th Floor,
New York, NY 10017, USA

ISBN: 978-1-63240-924-9

Cataloging-in-Publication Data

Contemporary electronics / edited by Norman Schultz.
        p. cm.
Includes bibliographical references and index.
ISBN 978-1-63240-924-9
1. Electronics. 2. Electronic apparatus and appliances. 3. Electronic systems.
4. Electrical engineering. I. Schultz, Norman.
TK7835 .C66 2020
621.381--dc23

For information on all Clanrye International publications
visit our website at www.clanryeinternational.com

CLANRYE
INTERNATIONAL

Contents

Preface

This book has been a concerted effort by a group of academicians, researchers and scientists, who have contributed their research works for the realization of the book. This book has materialized in the wake of emerging advancements and innovations in this field. Therefore, the need of the hour was to compile all the required researches and disseminate the knowledge to a broad spectrum of people comprising of students, researchers and specialists of the field.

Electronics is the science that is concerned with the emission, control and flow of electrons in vacuum and matter. In electronics, circuits are designed with active electrical components, associated passive electrical components and interconnection technologies. Active components consist of transistors, integrated circuits, vacuum tubes, diodes, sensors, etc. Some passive components are resistors, capacitors and inductors. Electric circuits can be analog or digital. Analog circuits use a continuous range of current or voltage whereas digital circuits are based on discrete values of current and voltage. Electronics is of tremendous use in signal processing, information processing and telecommunication. Analog electronics, optoelectronics, digital electronics, microelectronics and power electronics are some of the branches of electronics. This book strives to provide a fair idea about electronics and to help develop a better understanding of the latest advances within this field. It traces the progress of this field and highlights some of its key concepts and applications. With state-of-the-art inputs by acclaimed experts of this field, the book targets students and professionals.

At the end of the preface, I would like to thank the authors for their brilliant chapters and the publisher for guiding us all-through the making of the book till its final stage. Also, I would like to thank my family for providing the support and encouragement throughout my academic career and research projects.

Editor

Modeling and Fault Diagnosis of Interturn Short Circuit for Five-Phase Permanent Magnet Synchronous Motor

Jian-wei Yang, Man-feng Dou, and Zhi-yong Dai

School of Automation, Northwestern Polytechnical University, Xi'an 710072, China

Correspondence should be addressed to Jian-wei Yang; yangjianwei100@hotmail.com

Academic Editor: Ahmed F. Zobaa

Taking advantage of the high reliability, multiphase permanent magnet synchronous motors (PMSMs), such as five-phase PMSM and six-phase PMSM, are widely used in fault-tolerant control applications. And one of the important fault-tolerant control problems is fault diagnosis. In most existing literatures, the fault diagnosis problem focuses on the three-phase PMSM. In this paper, compared to the most existing fault diagnosis approaches, a fault diagnosis method for Interturn short circuit (ITSC) fault of five-phase PMSM based on the trust region algorithm is presented. This paper has two contributions. (1) Analyzing the physical parameters of the motor, such as resistances and inductances, a novel mathematic model for ITSC fault of five-phase PMSM is established. (2) Introducing an object function related to the Interturn short circuit ratio, the fault parameters identification problem is reformulated as the extreme seeking problem. A trust region algorithm based parameter estimation method is proposed for tracking the actual Interturn short circuit ratio. The simulation and experimental results have validated the effectiveness of the proposed parameter estimation method.

1. Introduction

Owing to high torque-to-current ratio, large power-to-weight ratio, high efficiency, high-power factor, high fault tolerance, robustness, and so forth, multiphase PMSMs have been paid more attention in high-power and high-reliability applications [1–3]. Compared with the traditional three-phase PMSM, with the added phase number, the fault tolerance of the multiphase PMSM is enhanced, and thus the reliability of the multiphase PMSM is improved. Therefore, multiphase PMSMs are widely used in fault-tolerant control systems [4, 5].

Fault diagnosis is the foundation of the fault-tolerant control of the electrical machines. In PMSMs, the usual faults include electrical faults, mechanical faults, and magnetic faults [6]. In electrical faults, short circuit faults form 21% of the faults occurring in electrical machines. The stator winding ITSC fault is the commonest short circuit fault in PMSMs. It always occurs due to insulation failures but develops into more serious faults very quickly [7]. So it is meaningful to research the effective fault diagnosis methods of stator winding interturn short circuit for PMSMs.

The current existing detection and diagnosis methods of ITSC fault can be commonly divided into off-line methods and on-line methods [8]. Compared to the off-line methods, in on-line methods, the PMSMs do not have to be taken out of service and predicting health condition and detecting faults at an incipient stage are made easier [9]. In recent years, with the application of neural network, fuzzy logic and particle swarm optimization (PSO), the artificial intelligence (AI) on-line fault detection, and diagnosis methods have drawn the attention of many authors [10]. The AI methods improve the robustness and efficiency of the fault diagnosis and have no need to interpret the collected data in relation to the occurring fault.

In some AI fault detection and diagnosis methods, such as literature [11], in order to detect and diagnose the severity of the stator winding interturn short circuit fault of PMSM, a mathematical model that can describe both healthy and fault conditions is needed first. Literature [12] built power losses model of five-phase PMSM with ITSC fault and analyzed the changes in power losses due to faults occurrence by finite elements simulations. However, this fault model is not suitable for AI fault diagnosis based on parameter optimization. Literature [13] and literature [14] proposed two mathematical models of PMSM with ITSC fault for fault diagnosis. Unfortunately, these models are all about three-phase PMSM and relatively complex. If the fault model of five-phase PMSM was built by the way shown in literature [13] and literature [14], the model would be more complex, and the calculation for the subsequent fault diagnosis based on parameter optimization would increase greatly. Thus, the efficiency of fault diagnosis would be affected. Therefore, it is meaningful to establish a relatively simple five-phase PMSM model with ITSC fault for fault diagnosis.

After the establishment of the fault model, in order to diagnose fault severity of the fault motor, the parameters associated with fault severity need to be identified. However, for the complex distribution of the parameters in the fault model, the identification problem is extremely difficult for nonlinear identification techniques. To overcome this difficulty, the fault diagnosis problem is transformed into a corresponding optimization problem and then solved by intelligent algorithm [15]. In recent years, many authors focus on PSO parameter optimization to deal with this problem, such as that shown in literature [16] and literature [17]. PSO is an evolution computation technique based on swarm intelligent methodology. PSO is initialized as a swarm of arbitrary particles (arbitrary solution), and then the optimal solution is discovered by iteration. However, the PSO algorithm creates the problems of partial convergence and precocious convergence when the particles' diversity is decreasing. Therefore, finding a better parameter optimization algorithm for five-phase PMSM fault diagnosis is essential.

In this paper, relatively simple mathematics models of the five-phase PMSM under both healthy and ITSC fault situations are established, respectively. Furthermore, a novel fault diagnosis method of ITSC based on the trust region algorithm is proposed for five-phase PMSM. With the aid of the trust region algorithm which is global convergence, the interturn short circuit ratio μ is estimated with a short time transient. The simulation and experimental results have validated both the correction of the established models and the effectiveness of the proposed parameter estimation method.

2. Model Analysis

2.1. Five-Phase PMSM Healthy Model. In order to establish the healthy model of five-phase PMSM, without loss of generality, the following assumptions are as follows:

(1) The magnetic circuit is linear. It is, in turn, that the magnetic circuit is not saturation.

(2) The stator winding current is sinusoidal, symmetrical, and without harmonics. The air gap magneto motive force (MMF) is sinusoidal.

(3) The rotor MMF is sinusoidal and the slot effect is neglected.

(4) The five-phase PMSM is nonsalient pole structure.

(5) Eddy currents and hysteresis losses are negligible.

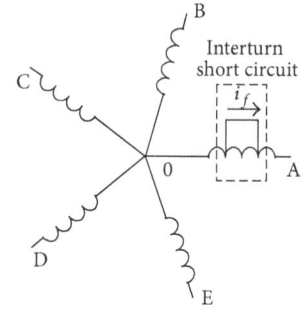

FIGURE 1: The schematic of five-phase PMSM with interturn short fault.

With these assumptions, the five-phase PMSM model can be provided by

$$U_s = R_s I_s + p\psi_s, \tag{1}$$

$$\psi_s = L_s I_s + \psi_m, \tag{2}$$

$$T_e = \frac{\partial W}{\partial \theta} = P\left[\frac{1}{2}I_s^T \frac{\partial L_s}{\partial \theta} I_s + I_s \frac{\partial \psi_m}{\partial \theta}\right]. \tag{3}$$

Equation (1) is the voltage balance equation, (2) is the flux equation, and (3) is the torque equation, where the stator phase voltage vector $U_s = [u_a\ u_b\ u_c\ u_d\ u_e]^T$; the stator phase current vector $I_s = [i_a\ i_b\ i_c\ i_d\ i_e]^T$; the stator winding resistance $R_s = r_s \times I_{5\times5}$; the stator flux vector $\psi_s = [\psi_a\ \psi_b\ \psi_c\ \psi_d\ \psi_e]^T$; the rotor flux vector $\psi_m = \psi_m[\cos\theta\ \cos(\theta-\alpha)\ \cos(\theta-2\alpha)\ \cos(\theta-3\alpha)\ \cos(\theta-4\alpha)]^T$ (θ is the rotor electrical angle and $\alpha = 72°$); L_s is the stator inductance matrix; $p = d/dt$ is the differential operator; P is the number of pole pairs; and T_e is the electromagnetic torque.

Because of adding two-phase windings, compared to traditional PMSM, the stator inductance matrix L_s of five-phase PMSM is more complex and it can be represented by

$$L_s = \begin{bmatrix} L_m & M_{ab} & M_{ac} & M_{ad} & M_{ae} \\ M_{ab} & L_m & M_{bc} & M_{bd} & M_{be} \\ M_{ac} & M_{bc} & L_m & M_{cd} & M_{ce} \\ M_{ad} & M_{bd} & M_{cd} & L_m & M_{de} \\ M_{ae} & M_{be} & M_{ce} & M_{de} & L_m \end{bmatrix}, \tag{4}$$

where L_m is the self-inductance of phase winding A (B, C, D, and E) and $M_{ab(c,d,e)}$ is the mutual-inductance between phase windings A and B (C, D, and E). Actually, the mutual-inductances can be expressed by $M_{ab} = L_m \cos\alpha$, $M_{ac} = L_m \cos 2\alpha$, $M_{ad} = L_m \cos 3\alpha$, and $M_{ae} = L_m \cos 4\alpha$.

2.2. Five-Phase PMSM Fault Model. Without loss of generality, assume that phase winding A causes ITSC fault and the rest of the phase windings is in healthy state. The five-phase PMSM with ITSC is shown in Figure 1. Note that a short circuit loop current i_f, which gives birth to braking torque, is produced in phase winding A. And thus the braking

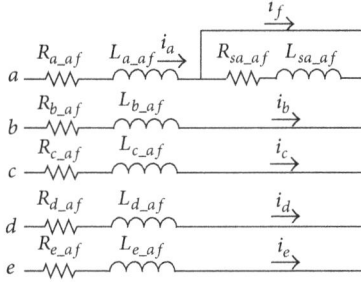

FIGURE 2: The equivalent circuit of five-phase PMSM with interturn short circuit fault.

torque affects the motor performance seriously. Besides, the effective turns number of phase winding A is reduced, and the values of the phase winding resistance, the self-inductance, the mutual-inductance, and the flux linkage are all changed accordingly.

In the ITSC fault model of five-phase PMSM, one of the most important parameters is the interturn short circuit ratio μ, which is defined as the ratio of the shorted turns number to the total turns number. When the ITSC fault occurs in phase winding A, depending on the physical relationship of the windings, the equivalent circuit of five-phase PMSM with ITSC fault is shown in Figure 2, where $R_{a(b,c,d,e)\_af}$ is the resistance of phase winding A (B, C, D, and E); R_{sa_af} is the short circuit winding resistance; $L_{a(b,c,d,e)\_af}$ is the self-inductance of phase winding A (B, C, D, and E); L_{sa_af} is the self-inductance of the short circuit winding; L_m is the self-inductance of phase winding under normal conditions. Actually, the resistance and the inductances can be expressed by

$$
\begin{aligned}
R_{a\_af} &= (1 - \mu) R_s, \\
R_{b\_af} &= R_{c\_af} = R_{d\_af} = R_{e\_af} = R_s, \\
R_{sa\_af} &= \mu R_s, \\
L_{a\_af} &= (1 - \mu)^2 L_m, \\
L_{sa\_af} &= \mu^2 L_m, \\
L_{b\_af} &= L_{c\_af} = L_{d\_af} = L_{e\_af} = L_m.
\end{aligned}
\tag{5}
$$

Besides, there are also three kinds of mutual-inductances existing. The first is the mutual- inductance $M_{ab(c,d,e)\_af}$ between the remaining normal winding of phase A and other phase (B, C, D, and E) windings, where $M_{ab(c,d,e)\_af} = (1 - \mu)M_{ab(c,d,e)}$. The second is the mutual-inductance M_{asa_af} between the remaining normal winding of phase A and the short circuit winding of phase A, where $M_{asa\_af} = (1-\mu)\mu L_m$. The third is the mutual-inductance $M_{sab(c,d,e)\_af}$ between the short circuit winding of phase A and other phase (B, C, D, and E) windings, where $M_{sab(c,d,e)\_af} = \mu M_{ab(c,d,e)}$. Since the phase windings B, C, D, and E are healthy, the mutual-inductances of the phase windings B, C, D, and E remain the values of the healthy model.

When phase winding A causes ITSC fault, the fluxes between the windings and the rotor can be derived as

$$
\begin{aligned}
\psi_{a\_af} &= (1 - \mu) \psi_m \cos \theta, \\
\psi_{b\_af} &= \psi_m \cos (\theta - \alpha), \\
\psi_{c\_af} &= \psi_m \cos (\theta - 2\alpha), \\
\psi_{d\_af} &= \psi_m \cos (\theta - 3\alpha), \\
\psi_{e\_af} &= \psi_m \cos (\theta - 4\alpha), \\
\psi_{sa\_af} &= \mu \psi_m \cos \theta.
\end{aligned}
\tag{6}
$$

According to the equivalent circuit and the analysis above, the voltage balance equation renders as

$$
\begin{aligned}
V_a ={}& R_{a\_af} i_a + L_{a\_af} \frac{di_a}{dt} + M_{ab\_af} \frac{di_b}{dt} + M_{ac\_af} \frac{di_c}{dt} \\
&+ M_{ad\_af} \frac{di_d}{dt} + M_{ae\_af} \frac{di_e}{dt} \\
&+ M_{asa\_af} \frac{d(i_a - i_f)}{dt} + \frac{d\psi_{a\_af}}{dt},
\end{aligned}
$$

$$
\begin{aligned}
V_b ={}& R_{b\_af} i_b + L_{b\_af} \frac{di_b}{dt} + M_{ab\_af} \frac{di_a}{dt} + M_{bc\_af} \frac{di_c}{dt} \\
&+ M_{bd\_af} \frac{di_d}{dt} + M_{be\_af} \frac{di_e}{dt} \\
&+ M_{bsa\_af} \frac{d(i_a - i_f)}{dt} + \frac{d\psi_{b\_af}}{dt},
\end{aligned}
$$

$$
\begin{aligned}
V_c ={}& R_{c\_af} i_c + L_{c\_af} \frac{di_c}{dt} + M_{ac\_af} \frac{di_a}{dt} + M_{bc\_af} \frac{di_b}{dt} \\
&+ M_{cd\_af} \frac{di_d}{dt} + M_{ce\_af} \frac{di_e}{dt} \\
&+ M_{csa\_af} \frac{d(i_a - i_f)}{dt} + \frac{d\psi_{c\_af}}{dt},
\end{aligned}
$$

$$
\begin{aligned}
V_d ={}& R_{d\_af} i_d + L_{d\_af} \frac{di_d}{dt} + M_{ad\_af} \frac{di_a}{dt} + M_{bd\_af} \frac{di_b}{dt} \\
&+ M_{cd\_af} \frac{di_c}{dt} + M_{de\_af} \frac{di_e}{dt} \\
&+ M_{dsa\_af} \frac{d(i_a - i_f)}{dt} + \frac{d\psi_{d\_af}}{dt},
\end{aligned}
$$

$$
\begin{aligned}
V_e ={}& R_{e\_af} i_e + L_{e\_af} \frac{di_e}{dt} + M_{ae\_af} \frac{di_a}{dt} + M_{be\_af} \frac{di_b}{dt} \\
&+ M_{ce\_af} \frac{di_c}{dt} + M_{de\_af} \frac{di_d}{dt} \\
&+ M_{esa\_af} \frac{d(i_a - i_f)}{dt} + \frac{d\psi_{e\_af}}{dt},
\end{aligned}
$$

$$0 = R_{sa\_af}\left(i_a - i_f\right) + L_{sa\_af}\frac{d\left(i_a - i_f\right)}{dt} + M_{asa\_af}\frac{di_a}{dt}$$

$$+ M_{sab\_af}\frac{di_b}{dt} + M_{sac\_af}\frac{di_c}{dt} + M_{sad\_af}\frac{di_d}{dt}$$

$$+ M_{sae\_af}\frac{di_e}{dt} + \frac{d\psi_{sa\_af}}{dt}. \tag{7}$$

The voltage balance equation (7) can be rewritten as

$$V_{af} = R_{af}I_{af} + L_{af}\frac{dI_{af}}{dt} + \frac{d\psi_{af}}{dt}, \tag{8}$$

where

$$V_{af} = \begin{bmatrix} V_a & V_b & V_c & V_d & V_e & 0 \end{bmatrix}^T,$$

$$I_{af} = \begin{bmatrix} i_a & i_b & i_c & i_d & i_e & i_f \end{bmatrix}^T,$$

$$R_{af} = \begin{bmatrix} R_s & 0 & 0 & 0 & 0 & -\mu R_s \\ 0 & R_s & 0 & 0 & 0 & 0 \\ 0 & 0 & R_s & 0 & 0 & 0 \\ 0 & 0 & 0 & R_s & 0 & 0 \\ 0 & 0 & 0 & 0 & R_s & 0 \\ \mu R_s & 0 & 0 & 0 & 0 & -\mu R_s \end{bmatrix},$$

$$L_{af} = \begin{bmatrix} L_m & M_{ab} & M_{ac} & M_{ad} & M_{ae} & -\mu L_m \\ M_{ab} & L_m & M_{ab} & M_{ac} & M_{ad} & -\mu M_{ab} \\ M_{ac} & M_{ab} & L_m & M_{ab} & M_{ac} & -\mu M_{ac} \\ M_{ad} & M_{ac} & M_{ab} & L_m & M_{ab} & -\mu M_{ad} \\ M_{ae} & M_{ad} & M_{ac} & M_{ab} & L_m & -\mu M_{ae} \\ \mu L_m & \mu M_{ab} & \mu M_{ac} & \mu M_{ad} & \mu M_{ae} & -\mu^2 L_m \end{bmatrix},$$

$$\psi_{af} = \psi_m \begin{bmatrix} \cos\theta & \cos(\theta - \alpha) & \cos(\theta - 2\alpha) & \cos(\theta - 3\alpha) & \cos(\theta - 4\alpha) & \mu\cos\theta \end{bmatrix}^T. \tag{9}$$

The electromagnetic torque equation of five-phase PMSM is

$$T_e = \frac{\partial W}{\partial \theta} = p\left[\frac{1}{2}I_{af}^T\frac{\partial L_{af}}{\partial \theta}I_{af} + I_{af}\frac{\partial \psi_{af}}{\partial \theta}\right]$$

$$= p\left[I_{af}\frac{\partial \psi_{af}}{\partial \theta}\right]. \tag{10}$$

The mechanical motion equation of five-phase PMSM is

$$T_e - T_L = \frac{Jd\omega}{dt} + B\omega, \tag{11}$$

where T_e is the electromagnetic torque; T_L is the load torque; J is the rotational inertia; B is the viscous friction coefficient; ω is the mechanical angular velocity.

3. Fault Diagnosis

3.1. Trust Region Algorithm. Trust region algorithm is a method for the extreme seeking. The method sets a trust region radius as the upper bound of the displacement length and, with the current iteration point as the center, determines a closed spherical region named trust region. By solving the optimal point of the quadratic approximation model to determine the candidate displacement, the nonlinear extreme problem is transformed into the extreme problem of solving the approximation quadratic model of the objective function within the trust region [18, 19]. If the candidate displacement enables the sufficient reduction to the objective function, the candidate displacement is adopted to be the new displacement and simultaneously maintains or expands the trust region radius for a new round of iteration. Otherwise, it indicates the approximate degree of the quadratic model and the objective function is unsatisfactory, the trust region radius should be reduced, and the extreme problem of the approximation quadratic model should be solved within the new trust region radius to obtain the new candidate displacement [20]. By the continuous iteration, the extreme optimization of nonlinear function is achieved.

The trust region methods can be understood by a typical unconstrained minimization problem:

$$\min_{x \in \mathbb{R}^n} f(x), \qquad (12)$$

where $f(x)$ is objective function to be minimized.

Suppose that x_k is the kth iteration, $f_k = f(x_k)$, $g_k = \nabla f(x_k)$, and B_k is the kth approximation of the Hesse Matrix $\nabla^2 f(x_k)$. So the trust region subproblem is

$$\min \quad q_k(d) = g_k^T d + \frac{1}{2} d^T B_k d \qquad (13)$$

$$\text{s.t.} \quad \|d\| \le \Delta_k,$$

where Δ_k is the trust region radius and $\|\cdot\|$ is the vector norm, usually $\|\cdot\|_2$ or $\|\cdot\|_\infty$. Suppose that the optimal solution of (13) is d_k; Δf_k is the decreasing of the kth iteration, and $\Delta f_k = f_k - f(x_k + d_k)$; Δq_k is the predicted decreasing correspondingly, and $\Delta q_k = q_k(0) - q_k(d_k)$. Define r_k as

$$r_k = \frac{\Delta f_k}{\Delta q_k}. \qquad (14)$$

Generally, $\Delta q_k > 0$. So if $r_k < 0$, $\Delta f_k < 0$, $x_k + d_k$ will not be the next iteration point. To solve the subproblem, the trust region radius needs to be reduced. If the value of r_k is closed to 1, it indicates that the quadratic model is a good approximation for the objective function within the trust region, $x_{k+1} := x_k + d_k$ can be used as the new iteration point, and meantime the trust region radius can be increased in the next iteration. For other cases, the trust region radius remains unchanged.

3.2. Fault Parameter Identification. In order to diagnose the fault severity of the fault motor, the interturn short circuit ratio μ needs to be identified based on the fault model. Because of the complex distribution of the parameter in the fault model, the identification problem is extremely difficult for nonlinear identification techniques. To overcome the difficulty, the fault diagnosis problem is first transformed into a corresponding optimization problem and then solved using trust region algorithm. Because of its simplicity, global convergence, and computational efficiency, trust region algorithm has been used extensively to solve a broad range of optimization problems. Numerous applications have applied trust region algorithm for parameter tuning and identification. The applications based on trust region algorithm are not limited by model structures, as are many traditional identification algorithms. As long as the model performs differently with different parameters, which is almost always true, trust region algorithm will be able to identify the unknown parameters in the models.

The principle of fault parameter (the interturn short circuit ratio μ) identification for five-phase PMSM under ITSC fault is as shown in Figure 3.

To begin with, give an arbitrary constant $\hat{\mu}(0) \in (0,1)$ as the initial estimation value of the interturn short circuit ratio for the five-phase fault PMSM model. Meanwhile, sample both the five-phase PMSM drive voltage V and the phase currents i from the actual PMSM. And then, a reference model of the five-phase PMSM with ITSC fault is established as (7), and its drive voltage is the same as the actual PMSM.

Afterwards, to diagnose the fault severity of the five-phase PMSM, or in other words, to identify the fault parameter (the interturn short circuit ratio μ) by trust region algorithm, a quadratic fitness function is introduced as

$$f(\hat{\mu}) = \sum_{n=1}^{5} (i_n - i_n^*)^2, \qquad (15)$$

where i and i^* are the phase current of the actual PMSM and reference model, respectively. In fact, the objective function is related to the parameters the interturn short circuit ratio u and its estimation $\hat{\mu}$. It can be proved that the objective function is equal to zero if and only if the actual interturn short circuit ratio is equal to its estimation.

At last, use the trust region algorithm to seek the interturn short circuit ratio estimation value $\hat{\mu}$ such that the fitness function achieves its minimum value zero. The specific steps of the trust region algorithm parameter identification can be realized as follows. In the process of parameter optimization, the gradient and the approximation of the Hesse Matrix for the quadratic fitness function are calculated, respectively, as (16) and (17).

Step 0. Select the initial parameters, where $0 \le \eta_1 < \eta_2 < 1$, $0 \le \tau_1 < 1 < \tau_2$, and $0 \le \varepsilon \ll 1$. Consider $\hat{\mu}(0) \in (0,1)$. The upper limit of the trust region radius is $\overline{\Delta}$, $\overline{\Delta} > 0$, and the initial trust region radius is Δ_0, $\Delta_0 \in (0, \overline{\Delta})$. Set $k := 0$.

Step 1. Calculate g_k as shown in (16); if $\|g_k\| \le \varepsilon$, stop the iteration

$$g_k = \nabla f(\hat{\mu})$$
$$= \left[\frac{\partial f(\hat{\mu})}{\partial i_1^*} \quad \frac{\partial f(\hat{\mu})}{\partial i_2^*} \quad \frac{\partial f(\hat{\mu})}{\partial i_3^*} \quad \frac{\partial f(\hat{\mu})}{\partial i_4^*} \quad \frac{\partial f(\hat{\mu})}{\partial i_5^*} \right]^T. \qquad (16)$$

Step 2. Solve the trust region subproblem of the objective function (15), and the solution is d_k.

Step 3. Calculate the value of r_k as (14), and in the calculation of r_k, q_k is calculated as shown in (14), where B_k is the kth approximation of the Hesse Matrix, as shown in

$$B_k = \nabla^2 f(\hat{\mu})$$

$$= \begin{bmatrix} \frac{\partial^2 f(\hat{\mu})}{\partial^2 i_1^*} & \frac{\partial^2 f(\hat{\mu})}{\partial i_1^* \partial i_2^*} & \frac{\partial^2 f(\hat{\mu})}{\partial i_1^* \partial i_3^*} & \frac{\partial^2 f(\hat{\mu})}{\partial i_1^* \partial i_4^*} & \frac{\partial^2 f(\hat{\mu})}{\partial i_1^* \partial i_5^*} \\[2mm] \frac{\partial^2 f(\hat{\mu})}{\partial i_2^* \partial i_1^*} & \frac{\partial^2 f(\hat{\mu})}{\partial^2 i_2^*} & \frac{\partial^2 f(\hat{\mu})}{\partial i_2^* \partial i_3^*} & \frac{\partial^2 f(\hat{\mu})}{\partial i_2^* \partial i_4^*} & \frac{\partial^2 f(\hat{\mu})}{\partial i_2^* \partial i_5^*} \\[2mm] \frac{\partial^2 f(\hat{\mu})}{\partial i_3^* \partial i_1^*} & \frac{\partial^2 f(\hat{\mu})}{\partial i_3^* \partial i_2^*} & \frac{\partial^2 f(\hat{\mu})}{\partial^2 i_3^*} & \frac{\partial^2 f(\hat{\mu})}{\partial i_3^* \partial i_4^*} & \frac{\partial^2 f(\hat{\mu})}{\partial i_3^* \partial i_5^*} \\[2mm] \frac{\partial^2 f(\hat{\mu})}{\partial i_4^* \partial i_1^*} & \frac{\partial^2 f(\hat{\mu})}{\partial i_4^* \partial i_2^*} & \frac{\partial^2 f(\hat{\mu})}{\partial i_4^* \partial i_3^*} & \frac{\partial^2 f(\hat{\mu})}{\partial^2 i_4^*} & \frac{\partial^2 f(\hat{\mu})}{\partial i_4^* \partial i_5^*} \\[2mm] \frac{\partial^2 f(\hat{\mu})}{\partial i_5^* \partial i_1^*} & \frac{\partial^2 f(\hat{\mu})}{\partial i_5^* \partial i_2^*} & \frac{\partial^2 f(\hat{\mu})}{\partial i_5^* \partial i_3^*} & \frac{\partial^2 f(\hat{\mu})}{\partial i_5^* \partial i_4^*} & \frac{\partial^2 f(\hat{\mu})}{\partial^2 i_5^*} \end{bmatrix}. \qquad (17)$$

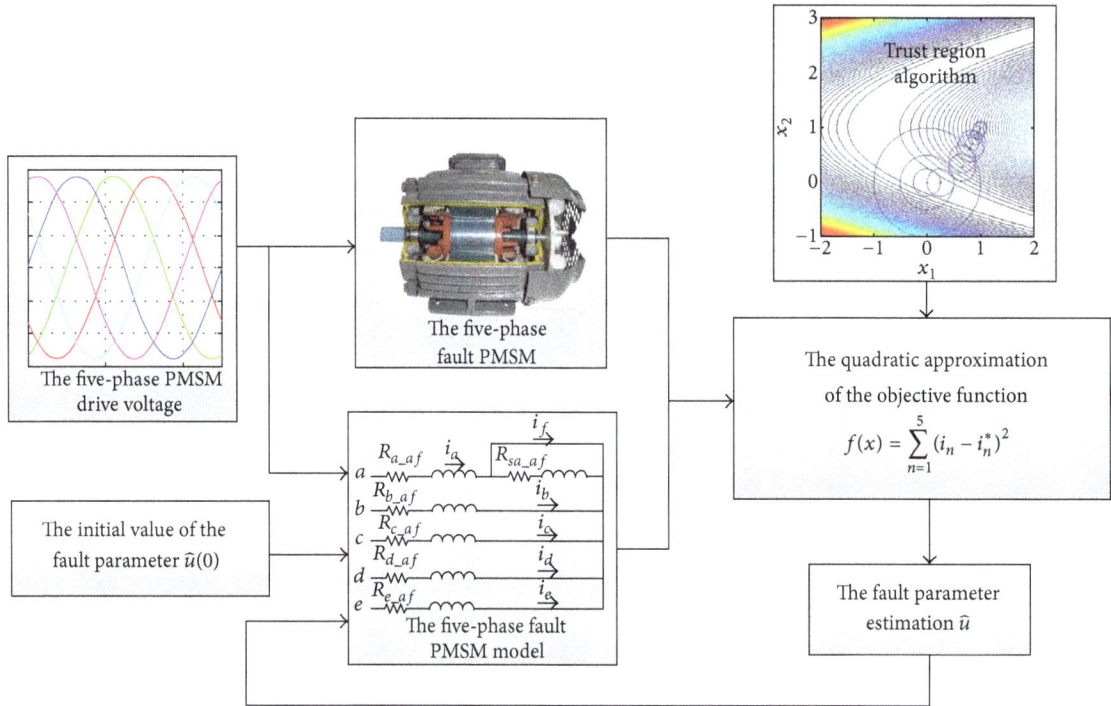

FIGURE 3: The block diagram of fault parameter identification based on trust region algorithm.

Step 4. Correct trust region radius as follows:

$$\Delta_{k+1} := \begin{cases} \tau_1 \Delta_k, & \text{if } r_k \leq \eta_1, \\ \Delta_k, & \text{if } \eta_1 < r_k \leq \eta_2, \\ \min\{\tau_2 \Delta_k, \overline{\Delta}\}, & \text{if } r_k \geq \eta_2, \|d_k\| = \Delta_k. \end{cases} \quad (18)$$

Step 5. If $r_k > \eta_1$, set $\hat{\mu}_{k+1} := \hat{\mu}_k + d_k$, $B_k := B_{k+1}$, $k := k + 1$, and go to Step 1; else set $\hat{\mu}_{k+1} := \hat{\mu}_k$, $k := k + 1$, and go to Step 2.

4. Simulation Analysis

In order to verify the model of the five-phase PMSM under ITSC fault and the trust region algorithm in the process of the fault parameter identification, the simulation has been done by MATLAB/Simulink. The parameters of five-phase PMSM are shown in Table 1.

Figure 4 shows the phase current waveforms in healthy state. Figures 5 and 6 show the phase current waveforms in fault state. Notice that the current of phase A (the yellow line) in fault state is significantly larger than the healthy ones. And with the increase of the fault parameter u, the current of phase A in fault state also increases. And meanwhile, the current waveforms of other phases are also affected by the fault phase A.

The parameters of five-phase PMSM for fault parameter identification based on trust region algorithm are shown in Table 2.

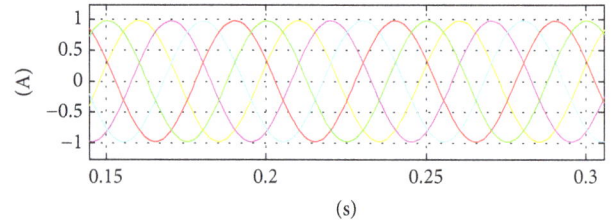

FIGURE 4: The phase current waveforms in normal state.

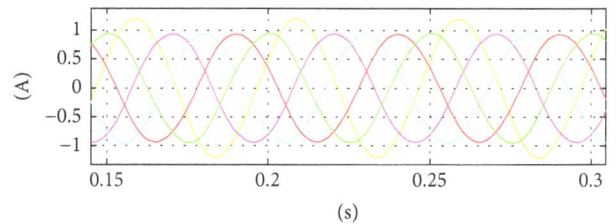

FIGURE 5: The phase current waveforms in fault state ($u = 0.2$).

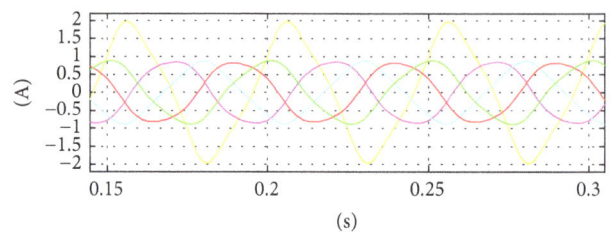

FIGURE 6: The phase current waveforms in fault state ($u = 0.5$).

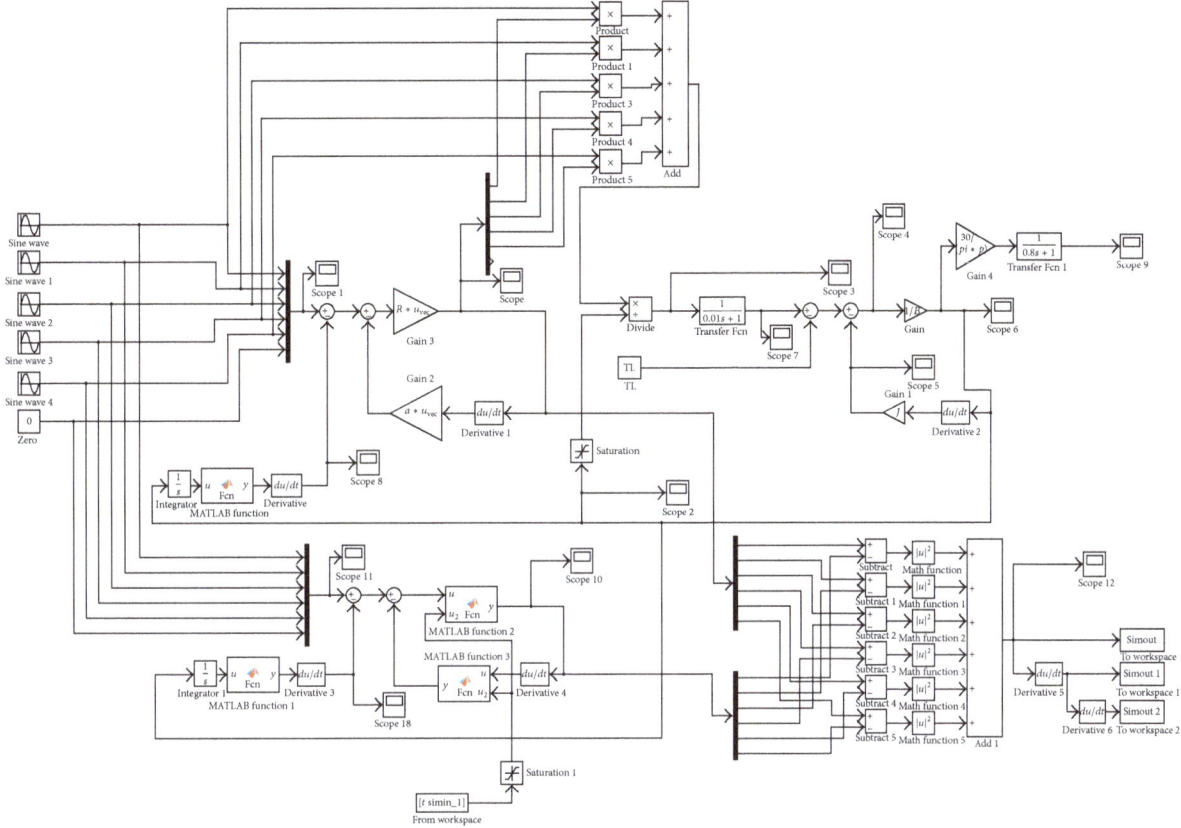

FIGURE 7: The simulation model of five-phase PMSM for fault parameter identification.

TABLE 1: Parameters of five-phase PMSM used for simulation.

Stator winding resistance	R_s	17.4 (Ω)
Self-inductance of phase winding	L_m	$4.5e-2$ (H)
Flux amplitude	ψ_m	0.1827 Wb
Moment of inertia	J	$0.8e-6$ (kgm^2)
Friction constant	B	$1e-3$ (Nms)
Number of pole pairs	p	4
Load torque	T_L	0.4 (Nm)
Input sinusoidal voltage amplitude	V	28 (V)
Input voltage frequency	f	20 (Hz)

FIGURE 8: Experimental devices.

5. Experimental Results

The Simulink simulation model of five-phase PMSM for fault parameter identification based on trust region algorithm is shown in Figure 7.

The results of five-phase PMSM for fault parameter identification based on trust region algorithm are shown in Table 3. The actual value of the interturn short circuit ratio μ is 0.2 (20% of the total turns occurring are shorted) and the initial estimation value of the interturn short circuit ratio u^* is 0.3. After six steps of the trust region seeking calculation, the estimation value of interturn short circuit ratio u^* converges to the actual value.

To verify the model of the five-phase PMSM under ITSC fault and the trust region algorithm in the process of the fault parameter identification, the experiment has been done. The parameters of five-phase PMSM are the same as shown in Table 1. The experimental devices are as shown in Figure 8. Figure 9 is the phase current waveforms of the five-phase PMSM under 20% ITSC fault in phase A. Figure 10 is the fitness function curve in optimization process under 20% ITSC fault in phase A.

Comparing the experimental results with the simulation results, it can be seen that the actual phase current waveforms

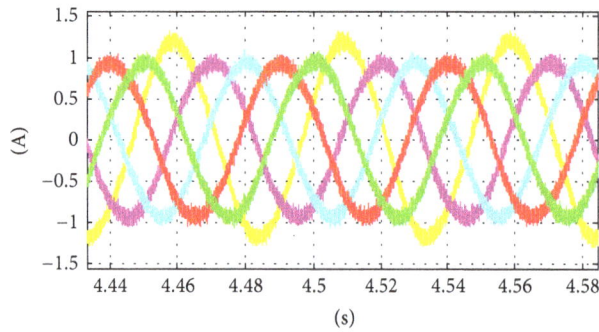

FIGURE 9: Phase current waveforms of the five-phase PMSM under 20% ITSC fault in phase A.

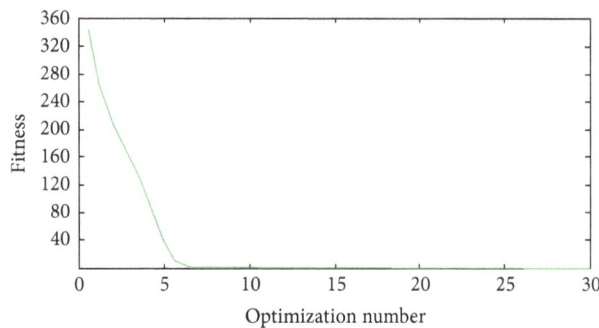

FIGURE 10: Fitness function curve in optimization process under 20% ITSC fault in phase A.

TABLE 2: Parameters of trust region algorithm.

η_1	0.1
η_2	0.75
$\bar{\Delta}$	2.0
τ_1	0.5
τ_2	2.0
ε	0.1

TABLE 3: Result of fault parameter identification.

k	Δ_0	r_k	u^*
1	1.0	0.1799	0.2834
2	1.0	0.2072	0.2668
3	1.0	0.2462	0.2502
4	1.0	0.3061	0.2335
5	1.0	0.4050	0.2167
6	1.0	0.5298	0.2001

are almost the same as the simulation analysis from Figures 5 and 9. The peak current values of the nonfault phases are all 1 A approximately in both the simulation results and the experimental results. And the peak current values of the fault phase are all about 1.2 A in both the simulation waveform and the experimental waveform. It proves the correction of the mathematics models built for the five-phase PMSM under both healthy and ITSC fault situations. The only difference between simulation results and experimental

results is that the experimental phase current waveforms are not so smooth as shown in the simulation results. The reason is that there may be interference during the actual current sampling process, which results in the current glitch.

Figure 10 is the fitness function curve in optimization process under 20% ITSC fault in phase A. It can be seen that the value of the fitness function is close to zero at the tenth optimization. And from Table 3, it can be seen that, in the simulation, after six steps of the trust region seeking calculation, the estimation value of interturn short circuit ratio u^* converges to the actual value. Compared to simulation results, the optimization number of the trust region seeking in the experiment is larger than the simulation shown, for the reason that the actual phase voltage and current of the motor are sampled by the voltage sensor and current sensor, and there are errors during the sampling procedure. So the optimization number is added. However, it is obvious that both the simulation and the experiment are all successful implementations of the parameter identification and prove the correctness of the proposed parameter estimation method for the fault diagnosis.

6. Conclusions

In this paper, the mathematics models for the five-phase PMSM under both healthy and ITSC fault situations are established, respectively. Furthermore, a novel fault diagnosis method of ITSC based on the trust region algorithm is proposed for five-phase PMSM. With the aid of the trust region algorithm, the interturn short circuit ratio μ is estimated with a short time transient. The simulation and experimental results have validated both the correction of the established models and the effectiveness of the proposed parameter estimation method.

Conflict of Interests

The authors declare that there is no conflict of interests regarding the publication of this paper.

Acknowledgment

This research was supported by the National Natural Science Foundation of China (Grant no. 51407143).

References

[1] K. D. Hoang, Y. Ren, Z.-Q. Zhu, and M. Foster, "Modified switching-table strategy for reduction of current harmonics in direct torque controlled dual-three-phase permanent magnet synchronous machine drives," *IET Electric Power Applications*, vol. 9, no. 1, pp. 10–19, 2015.

[2] R. Islam, M. Islam, J. Tersigni, and T. Sebastian, "Inter winding short circuit faults in permanent magnet synchronous motors used for high performance applications," in *Proceedings of the 4th Annual IEEE Energy Conversion Congress and Exposition (ECCE '12)*, pp. 1291–1298, September 2012.

[3] G. Vinson, M. Combacau, T. Prado, and P. Ribot, "Permanent magnets synchronous machines faults detection and identification," in *Proceedings of the 38th Annual Conference on IEEE Industrial Electronics Society (IECON '12)*, pp. 3925–3930, Montreal, Canada, October 2012.

[4] B. G. Gu, J. H. Choi, and I. S. Jung, "A dynamic modeling and a fault detection scheme of a PMSM under an inter turn short," in *Proceedings of the IEEE Vehicle Power and Propulsion Conference (VPPC '12)*, pp. 1074–1080, Seoul, Republic of Korea, October 2012.

[5] N. Leboeuf, T. Boileau, B. Nahid-Mobarakeh, N. Takorabet, F. Meibody-Tabar, and G. Clerc, "Effects of imperfect manufacturing process on electromagnetic performance and online interturn fault detection in PMSMs," *IEEE Transactions on Industrial Electronics*, vol. 62, no. 6, pp. 3388–3398, 2015.

[6] Z. Wang, J. Yang, H. Ye, and W. Zhou, "A review of permanent magnet synchronous motor fault diagnosis," in *Proceedings of the IEEE Transportation Electrification Conference and Expo (ITEC Asia-Pacific '14)*, Beijing, China, September 2014.

[7] A. Gandhi, T. Corrigan, and L. Parsa, "Recent advances in modeling and online detection of stator interturn faults in electrical motors," *IEEE Transactions on Industrial Electronics*, vol. 58, no. 5, pp. 1564–1575, 2011.

[8] M. Eftekhari, M. Moallem, S. Sadri, and A. Shojaei, "Review of induction motor testing and monitoring methods for interturn stator winding faults," in *Proceedings of the 21st Iranian Conference on Electrical Engineering (ICEE '13)*, pp. 1–6, IEEE, Mashhad, Iran, May 2013.

[9] D. Yao and H. Toliyat, "A review of condition monitoring and fault diagnosis for permanent magnet machines," in *Proceedings of the IEEE Power and Energy Society General Meeting (PES '12)*, pp. 1–4, San Diego, Calif, USA, July 2012.

[10] M. A. Shamsi Nejad and M. Taghipour, "Inter-turn stator winding fault diagnosis and determination of fault percent in PMSM," in *Proceedings of the IEEE Applied Power Electronics Colloquium (IAPEC '11)*, pp. 128–131, Johor Bahru, Malaysia, April 2011.

[11] F. Grouz, L. Sbita, and M. Boussak, "Particle swarm optimization based fault diagnosis for non-salient PMSM with multi-phase inter-turn short circuit," in *Proceedings of the 2nd International Conference on Communications, Computing and Control Applications (CCCA '12)*, pp. 1–6, IEEE, Marseille, France, December 2012.

[12] H. Saavedra, J.-R. Riba, and L. Romeral, "Inter-turn fault detection in five-phase pmsms. Effects of the fault severity," in *Proceedings of the 9th IEEE International Symposium on Diagnostics for Electric Machines, Power Electronics and Drives (SDEMPED '13)*, pp. 520–526, Valencia, Spain, August 2013.

[13] L. Romeral, J. C. Urresty, J.-R. R. Ruiz, and A. G. Espinosa, "Modeling of surface-mounted permanent magnet synchronous motors with stator winding interturn faults," *IEEE Transactions on Industrial Electronics*, vol. 58, no. 5, pp. 1576–1585, 2011.

[14] N. H. Obeid, T. Boileau, and B. Nahid-Mobarakeh, "Modeling and diagnostic of incipient inter-turn faults for a three phase permanent magnet synchronous motor," in *Proceedings of the IEEE Industry Applications Society Annual Meeting*, pp. 1–8, IEEE, Vancouver, Canada, October 2014.

[15] B.-G. Park, R.-Y. Kim, and D.-S. Hyun, "Fault diagnosis using recursive least square algorithm for permanent magnet synchronous motor drives," in *Proceedings of the 8th IEEE International Conference on Power Electronics and ECCE Asia (ICPE-ECCE '11)*, pp. 2506–2510, IEEE, Jeju, Republic of Korea, May-June 2011.

[16] F. Grouz, L. Sbita, and M. Boussak, "Modelling for non-salient PMSM with multi-phase inter-turn short circuit," in *Proceedings of the 10th IEEE International Multi-Conference on Systems, Signals & Devices (SSD '13)*, pp. 1–6, IEEE, Hammamet, Tunisia, March 2013.

[17] W. Liu, L. Liu, I.-Y. Chung, D. A. Cartes, and W. Zhang, "Modeling and detecting the stator winding fault of permanent magnet synchronous motors," *Simulation Modelling Practice and Theory*, vol. 27, pp. 1–16, 2012.

[18] Z. Min and S. Pingping, "Nonlinear model predictive control algorithm based on filter-trust-region method," in *Proceedings of the 31st Chinese Control Conference (CCC 2012)*, pp. 4069–4074, IEEE, Hefei, China, July 2012.

[19] P. M. Moubarak, "Trust-region reflective adaptive controller for time varying systems," *IET Control Theory & Applications*, vol. 9, no. 2, pp. 240–247, 2015.

[20] Y. Shu-Ping, Y. Xiu-Gui, and L. Zai-Ming, "A trust region algorithm based on general curve-linear searching direction for unconstrained optimization," in *Proceedings of the 4th International Conference on Intelligent Computation Technology and Automation (ICICTA '11)*, pp. 328–332, Shenzhen, China, March 2011.

2

A Miniaturized Reconfigurable CRLH Leaky-Wave Antenna Using Complementary Split-Ring Resonators

Damiano Patron ⓘ, Yuqiao Liu, and Kapil R. Dandekar

Department of Electrical and Compnuter Engineering, Drexel University, 3141 Chestnut Street, Philadelphia, PA 19104, USA

Correspondence should be addressed to Damiano Patron; damiano.patron@gmail.com

Academic Editor: John N. Sahalos

Composite Right-/Left-Handed (CRLH) Leaky-Wave Antennas (LWAs) are a class of radiating elements characterized by an electronically steerable radiation pattern. The design is comprised of a cascade of CRLH unit cells populated with varactor diodes. By varying the voltage across the varactor diodes, the antenna can steer its directional beam from broadside to backward and forward end-fire directions. In this paper, we discuss the design and experimental analysis of a miniaturized CRLH Leaky-Wave Antenna for the 2.4 GHz WiFi band. The miniaturization is achieved by etching Complementary Split-Ring Resonator (CSRR) underneath each CRLH unit cell. As opposed to the conventional LWA designs, we take advantage of a LWA layout that does not require thin interdigital capacitors; thus we significantly reduce the PCB manufacturing constraints required to achieve size reduction. The experimental results were compared with a nonminiaturized prototype in order to evaluate the differences in impedance and radiation characteristics. The proposed antenna is a significant achievement because it will enable CRLH LWAs to be a viable technology not only for wireless access points, but also potentially for mobile devices.

1. Introduction

Reconfigurable antennas have received significant attention in the literature with respect to static antennas (antennas with a fixed radiation pattern) thanks to their capability of dynamically changing their radiation properties. These antennas can adapt their characteristics in response to the behavior of the wireless channel and be used for a variety of applications including throughput maximization [1, 2], interference management [3], directional networking [4], and security [5].

Adaptive antennas can be divided into two subclasses: phased arrays and reconfigurable antennas. While the former subclass requires multiple radiating elements and phase shifting networks [6, 7], the latter subclass of reconfigurable antennas consists of a single radiating element, capable of generating different patterns or polarization [8, 9]. The reconfigurable antenna solution is thus preferable with respect to a phased array antenna mainly because: (i) it employs a single active element and therefore it occupies a small space and (ii) it allows for high radiation efficiency since it does not employ phase shifters and power dividers.

Various types of reconfigurable antennas capable of changing pattern and polarization have been proposed in the literature. These antennas may employ embedded switches or variable capacitors to change the current distribution on the metallization of the active element [10, 11] or may employ an active antenna element surrounded by passive elements (i.e., parasitic elements) loaded with variable capacitors or connected to switches [12, 13]. Particularly interesting is the design of Composite Right-/Left-Handed (CRLH) Reconfigurable Leaky-Wave Antennas (LWAs), a two-port metamaterial-based design that is able to steer its directive beam from broadside to backward and forward angles [14].

Leaky-wave antennas are based on the concept of traveling wave, as opposed to conventional resonating-wave behavior. When an RF signal is applied to the input port, the traveling wave progressively "leaks" power as it travels along the waveguide structure. LWAs can also be seen as a phased array traveling wave antenna with amplitude decaying excitation and progressive phase shift as a result of the wave traveling along each unit cell. This leakage phenomenon is directly related to the directivity of the radiated beam. Besides conventional PCB substrates, recent developments in LWA

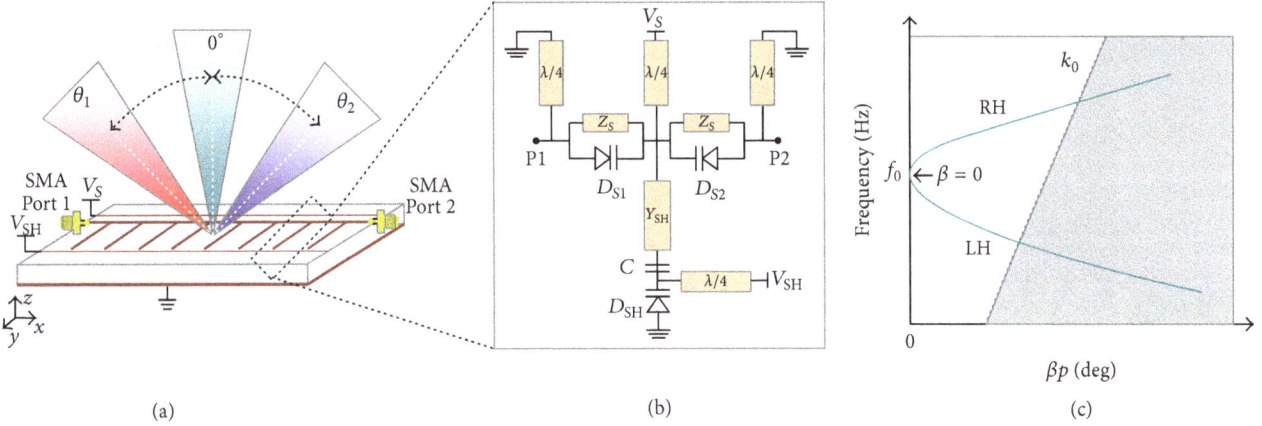

FIGURE 1: (a) Sketch of a CRLH Leaky-Wave Antenna with beam steering from broadside $\theta = 0°$ to backward θ_1 and forward θ_2 angles. (b) Schematic of a conventional CRLH unit cell. (c) Dispersion diagram used to evaluate the propagation constant β and estimate the main beam angle θ.

designs have shown the possibility of designing CLRH LWA by using liquid crystals. For example, in [15, 16], the authors present CRLH LWA made of injected liquid crystals with the beam steering achieved through an external electric field. As opposed to the proposed LWA, the use of liquid crystals requires very high bias voltage and the beam steering is limited to 20–30 degrees.

Although the planar and compact form factors of the LWAs make them suitable for wireless base stations, they cannot be exploited on mobile devices due to size constrains. In this paper, we will address this limitation by presenting an approach that will make LWAs more suitable for mobile devices.

Current attempts to miniaturize antenna dimensions involve the use of nonconventional substrates with high or enhanced dielectric constant [17, 18]. Other techniques were developed where the substrate is made by stacking reactive layers [19, 20]. Unfortunately, these techniques introduce more manufacturing complexity and bulk. In [21], the authors propose the design of a miniaturized CRLH LWA using metallic vias and interdigital lines. Even though the results show a wide beam scanning from −60 to +67 degrees, the beams are characterized by a poor front-to-back ratio and the patterns are not electrically reconfigurable. On the other hand, recent developments in defected ground structures have shown the possibility of properly etching the ground plane of transmission lines or antennas in order to change their cut-off and resonant frequencies [22, 23]. As a result, devices with small dimension can be loaded with complementary resonators on the ground plane to resonate at lower frequencies, achieving miniaturization [24, 25]. We are unaware of any previous work that has achieved reconfigurable antenna miniaturization through defected ground structures (e.g., designing complementary resonators).

In this paper, we apply a defected ground technique to achieve miniaturization of reconfigurable antennas. In particular, we build upon the LWA design introduced in [26] in which, as opposed to the conventional LWAs, we greatly reduce PCB manufacturing constraints by avoiding the use

of thin interdigital capacitors. In this paper, we designed the miniaturized LWA by applying a CSRR underneath each unit cell to achieve miniaturization of the top layer radiating layout. The unit cell is designed and characterized to resonate at 2.4 GHz and provide the largest possible beam steering. Relative to a conventional 2.4 GHz LWA, the overall dimension can be halved while maintaining good impedance matching, relatively high front-to-back ratio, and good beam steering performance.

The miniaturized LWA is designed to exhibit good impedance matching within the 2.41–2.48 GHz band, for WiFi operations on mobile devices such as laptops or tablets.

The paper is organized as follows: Section 2 provides a brief overview on LWAs and the metamaterial unit cell. Section 3 describes the design of the miniaturized LWA unit cell loaded with CSRR. Section 4 presents the LWA design along with experimental analysis of impedance and radiation patterns. These characteristics are then compared with a nonminiaturized prototype. Finally, conclusions are drawn in Section 5.

2. Background

The reconfigurable CRLH LWA can be realized as a 2-port radiating element with tunable radiation properties. The layout is made by a series of N metamaterial unit cells [27], cascaded in order to create a periodic structure from port 1 to port 2, as shown in Figure 1(a). Unlike conventional resonating-wave antennas, the LWA is based on the concept of a traveling wave. When a radio-frequency signal is applied to one of the input ports, the traveling wave leaks out energy as it progressively travels toward the second port. This energy leakage determines the directivity of the radiated beam and is a function of the propagation constant along the structure.

In LWAs, the radiation properties are determined by the complex propagation constant $\gamma = \alpha - j\beta$, where α is the attenuation constant and β is the phase constant. While the former corresponds to a loss due to the leakage of energy, the latter determines the radiation angle of the main beam.

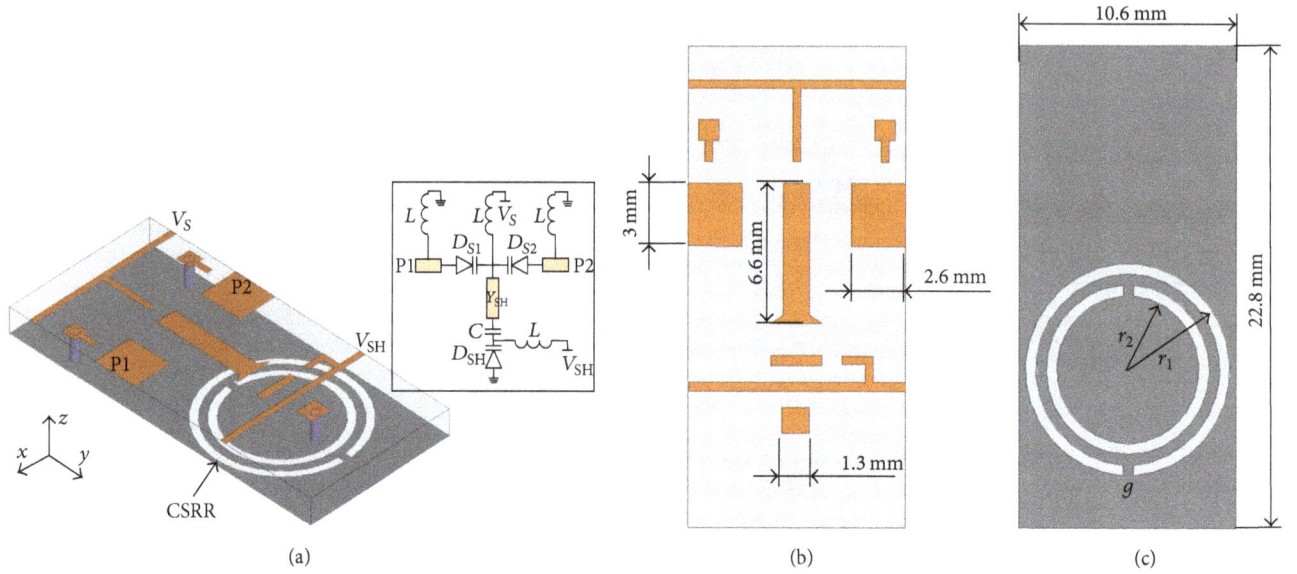

FIGURE 2: (a) 3D HFSS model of the LWA unit cell, with CSRR etched on the ground plane. (b) 2D top layer layout and dimensions. (c) Bottom layer with ground plane and CSRR design. r_1 = 5 mm and r_2 = 4 mm. The gap g and the distance between the two rings are 0.5 mm.

Additionally, the relationship between β and the wavenumber k_0 defines the regions of operation.

The dispersion diagram in Figure 1(c) depicts the absolute value of β and the two regions of operation. The darker area where $|\beta| > k_0$ represents the guided wave, where the energy is propagated from port 1 to port 2, whereas the area where $|\beta| < k_0$ represents the radiated region. The angle of the main beam can be determined by the following:

$$\theta = \sin^{-1}\left(\frac{\beta}{k_0}\right). \qquad (1)$$

If we assume that port 2 is fed an input signal and port 1 is terminated with a 50 Ω load, at frequency $f0$ where $\beta = 0$, the antenna radiates a main lobe in broadside direction $\theta = 0°$, which is perpendicular to the antenna's plane. For frequencies where $\beta > 0$ (positive slope of $|\beta|$) the antenna operates in RH region, steering the beam around the left semiplane θ_1. On the other hand, when $\beta < 0$ (negative slope of $|\beta|$) it operates in LH region, and radiation occurs within the symmetric half-plane θ_2. This frequency-dependent behavior allows for the scanning of the main beam from back-fire to end-fire directions. The introduction of tunable capacitances in the unit cell can turn the antenna from a frequency-controlled to a voltage-controlled beam steering radiator.

Several voltage-controlled LWAs have been developed in the literature [14, 28] and the circuit model of the conventional metamaterial unit cell can be described as in Figure 1(b). The structure is comprised of both series and shunt components. The series portion is designed with two interdigital capacitors and two varactor diodes D_{S1} and D_{S2} connected in parallel. The shunt portion is composed of a stub and a varactor diode D_{SH} in series. By adding a varactor-loaded shunt stub, the shunt admittance Y_{SH} of the unit cell can also be tuned. In addition, the independent control

through V_{SH} provides an additional degree of freedom, leading to improved tunability of scanning range and impedance matching [14, 28]. The capacitor C acts as DC-block for the two bias lines V_S and V_{SH}. Three $\lambda/4$ microstrip transformers provide the DC bias lines to the diodes. The introduction of varactor diodes allows for a change in capacitance through the reverse bias voltage, and the propagation constant β becomes a function of the diode's voltage. As a result, the curve depicted in Figure 1(c) can be varied along the vertical axis, and the radiator can steer the main beam from backward to forward directions at a given frequency.

Unlike the aforementioned designs, for the miniaturization of the CRLH LWA, we took advantage of an improved design presented in [26] which avoids the use of interdigital capacitors as part of the unit cell model. Therefore, we avoid the manufacturing challenges that may be introduced by etching the very thin fingers that constitute the interdigital capacitors.

From a manufacturing and potential commercialization perspective, this is a significant advantage, especially as research is conducted to miniaturize the layout.

In the next section, we discuss the design of the miniaturized CRLH unit cell along with the CSRR. Through experimental analysis of the scattering parameters (S-parameters) we will evaluate the impedance characteristics and the expected radiation angles from the dispersion diagram.

3. Design of the CRLH Unit Cell

The LWA unit cell shown in Figure 2(a) was etched on a conventional FR-4 substrate having dielectric constant s and thickness t = 1.6 mm. The top layout and the CSRR were tuned to operate within the entire 2.4 GHz band.

The unit cell was designed and tuned using the full-wave electromagnetic simulator Ansoft HFSS [29]. In order to

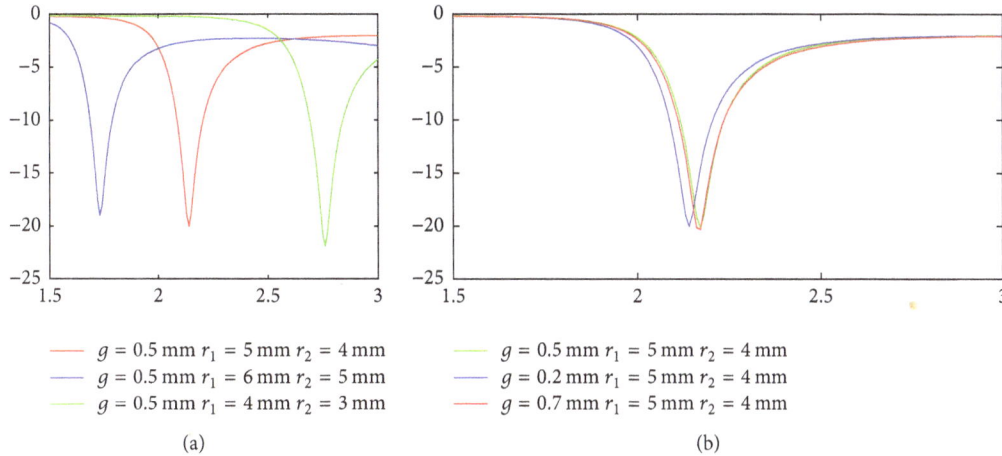

FIGURE 3: (a) Parametric simulation of change in inner and outer radii r_1 and r_2. (b) Parametric simulation of change in the gap g.

perform more realistic simulations, each lumped component was measured through a 2-port fixture and a Vector Network Analyzer (VNA). Then, the S-parameters (S2P) were loaded into the circuit simulator Ansoft Designer [30]. The cosimulation between HFSS and Designer allowed for evaluation of the 3D model using the actual S2P parameters. As a varactor diode we selected an Infineon BB833 in SOD323 package, designed to operate up to 2.5 GHz [31]. We chose to use the BB833 because it provided a large dynamic range at low voltages, which reduced the power consumption and the complexity of the control board. In order to get a qualitative evaluation of the capacitance range and loss under reverse bias voltage, we extracted the junction capacitance C_J and the series resistance R_S from the measured S2P. The plot in Figure 3 shows that the series resistance falls within the range $1.7\,\Omega \leq R_S \leq 1.85\,\Omega$ within the entire reverse voltage sweep. However, when $V_R \leq 10\,\text{V}$, the junction capacitance exhibits larger dynamic range: $18\,\text{pF} \leq C_J \leq 3\,\text{pF}$ (Figure 4). The final unit cell layout has been optimized to take advantage of this large C_J variation under low bias voltages, achieving the largest possible tunability of the phase constant β.

The CRLH behavior is determined by designing the unit cell to have proper series capacitance and a shunt inductive component. The series capacitance is achieved by placing two varactor diodes in series with a common cathode (D_{S1} and D_{S2}). The inductive part is designed by means of a shunt stub with a varactor diode (D_{SH}) placed in series. The dynamic tuning is accomplished by changing the reverse voltage V_R of the two bias line V_S and V_{SH}. A $C = 0.5\,\text{pF}$ capacitor was added to the shunt stub in order to decouple the two bias voltages. To further reduce manufacturing complexity and form factor, we have used $L = 220\,\text{nH}$ inductors that act as RF chokes to provide the two bias voltages. The inset in Figure 2(a) depicts the resulting schematic of the LWA unit cell. The dimensions are shown in Figure 2(b), and the gaps are properly designed to include the lumped components. The series microstrips that connect D_{S1} and D_{S2} were calculated and optimized to achieve a characteristic impedance of $50\,\Omega$, whereas the shunt microstrip that connect D_{SH} was scaled down from the reference dimension in [26], due to the

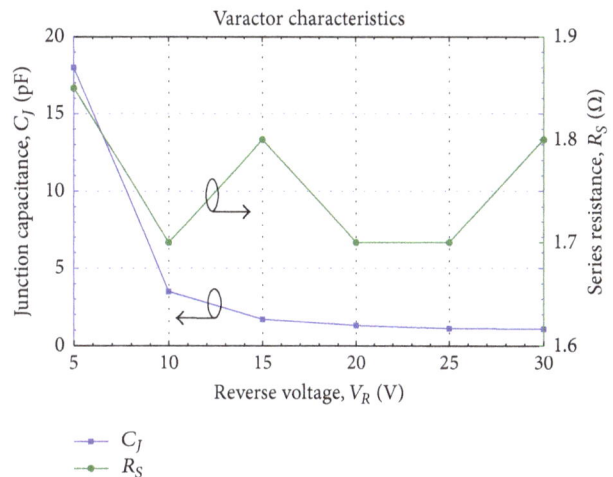

FIGURE 4: Junction capacitance C_J and series resistance R_S as function of the reverse voltage V_R. The values were extracted from the measured S-parameters. While R_S maintains a relatively constant value within the entire voltage sweep, the capacitance C_J exhibits a larger dynamic range when $V_R \leq 10\,\text{V}$.

contribution of the CSRR. The thinner bias lines that connect the RF chokes were kept as short as possible and the width matches the pads of the SMT chokes.

For lab characterization purpose, the varactor diodes were biased using bench-top power supplies. On the other hand, for real-time dynamic biasing of the LWA, our group developed a control board comprising voltage boost circuits (0–30 V) and an FPGA. The lookup table contained within the FPGA allows switching between the different pattern configurations and it can be controlled through an external SPI or I2C command.

Simulations have shown that, by using a standard ground plane, the proposed unit cell operates in the frequency region of 5 GHz. In order to reduce the operating band, a single CSRR was etched underneath the unit cell. In [32], it has been shown that when a CSRR is etched on the ground plane

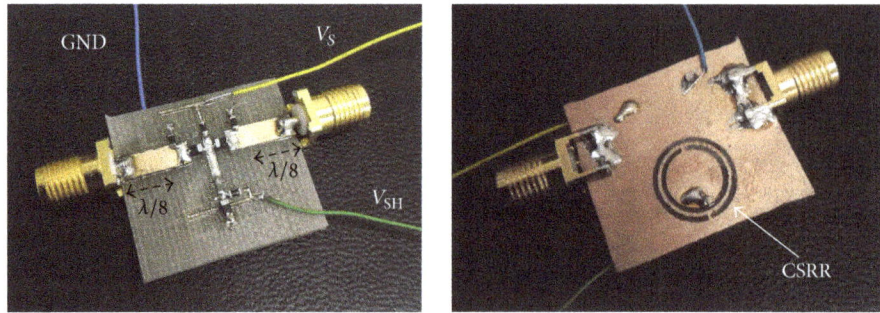

FIGURE 5: Top and bottom layer pictures of the miniaturized LWA unit cell. The design is etched between two $\lambda/8$ microstrip lines for S-parameter measurements.

of a $50\,\Omega$ microstrip line, due to the Babinet principle and, complementarity, the microstrip loaded with CSRR behaves like a one-dimensional effective medium with a negative permittivity within a region around the CSRR resonance. Thanks to this change of permittivity, the top layer antenna structure resonates at lower frequencies.

In order to find the optimal CSRR radii and gap dimensions, we started by simulating the insertion loss of a $50\,\Omega$ microstrip line above a parametric CSRR. The simulations have shown a direct proportion between the CSRR radii and the resonant frequency. In fact when the radii are reduced, the resonance shifts down in frequency as shown in Figure 3. The gap g acts as fine-tuning element to optimize the resonance point. In our case, after finding the suboptimal dimension for resonance around 2.4 GHz, we choose to fine-tune the CSRR by looking at the frequency reduction of the LWA unit cell down to 2.45 GHz. The optimal CSRR layout is shown in Figure 2(c).

The outer radius is $r_1 = 5$ mm, while the inner radius is $r_2 = 4$ mm. The gap g on both rings, as well as the distance between them, is 0.5 mm. From the simulations, we noticed that when the CSRR is positioned at the center of the unit cell, the miniaturization effect is reduced and the resulting radiation patterns exhibit a pronounced back lobe due to radiation leakage from the CSRR apertures on the ground plane. For this reason, the CSRR was slightly moved from the center to the shunt part of the unit cell, in order to reduce the radiation from the ground plane and enhance the front-to-back ratio. As we will see in Section 3.1, the effects of the CSRR on the unit cell characteristics are to extend the S_{11} bandwidth, while the dispersion curve βp is intentionally tuned in the RH region through the varactor C_j operating point and the shunt stub dimension.

The CRLH unit cell can exhibit balanced or unbalanced resonances, based on the series and shunt resonant frequencies ω_{se}, ω_{sh}. While the unbalanced unit cell ($\omega_{se} \neq \omega_{sh}$) supports two different frequencies, the lower for the LH and the higher for the RH region, we used balanced unit cell ($\omega_{se} = \omega_{sh}$) in order to avoid the gap between the RH and LH regions and match the structure over a broad bandwidth. In terms of radiating regions, a CRLH unit cell can typically operate in either RH or LH regimes. However, in order to achieve the maximum beam coverage by switching between the two input ports, we have optimized the design within the

RH region ($|\beta| > 0$). Similar to [33], port 1 is used and the beam can be steered from $0°$ to max$\{\theta_2\}$, while by switching to port 2 the beam covers the symmetrical quadrant from $0°$ to max$\{\theta_1\}$. This design choice enables full-space beam steering, while taking advantage of the high C_J variation under low voltage regimes. Due to the 2-port switching, a similar beam steering mechanism can be achieved using unbalanced CRLH unit cells.

The next subsection describes the experimental analysis conducted on a miniaturized unit cell prototype. We evaluated the impedance characteristics and the expected radiation angles from the dispersion diagram.

3.1. Characterization Results. S-parameter measurements were carried out to assess the performance of a miniaturized unit cell prototype and validate the simulation results. The unit cell was etched between two $\lambda/8$ transmission lines for soldering the SMA connectors. An Agilent N5230A Vector Network Analyzer was calibrated with the port extension function for deembedding the two extra lengths. Top and bottom layers of the manufactured unit cell are shown in Figure 5.

Figure 6 shows measured and simulated S-parameters for four arbitrary configurations. Due to port symmetry, in this plot, we assume $S_{11} = S_{22}$ and $S_{12} = S_{21}$ for greater visual clarity. By observing the S_{11} curves, we can note that the proposed miniaturized unit cell maintains good impedance matching within the bandwidth of interest from 2.41 to 2.48 GHz. The 10 dB bandwidths are between 220 MHz $\leq$ BW $\leq$ 650 MHz. The measured and simulated S-Parameters of the unit cell are in good agreement around the bandwidth of interest: 2.4 GHz and 2.6 GHz.

Outside the desired bandwidth, the traces start to differ because of the narrow-band S-parameter fixture used for testing. However, we chose to keep a large x-axis range in order to highlight the 10 dB bandwidth.

If we define the phase constant β as $\beta = d^{-1}\cos^{-1}(1 + Z(\omega)Y(\omega))$ [27], where $Z(\omega)$ is the series impedance; $Y(\omega)$ is the shunt admittance; the two series varactors D_{S1}, D_{S2} vary with $Z(\omega)$ while the shunt varactor D_{SH} varies with $Y(\omega)$. Furthermore, from our measurements we noticed that the series voltage V_S has major control in changing configurations, while the voltage V_{SH} allows for fine-tuning the S-parameters, maintaining the Bloch impedance relatively

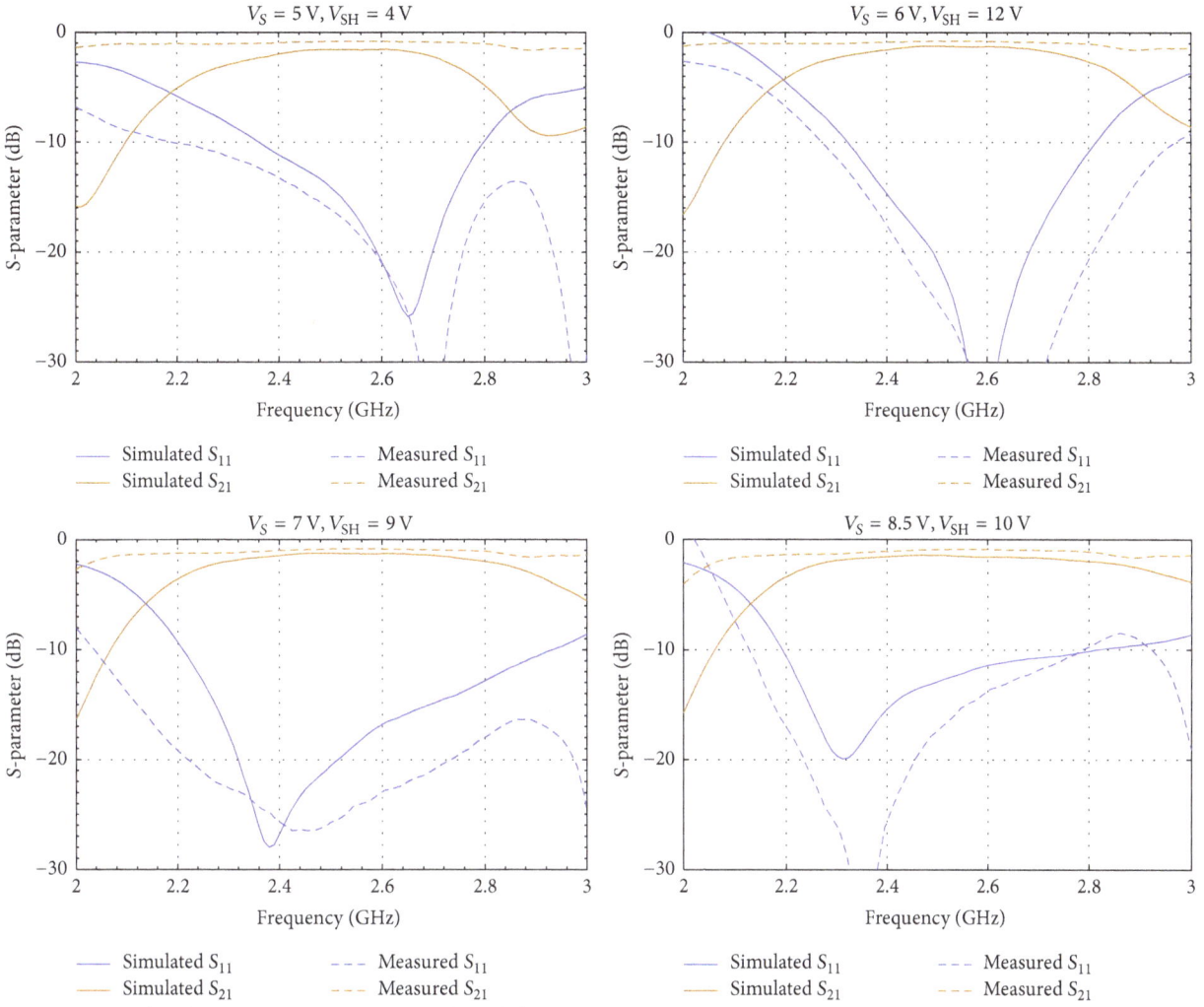

FIGURE 6: Simulated and measured S-parameters for four different configurations. While V_S acts as major controller for the center frequency, V_{SH} allows for fine-tuning and improvement of the impedance.

constant and close to 50 Ω. The insertion loss, which includes both actual losses and radiation leakage, is between $0.8\,\text{dB} \leq S_{21} \leq 1.5\,\text{dB}$ among the different configurations. The higher deviation between simulation and measurement at the two sides of the bandwidth is potentially due to the S-parameters fixture used to extract the S2P of each lumped component.

In order to evaluate the beam steering capabilities, the dispersion diagram was created using the following equation [34] and the measured S-parameters:

$$\beta p = \cos^{-1}\left(\frac{1 - S_{22}S_{22} + S_{21}S_{12}}{2S_{21}}\right). \qquad (2)$$

The dispersion diagram in Figure 7 shows the result for four different configurations. We can note that, within the bandwidth of interest, the curves are upward sloping, denoting operation in RH regime. The expected radiation angles θ can be estimated through the equation shown in the inset of Figure 7 and computed at the desired frequency. By assuming WiFi operation at 2.46 GHz (channel 11), the miniaturized unit cells allow for steering the radiated beam approximately from $\theta = 21°$ to $\theta = 55°$ with respect to

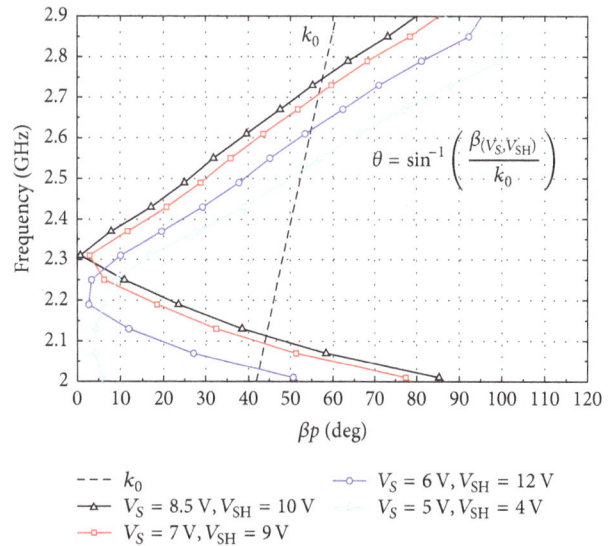

FIGURE 7: Dispersion diagram of the proposed miniaturized CRLH unit cell. The four different states were taken for incremental values of bias voltages V_S V_{SH}. The desired frequency bandwidth, 2.41–2.48 GHz, falls within the RH radiated region.

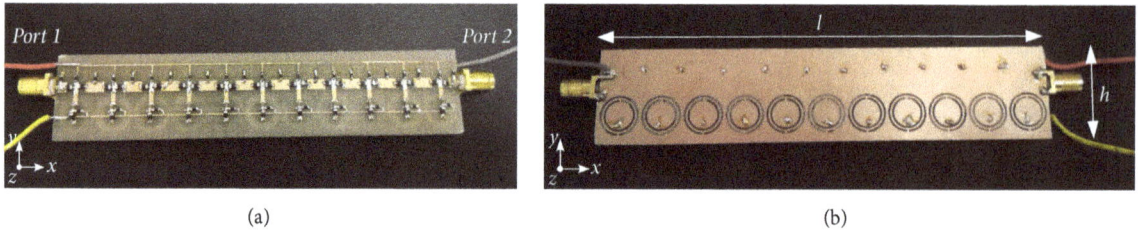

(a)   (b)

FIGURE 8: Picture of the miniaturized CRLH LWA. (a) Top layer with cascade or $N = 11$ unit cells. (b) Bottom layer with CSRRs. The overall dimensions are $l = 11.5$ cm, $h = 2.3$ cm.

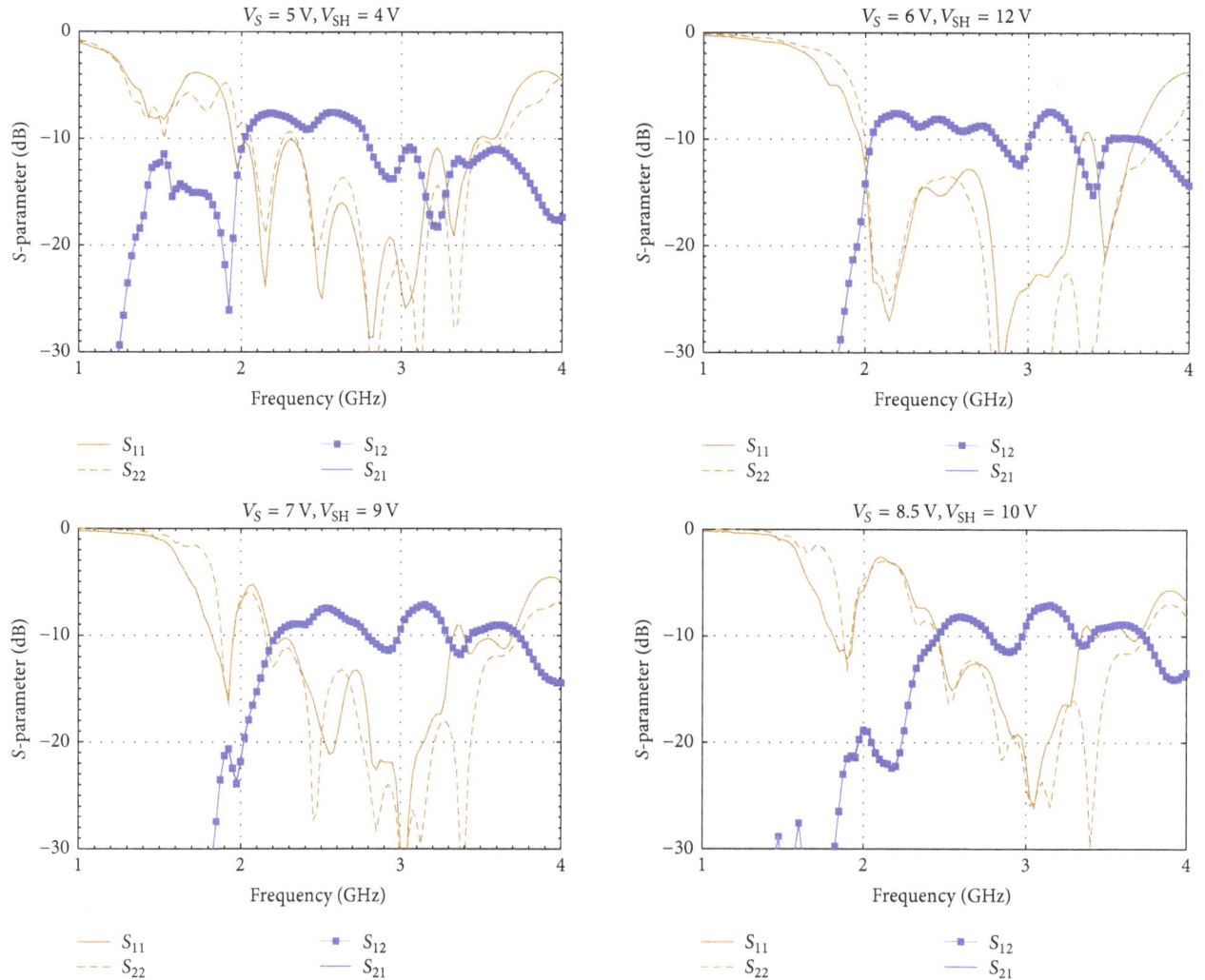

FIGURE 9: Measured S-parameters of the miniaturized CRLH LWA.

broadside direction. The beginning of the flip observed in the $V_S = 5$ V, $V_{SH} = 4$ V curve is due to the space harmonics periodicity of β, which is given by

$$\beta = \beta_0 + \frac{2n\pi}{d}, \tag{3}$$

where β_0 is the lowest order mode phase constant, n is the space harmonics $(0, \pm 1, \pm 2, \ldots)$, and d is the period. Although the continuous biasing of varactor diodes allows for a theoretically infinite number of configurations, in

Table 1, we summarize four significant configurations chosen to achieve uniform beam steering based on the HPBW of each beam. The relative Bloch impedance Z_b and expected beam angle _ are also reported.

The aforementioned results enable the cascading of the miniaturized unit cell to create a leaky-wave antenna for the 2.4 GHz WiFi band. In the next section, we discuss the design of this reconfigurable CRLH LWA made by cascading 11 miniaturized unit cells, with experimental analysis of impedance and radiation characteristics.

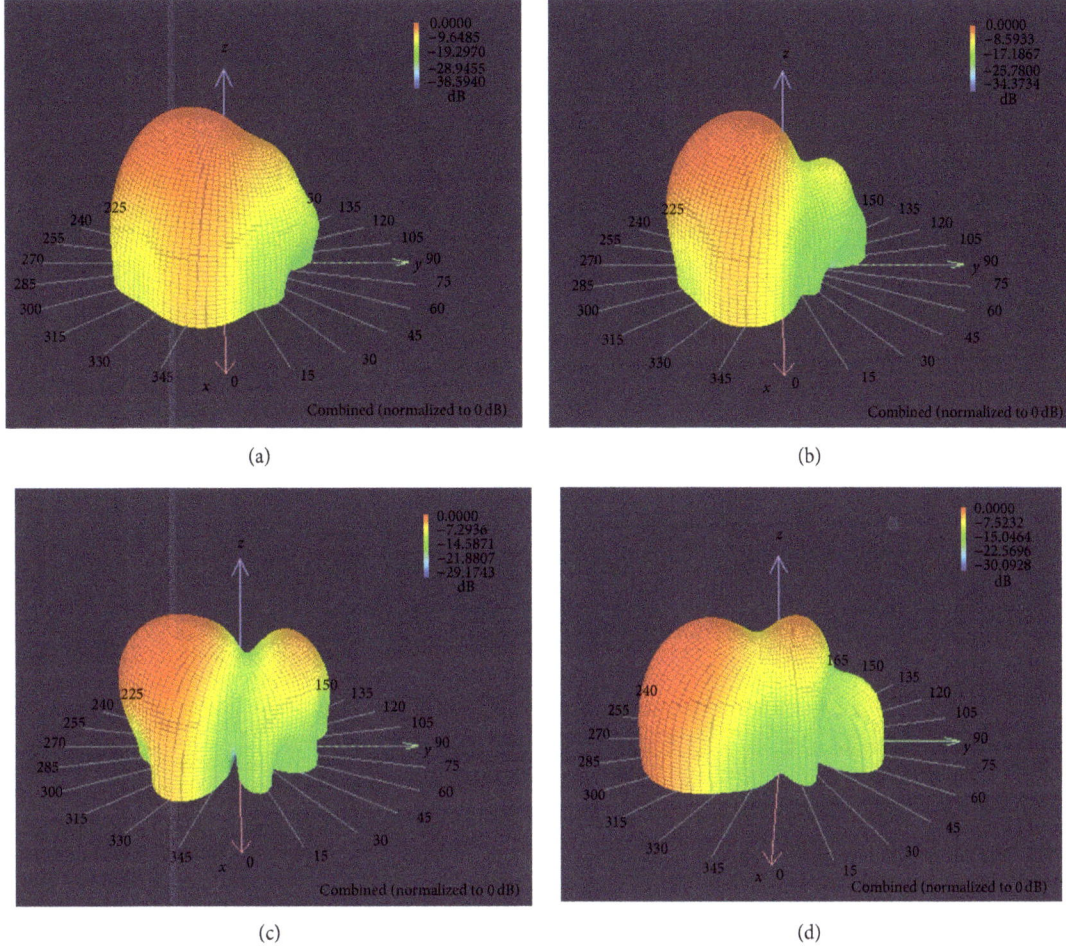

(a)   (b)

(c)   (d)

FIGURE 10: Measured 3D radiation patterns at 2.46 GHz for four configurations. Port 2 oriented in $+y$ direction was connected to a signal generator, while port 1 was terminated to a 50 Ω matched load. (a) $V_S = 8.5$ V, $V_{SH} = 10$ V; (b) $V_S = 7$ V, $V_{SH} = 9$ V; (c) $V_S = 6$ V, $V_{SH} = 12$ V; (d) $V_S = 5$ V, $V_{SH} = 4$ V.

TABLE 1: Summary of four different configurations at the frequency of 2.46 GHz.

Configuration $\{V_S, V_{SH}\}$	Block impedance Z_b	Beam angle θ
{8.5 V, 10 V}	$42 + j8\ \Omega$	21°
{7 V, 9 V}	$37 + j7\ \Omega$	28°
{6 V, 12 V}	$47 + j10\ \Omega$	38°
{5 V, 4 V}	$56 + j9\ \Omega$	55°

4. Miniaturized CRLH Leaky-Wave Antenna

The periodic structure of the miniaturized CRLH LWA was designed by cascading a series of unit cells described in Section 3. As illustrated in Figure 8, the antenna consists of $N = 11$ unit cells and has overall dimension $l = 11.5$ cm and $h = 2.3$ cm. The number of unit cells was selected to achieve positive gain and obtain a fair comparison with the earlier LWA presented in [26]. By switching between the two input ports, the antenna allows for the generation of two independent beams that can be steered from back-fire to

end-fire, with expected beam angles θ estimated during the unit cell analysis.

4.1. Input Impedance. The return loss and the isolation of the two input ports have been measured through a VNA. The S_{11} and S_{22} scattering parameters describe the impedance integrity between the antenna's ports and 50 feedlines, whereas S_{12} and S_{21} render the isolation achievable between them. Figure 9 shows the measured scattering parameters for the four configurations listed in Table 1.

Both input ports exhibit good impedance matching within the 2.41–2.48 GHz band, the small discrepancies between the S_{11} and S_{22} curves are potentially due to the manufacturing process and, in particular, the manual population of the board. We also note that the 10 dB bandwidth is relatively large, between 1 GHz $\leq$ BW $\leq$ 1.3 GHz. In terms of decoupling between the two ports, at 2.46 GHz, the antenna's isolation is within the range of 8 dB $\leq S_{21} \leq$ 10 dB.

4.2. Radiation Pattern. In order to evaluate the radiation characteristics of the proposed antenna and the agreement with the expected angles, we have measured the radiation

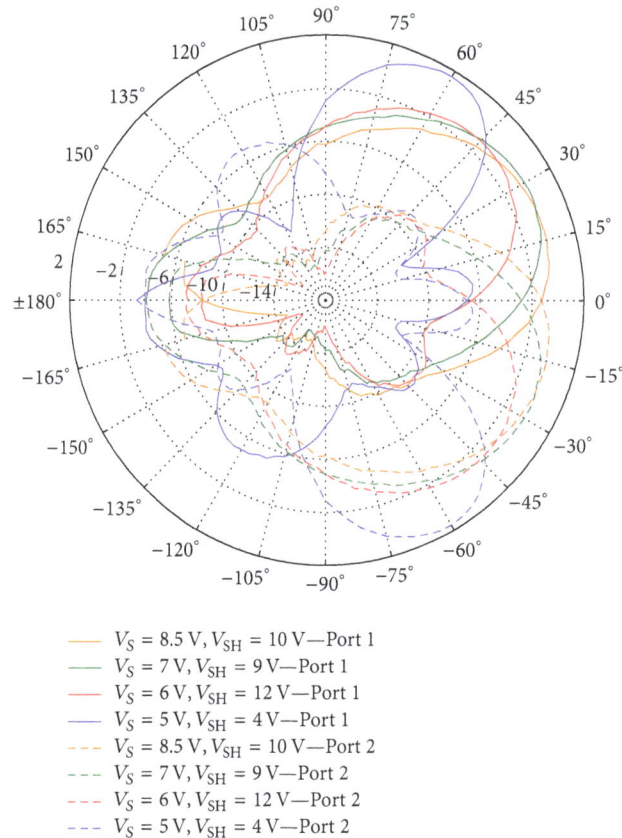

$V_S = 8.5\,\mathrm{V}, V_{SH} = 10\,\mathrm{V}$—Port 1
$V_S = 7\,\mathrm{V}, V_{SH} = 9\,\mathrm{V}$—Port 1
$V_S = 6\,\mathrm{V}, V_{SH} = 12\,\mathrm{V}$—Port 1
$V_S = 5\,\mathrm{V}, V_{SH} = 4\,\mathrm{V}$—Port 1
$V_S = 8.5\,\mathrm{V}, V_{SH} = 10\,\mathrm{V}$—Port 2
$V_S = 7\,\mathrm{V}, V_{SH} = 9\,\mathrm{V}$—Port 2
$V_S = 6\,\mathrm{V}, V_{SH} = 12\,\mathrm{V}$—Port 2
$V_S = 5\,\mathrm{V}, V_{SH} = 4\,\mathrm{V}$—Port 2

Figure 11: Azimuth (x-z) view of the total beam steering capabilities. The solid lines depict beams generated by exciting port 1 (and port 2 terminated to a 50 Ω load), whereas the dashed lines depict beams generated by exciting port 2 (and port 1 terminated to a 50 Ω load).

patterns for the four configurations listed in Table 1. For this purpose, we used the tool EMSCAN RFxpert [35], which is a bench-top measurement system that enables us to get 3D and 2D antenna pattern measurements in real time. Figure 10 shows the 3D antenna directivity graphs measured at 2.46 GHz by exciting port 2 and terminating port 1 to a 50 Ω load. The steering angles are in good agreement with the expected values. The minimum gain is 0 dBi while the peak is about 2 dBi, with front-to-back ratio between 5 and 8 dB, depending on the adopted configuration. The front-to-back ratio of the proposed antenna is suboptimal due to the presence of the defected ground plane. In fact a percentage of the total radiation leaks out from the CSRR openings, resulting in some radiation from the back of the PCB. In general, directional antennas that employ CSRR suffer from poor front-to-back ratio [25], but, in the case of the proposed antenna, the patterns maintain a higher level of directivity. The major losses that limit the gain are the series resistance of the varactor diode R_S and the lossy FR-4 substrate. More expensive substrates can provide much lower loss factors, while the series resistance of varactor diodes could be improved by choosing a smaller package or a more expensive model. Further measurements were conducted in an anechoic chamber and Figure 11 illustrates the azimuth cut (x-z) with the complete set of radiation patterns accomplished by switching between the two input ports. The total steering angle is about 120° and the half-power beamwidth

(HPWB) of each beam, between 40° and 60°, allows for nearly uniform coverage. The measurements in Figure 11 denote good agreement with the expected beam angles listed in Table 1. By comparing the same voltage configurations, the error between the estimated and the measured beam angles is between 0°–5° across all the configurations.

In terms of beams polarization, we observed that the miniaturized CRLH LWA maintains linear polarization across all the configurations, similar to a conventional LWA. In Figure 12 we show the normalized plots of co-pol and cross-pol for four beams at ±60° and ±30°. For all the configurations, the cross-pol is at least 5 dB lower than the co-pol confirming that the radiated fields are linearly polarized.

In terms of radiation efficiency, we have estimated that the average components' losses on each LWA unit cell are approximately 0.6 dB total. Therefore, for the 11-cell LWA presented in this manuscript, the expected radiation efficiency is about 21% including the ports' return loss.

4.3. Comparison with Conventional LWA. In Figure 13, we compare the size of the proposed miniaturized LWA with an earlier conventional design of a LWA [26]. Both antennas were designed by cascading 11 unit cells; however, the miniaturized LWA is about 53% smaller than the conventional LWA. We then conducted a qualitative comparison of the electrical and radiation characteristics to evaluate the

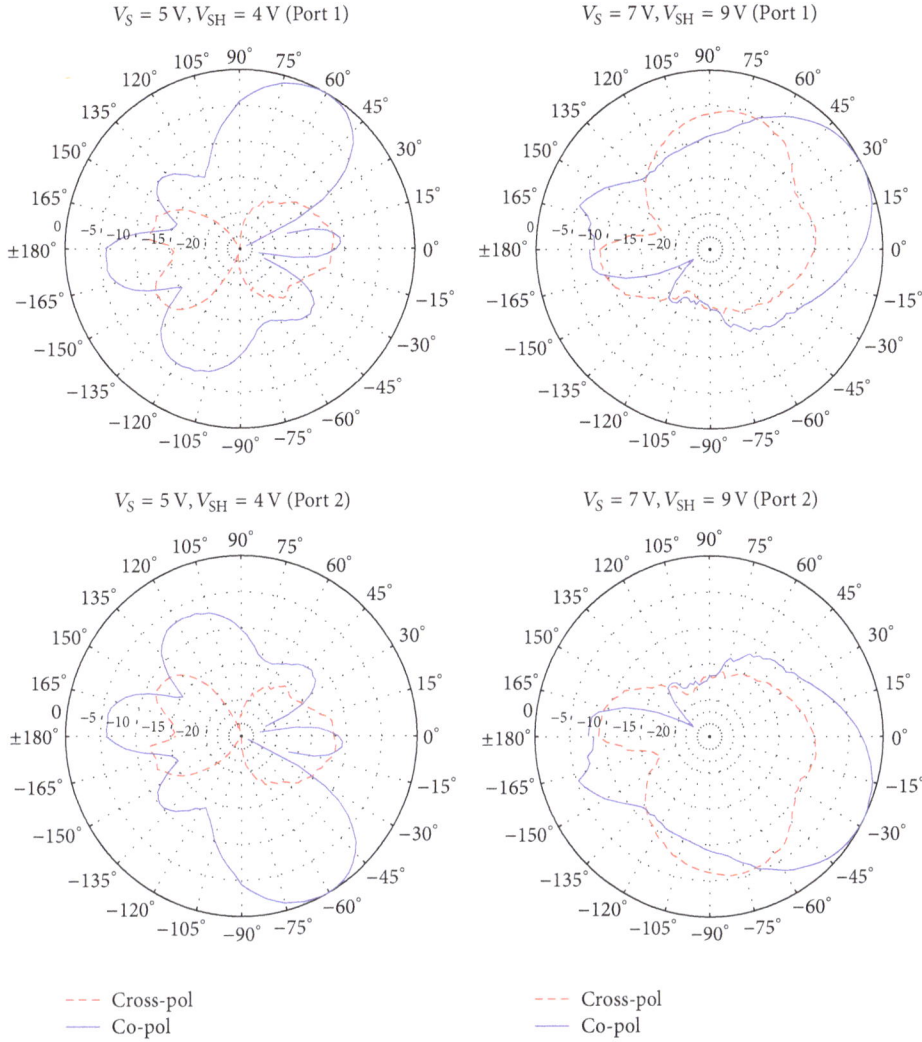

$V_S = 5\,\mathrm{V}, V_{SH} = 4\,\mathrm{V}$ (Port 1)

$V_S = 7\,\mathrm{V}, V_{SH} = 9\,\mathrm{V}$ (Port 1)

$V_S = 5\,\mathrm{V}, V_{SH} = 4\,\mathrm{V}$ (Port 2)

$V_S = 7\,\mathrm{V}, V_{SH} = 9\,\mathrm{V}$ (Port 2)

--- Cross-pol
— Co-pol

--- Cross-pol
— Co-pol

FIGURE 12: Co-pol and cross-pol of the beams at ±60° and ±30°. The miniaturization of the CRLH LWA maintains the linear polarization, with the cross-pol at least 5 dB lower than the co-pol.

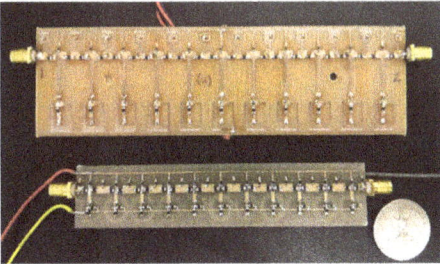

FIGURE 13: Comparison between the conventional and the proposed miniaturized reconfigurable LWA. Both antennas were designed by cascading $N = 11$ unit cells. The former occupies an area of 56 cm^2 while the latter occupies an area of 26.5 cm^2.

TABLE 2: Comparison between conventional and miniaturized LWAs.

	Conventional LWA	Miniaturized LWA
Dimension	56 cm^2	26.5 cm^2
10 dB bandwidth (Max)	30 MHz	1.3 GHz
Isolation (min)	10 dB	8 dB
Peak gain	4 dBi	2 dBi
Front-to-back ratio (avg)	8 dB	7 dB
Beam steering coverage	120°	120°

performance of the proposed miniaturized LWA. A summary is shown in Table 2.

The 10 dB bandwidth of the miniaturized LWA is significantly larger than the conventional model. However, it is important to recall that, due to the frequency dependency, different frequency regions will exhibit different handedness regions (i.e., RH or LH) and thus different steering angles. Moreover, when the dispersion curve approaches the propagation regime, the beam's directivity, and gain degrade.

Due to the smaller dimension, the isolation between the two input ports is lower with respect to the standard model. Although more than the 85% of the energy is radiated and attenuated through the structure, the employment of a single-pole-double-throw (SPDT) switch would allow further decoupling the two ports and switching between them to generate the desired back-fire and end-fire beams. Furthermore, the cascade of additional unit cells can also lead to higher isolation between the ports and increases the radiated gain.

In terms of radiation characteristics, the miniaturized LWA allows for beam steering of about 120° around the azimuth plane, similar to the earlier version. The peak gain is 2 dB lower, but sufficient to utilize the antenna for mobile applications [36]. The front-to-back ratios are comparable, with both antennas performing between 4 and 8 dB, depending on the adopted configuration.

5. Conclusion

In this paper, we presented the design of a miniaturized reconfigurable leaky-wave antenna, where the size reduction was accomplished by etching a Complementary Split-Ring Resonator (CSRR) underneath each unit cell.

The CSRR was designed to decrease the size of an improved design of CRLH unit cell, covering the whole WiFi band from 2.41 GHz to 2.48 GHz. The absence of interdigital capacitors greatly reduces manufacturing constraints for size reduction while also allowing the application of the CSRR miniaturization technique.

Numerical and experimental analyses of the miniaturized unit cell have shown good impedance performance and relatively large variations of the dispersion curves, which leads to large beam steering.

After fine-tuning the unit cell for the desired radiating region and steering angles, the miniaturized leaky-wave antenna has been designed by cascading 11 unit cells. With respect to an equivalent conventional LWA model, the miniaturized antenna is 53% smaller and exhibits a larger 10 dB bandwidth. The radiation patterns were in good agreement with the expected angles, and the total azimuth coverage is about 120° with gains between 0 and 2 dBi.

In conclusion, the technique of etching CSRR on reconfigurable leaky-wave antennas has been shown to be successful for size reduction and maintenance of good radiating performance. The proposed solution enables the development of miniaturized reconfigurable antennas that do not require expensive and customized substrates. In future work, the antenna will be applied on software-defined radios to realize new wireless networking applications exploiting directionality on mobile device platforms.

Conflicts of Interest

The authors declare that there are no conflicts of interest regarding the publication of this paper.

Acknowledgments

This material is based upon work supported by the National Science Foundation under Grant CNS-1422964.

References

[1] J. Kountouriotis, D. Piazza, K. R. Dandekar, M. D'Amico, and C. Guardiani, "Performance analysis of a reconfigurable antenna system for MIMO communications," in *Proceedings of the 5th European Conference on Antennas and Propagation, EUCAP 2011*, pp. 543–547, ita, April 2011.

[2] N. Gulati and K. R. Dandekar, "Learning state selection for reconfigurable antennas: A multi-armed bandit approach," *IEEE Transactions on Antennas and Propagation*, vol. 62, no. 3, pp. 1027–1038, 2014.

[3] R. Bahl, N. Gulati, K. R. Dandekar, and D. Jaggard, "Impact of pattern reconfigurable antennas on Interference Alignment over measured channels," in *Proceedings of the 2012 IEEE Globecom Workshops, GC Wkshps 2012*, pp. 557–562, USA, December 2012.

[4] R. Wang, X. Wang, T. Chow et al., "Capacity and performance analysis for adaptive multi-beam directional networking," in *Proceedings of the Military Communications Conference 2006, MILCOM 2006*, USA, October 2006.

[5] P. Mookiah, J. Kountouriotis, R. Dorsey, B. Shishkin, and K. R. Dandekar, "Securing wireless links at the physical layer through reconfigurable antennas," in *Proceedings of the 2010 IEEE International Symposium on Antennas and Propagation and CNC-USNC/URSI Radio Science Meeting - Leading the Wave, AP-S/URSI 2010*, Canada, July 2010.

[6] R. F. Harrington, "Reactively Controlled Directive Arrays," *IEEE Transactions on Antennas and Propagation*, vol. 26, no. 3, pp. 390–395, 1978.

[7] M. A. Y. Abdalla, K. Phang, and G. V. Eleftheriades, "A planar electronically steerable patch array using tunable PRI/NRI phase shifters," *IEEE Transactions on Microwave Theory and Techniques*, vol. 57, no. 3, pp. 531–541, 2009.

[8] D. Piazza, P. Mookiah, D. Michele, and K. R. Dandekar, "Pattern and polarization reconfigurable circular patch for MIMO systems," in *Proceedings of the 3rd European Conference on Antennas and Propagation, EuCAP 2009*, pp. 1047–1051, deu, March 2009.

[9] Y. J. Sung, T. U. Jang, and Y.-S. Kim, "A reconfigurable microstrip antenna for switchable polarization," *IEEE Microwave and Wireless Components Letters*, vol. 14, no. 11, pp. 534–536, 2004.

[10] D. Patron, K. R. Dandekar, and A. S. Daryoush, "Optical control of reconfigurable antennas and application to a novel pattern-reconfigurable planar design," *IEEE Journal of Lightwave Technology*, no. 99, 2014.

[11] Q. Liu and P. S. Hall, "Varactor-loaded left handed loop antenna with reconfigurable radiation patterns," in *Proceedings of the 2009 IEEE International Symposium on Antennas and Propagation and USNC/URSI National Radio Science Meeting, APSURSI 2009*, USA, June 2009.

[12] D. V. Thiel, "Switched parasitic antennas and controlled reactance parasitic antennas: A systems comparison," in *Proceedings of the IEEE Antennas and Propagation Society Symposium 2004 Digest held in Conjunction with: USNC/URSI National Radio Science Meeting*, pp. 3211–3214, usa, June 2004.

[13] D. V. Thiel, "Impedance variations in controlled reactance parasitic antennas," in *Proceedings of the 2005 IEEE Antennas and Propagation Society International Symposium and USNC/URSI Meeting*, pp. 671–674, USA, July 2005.

[14] S. Lim, C. Caloz, and T. Itoh, "Electronically-controlled metamaterial-based transmission line as a continuous-scanning

leaky-wave antenna," in *Proceedings of the 2004 IEEE MITT-S International Microwave Symposium Digest*, pp. 313–316, usa, June 2004.

[15] B.-J. Che, F.-Y. Meng, J.-H. Fu, K. Zhang, G.-H. Yang, and Q. Wu, "A dual band CRLH leaky wave antenna with electrically steerable beam based on liquid crystals," in *Proceedings of the 17th Biennial IEEE Conference on Electromagnetic Field Computation, IEEE CEFC 2016*, USA, November 2016.

[16] M. Roig, M. Maasch, C. Damm, and R. Jakoby, "Dynamic beam steering properties of an electrically tuned liquid crystal based CRLH leaky wave antenna," in *Proceedings of the 2014 8th International Congress on Advanced Electromagnetic Materials in Microwaves and Optics, METAMATERIALS 2014*, pp. 253–255, Denmark, August 2014.

[17] J. S. Kula, D. Psychoudakis, W.-J. Liao, C.-C. Chen, J. L. Volakis, and J. W. Halloran, "Patch-antenna miniaturization using recently available ceramic substrates," *IEEE Antennas and Propagation Magazine*, vol. 48, no. 6, pp. 13–20, 2006.

[18] P. M. T. Ikonen, K. N. Rozanov, A. V. Osipov, P. Alitalo, and S. A. Tretyakov, "Magnetodielectric substrates in antenna miniaturization: Potential and limitations," *IEEE Transactions on Antennas and Propagation*, vol. 54, no. 11, pp. 3391–3399, 2006.

[19] H. Mosallaei and K. Sarabandi, "Design and modeling of patch antenna printed on magneto-dielectric embedded-circuit meta-substrate," *IEEE Transactions on Antennas and Propagation*, vol. 55, no. 1, pp. 45–52, 2007.

[20] P. Mookiah and K. R. Dandekar, "Metamaterial-substrate antenna array for MIMO communication system," *IEEE Transactions on Antennas and Propagation*, vol. 57, no. 10, pp. 3283–3292, 2009.

[21] G.-C. Wu, G.-M. Wang, H.-X. Peng, X.-J. Gao, and J.-G. Liang, "Design of leaky-wave antenna with wide beam-scanning angle and low cross-polarisation using novel miniaturised composite right/left-handed transmission line," *IET Microwaves, Antennas & Propagation*, vol. 10, no. 7, pp. 777–783, 2016.

[22] D. Ahn, J. Park, C. Kim, J. Kim, Y. Qian, and T. Itoh, "A design of the low-pass filter using the novel microstrip defected ground structure," *IEEE Transactions on Microwave Theory and Techniques*, vol. 49, no. 1, pp. 86–93, 2001.

[23] A. Anand, A. Bansal, K. Shambavi, and Z. C. Alex, "Design and analysis of microstrip line with novel defected ground structure," in *Proceedings of the 2013 International Conference on Advanced Nanomaterials and Emerging Engineering Technologies, ICANMEET 2013*, pp. 670–673, India, July 2013.

[24] Y. Xie, L. Li, C. Zhu, and C. Liang, "A novel dual-band patch antenna with complementary split ring resonators embedded in the ground plane," in *Progress In Electromagnetics Research Letters*, vol. 25, pp. 117–126, 2011.

[25] M. S. Sharawi, M. U. Khan, A. B. Numan, and D. N. Aloi, "A CSRR loaded MIMO antenna system for ISM band operation," *IEEE Transactions on Antennas and Propagation*, vol. 61, no. 8, pp. 4265–4274, 2013.

[26] D. Patron, H. Paaso, A. Mammela, D. Piazza, and K. R. Dandekar, "Improved design of a CRLH leaky-wave antenna and its application for DoA estimation," in *Proceedings of the 2013 3rd IEEE-APS Topical Conference on Antennas and Propagation in Wireless Communications, IEEE APWC 2013*, pp. 1343–1346, September 2013.

[27] C. Caloz and T. Itoh, *Electromagnetic metamaterials transmission line theory and application*, John Wiley Sons, Hoboken, NJ, 2006.

[28] D. Piazza, D. Michele, and K. R. Dandekar, "Two port reconfigurable crlh leaky wave antenna with improved impedance matching and beam tuning," in *Proceedings of the 3rd European Conference on Antennas and Propagation, EuCAP 2009*, pp. 2046–2049, deu, March 2009.

[29] *Ansoft HFSS User Manual*, Ansoft Corp., Pittsburg, PA, USA, 2005.

[30] *Getting Started with Ansoft Designer*, Ansoft Corp., Pittsburg, Pa, USA, 2003.

[31] N. Stollon, *On-Chip Instrumentation*, Springer US, Boston, MA, 2011.

[32] F. Falcone, T. Lopetegi, J. D. Baena, R. Marqués, F. Martín, and M. Sorolla, "Effective negative −/spl epsiv/stopband microstrip lines based on complementary split ring resonators," *IEEE Microwave and Wireless Components Letters*, vol. 14, no. 6, pp. 280–282, 2004.

[33] H. V. Nguyen, S. Abielmona, and C. Caloz, "Performance-enhanced and symmetric full-space scanning end-switched CRLH LWA," *IEEE Antennas and Wireless Propagation Letters*, vol. 10, pp. 709–712, 2011.

[34] A. Mehdipour, T. A. Denidni, and A. Sebak, "Multi-band miniaturized antenna loaded by ZOR and CSRR metamaterial structures with monopolar radiation pattern," *IEEE Transactions on Antennas and Propagation*, vol. 62, no. 2, pp. 555–562, 2014.

[35] "RFxpert User Manual," EMSCAN, 2013, http://www.emscan.com.

[36] K.-L. Wong and P.-J. Ma, "Small-size internal antenna for LTE/WWAN operation in the laptop computer," in *Proceedings of the International Conference on Applications of Electromagnetism and Student Innovation Competition Awards, AEM2C 2010*, pp. 152–156, Taiwan, August 2010.

3

A New Digital to Analog Converter based on Low-Offset Bandgap Reference

Jinpeng Qiu, Tong Liu, Xubin Chen, Yongheng Shang, Jiongjiong Mo, Zhiyu Wang, Hua Chen, Jiarui Liu, Jingjing Lv, and Faxin Yu

School of Aeronautics and Astronautics, Zhejiang University, No. 38 Zheda Road, Hangzhou 310027, China

Correspondence should be addressed to Yongheng Shang; yh_shang@zju.edu.cn

Academic Editor: Muhammad Taher Abuelma'atti

This paper presents a new 12-bit digital to analog converter (DAC) circuit based on a low-offset bandgap reference (BGR) circuit with two cascade transistor structure and two self-contained feedback low-offset operational amplifiers to reduce the effects of offset operational amplifier voltage effect on the reference voltage, PMOS current-mirror mismatch, and its channel modulation. A Start-Up circuit with self-bias current architecture and multipoint voltage monitoring is employed to keep the BGR circuit working properly. Finally, a dual-resistor ladder DAC-Core circuit is used to generate an accuracy DAC output signal to the buffer operational amplifier. The proposed circuit was fabricated in CSMC 0.5 μm 5 V 1P4M process. The measured differential nonlinearity (DNL) of the output voltages is less than 0.45 LSB and integral nonlinearity (INL) less than 1.5 LSB at room temperature, consuming only 3.5 mW from a 5 V supply voltage. The DNL and INL at −55°C and 125°C are presented as well together with the discussion of possibility of improving the DNL and INL accuracy in future design.

1. Introduction

Along with the development of the semiconductor technology, phased array radar system has become more and more attractive in the area of space communication. However, due to the source limitation in space environment, it constrains that the design of the transmit/receive (TR) module has to have many control ability to switch between transmit and receive mode. This has motivated the design of high performance control chip with respect to accuracy, linearity, and stability under harsh space environment [1–4].

The duty of the control chip is to deliver different bias voltage to corresponding chip by following instructions from command computer. Therefore, its performance is highly relay on the design of its digital to analogous converter (DAC). There are many types of DAC structures that have been proposed in the literature [5–12]. Resistor based DAC architecture is one of the most attractive due to its simplicity structure, high stability, high monotonic, and low power consumption [12] and is a good choice for space application. However, for resistor type of DAC architecture, the DAC's number of bits is proportional to the number of resistor used

in the circuit. The higher number of resistors further leads to larger of chip area. It also faces the increasing of instability due to the effectiveness of processing technology on the larger number of resistors. In order to solve such problems, the researchers have developed many types of architecture such as subranging [5], dual-ladder resistor [6], resistor-floating-resistor-string [7], embedded operational amplifier [8], and multibits calibration [9]. Nonetheless, all those methods are able to improve some the DAC circuit in different ways; there are still more space that can be improved, especially in the application of space environment. This has motivated the introduction of low-offset bandgap reference (BGR) structure [10, 11]. The good side is that the BGR circuit is able to provide a reference voltage to the DAC circuit for improving the stability of DAC circuit for a normal application, but for the harsh environmental application such as space application, the BGR circuit needs to be improved further to cope with the application.

In this work, we have introduced a new development of a 12-bit high precision, low power consumption, and high stability DAC circuit based on an improved low-offset BGR

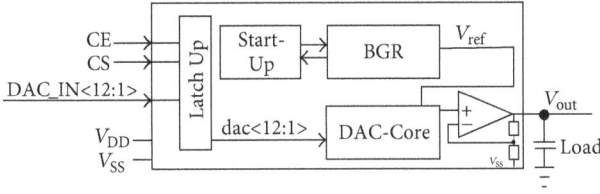

FIGURE 1: Block diagram of the proposed DAC circuit.

and dual-resistor ladder circuit. Three main blocks have been optimized to achieve the design constrains.

2. DAC Overall Circuit Design

The proposed circuit design mainly includes three parts: BGR block, Start-Up Block, and DAC-Core Block as illustrated in Figure 1. The input signal of the proposed DAC includes a pair of latch up signal "CE" and "CS"; it decides which input DAC instruction is stored into the "Latch Up" register. The "DAC IN<12:1>" is a 12-bit input instruction from an upper-computer which will be transferred to a set of bias voltages "V_{out}" and act on the following chipset in a TR module. "V_{DD}" and "V_{SS}" correspond to "+5 V" and "0 V" and are the power supply of the DAC. The reference voltage from BGR circuit is $V_{ref} = 2.5$ V. The end stage operational amplifier is a two-stage PMOS class-AB output buffer with a 0.2 V to 4.8 V output voltages.

3. The BGR Circuit Design

The aim of the BGR circuit is to provide a stable reference voltage "V_{ref}" to the "DAC-Core" circuit without the effects from variation of power supply and environment condition such as temperature and radiation. There are two ways that have been employed in this paper to improve the stability of "V_{ref}." One is the use of two-cascade transistor structure; the second is the use of two low-offset operational amplifier with self-contained feedback circuit. They are detailed as follows.

First, let us start with analysis of the traditional BGR circuit illustrated in Figure 2 [12]; the output voltage "V_{ref}" can be written as

$$V_{ref} = V_{be\text{-}Q_2} + \left(1 + \frac{R_2}{R_1}\right)\left(V_t \ln(n) - V_{OS}\right), \qquad (1)$$

where "$V_{be\text{-}Q_2}$" is base-emitter voltage for Q_2 (≈ 0.7 V), "V_t" is the electron thermal voltage (≈ 0.026 V), "V_{OS}" is the input offset voltage for operational amplifier, "n" is the ratio between number of Q_2 and number of Q_1. From (1), we know that the most troublesome part is "V_{OS}." If we can reduce the effect of "V_{OS}," then we have found a way of stabilize the overall DAC circuit. This is done by using a low-offset BGR circuit which is shown in Figure 3. This architecture employs two-cascade transistors to minimize the contribution of "V_{OS}" on "V_{ref}."

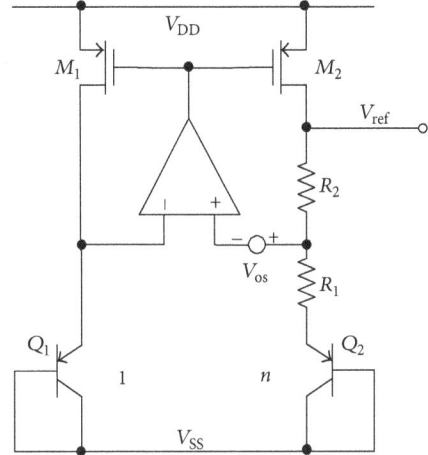

FIGURE 2: The traditional BGR circuit.

From Figure 3, the output voltage of the improved low-offset BGR circuit is written as

$$V_{ref} = 2V_{be\text{-}Q_3Q_4} + \left(1 + \frac{R_2}{R_1}\right)\left(2V_t \ln(n) - V_{OS}\right), \qquad (2)$$

where "$V_{be\text{-}Q_3Q_4}$" is the base-emitter voltage for transistor "Q_3" and "Q_4". Comparing (1) and (2), the effectiveness of "V_{OS}" to the "V_{ref}" is reduced by increasing the value of "$2V_{be\text{-}Q_3Q_4}$" and "$2V_t \ln(n)$." However, the value of "V_{OS}" is untouched from (2), which still causes problems. Therefore, in order to reduce the effect of "V_{OS}" further, two low-offset operational amplifiers (see Figure 4) with self-contained feedback circuit are employed in here to reduce the effects of PMOS current-mirror mismatch and its channel modulation without the decrease of supply power redundancy. It further leads to the reducing of "V_{OS}" value of the overall BGR circuit.

4. Start-Up Circuit Design

The Start-Up circuit has a self-bias and multipoint monitor structure which provides a self-bias current to the BGR circuit. As demonstrated in Figure 5, when the power of the DAC is switched on, points A and B will have a high voltage level; "V_{ref1}" and "V_{ref2}" are at low voltage level. The current at BGR circuit is "0." The MOSFET M7–M10 are "off." M5 and M6 are "on." Current is injected into BGR circuit and the voltage level starts to drop at points A and B at the same time. Following this process, M1 and M2 are switched on which indicates the BGR circuit is fully functioning. M7 and M8 mirror the BGR circuit current "ib." At this point, M5 and M6 are switched off, and the Start-Up circuit changes to low power consumption state. The voltages at "V_{ref1}," "V_{ref2}," and point A are monitored during the DAC working state. Once the voltage at those three points is affected by the environment condition, they will wake up the Start-Up circuit to keep the BGR circuit working properly.

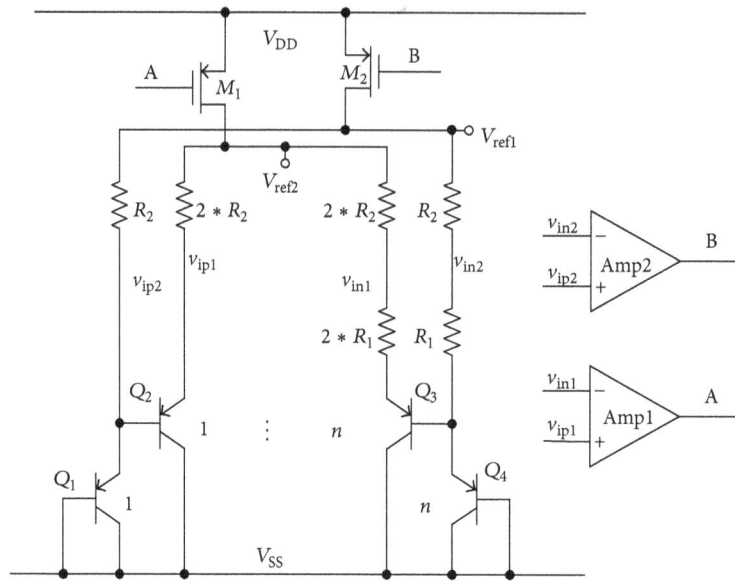

FIGURE 3: Improved low-offset BGR circuit.

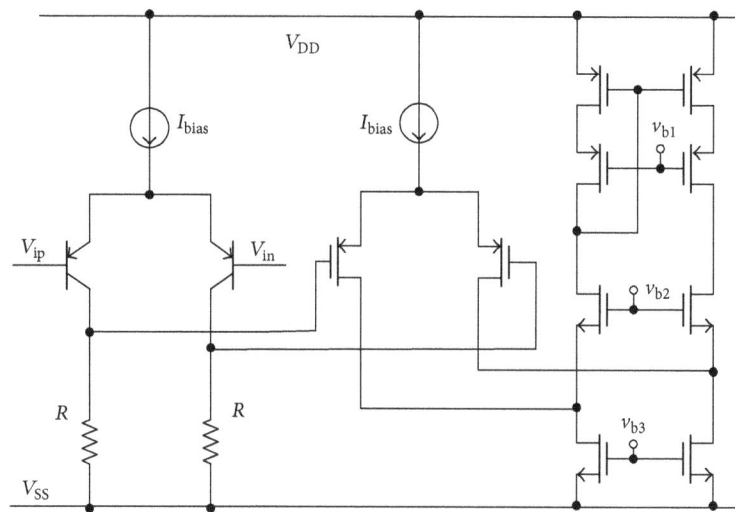

FIGURE 4: Low-offset operational amplifier circuit.

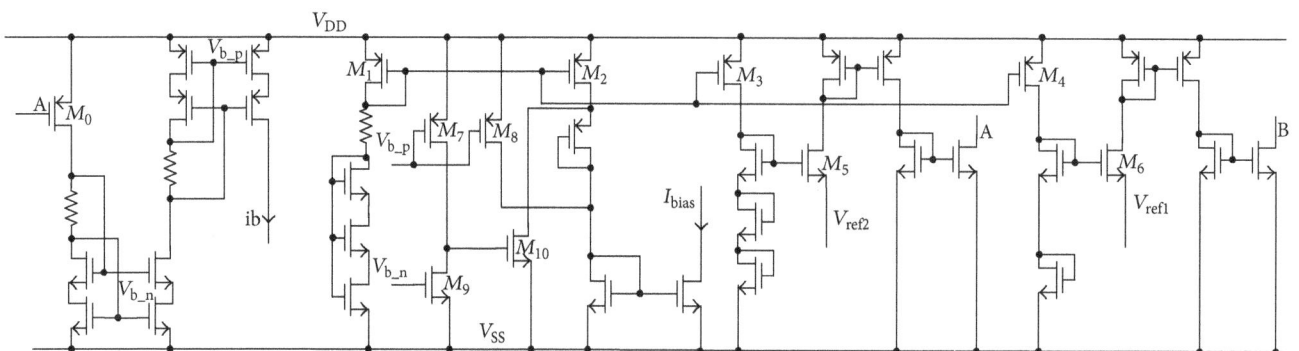

FIGURE 5: Start-Up circuit with self-bias current source.

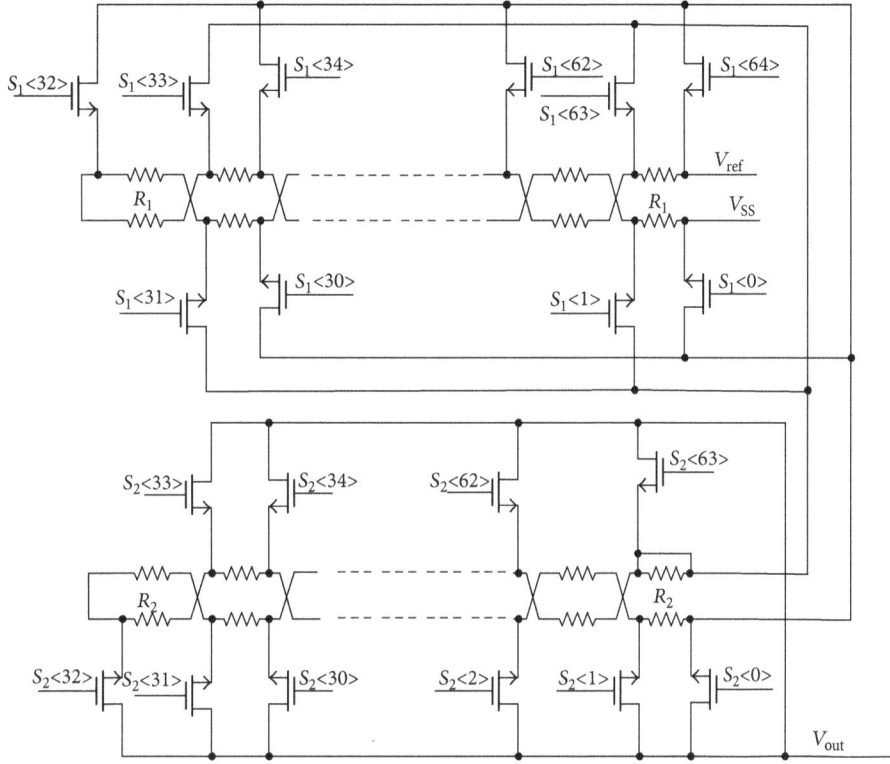

FIGURE 6: Dual ladder resistor string circuit.

5. DAC-Core Circuit Design

The design of the DAC-Core circuit is based on resistor structure to achieve high output monotonic and stability with simply layout and low power dissipation. A two stage dual-ladder resistor string circuit structure is used in here to trade off between the accuracy and chip size. R_1 and R_2 are the voltage splitter at the first and second stage. There are two switches S_1 and S_2. They have been controlled by DAC IN<12:7> for S_1 and DAC IN<6:1> for S_2.

The main problem of such design is that when S_1 is in action, the fluctuation of its switch on resistance (R_{on}) will result in the voltage value change for the second stage resistors. This leads to the lower of the DAC output accuracy. In order to solve this problem, we have adjusted the size of the 65 switches to achieve a full match, low variation of switch on resistance. This is done as follows. First, when the switches in the first stage are switched on in pair, the connection point voltage can be written as "V_r" and "V_p" (see Figure 7); then the effective resistance between those two points is written as

$$R_{Parallel} = R_1 \parallel (2R_{on} + 63R_2) = \frac{R_1 (63R_2 + 2R_{on})}{R_1 + 63R_2 + 2R_{on}}. \quad (3)$$

Before switching,

$$V_1 = \frac{R_{par} V_{ref}}{63R_1 + R_{par}} \times \frac{63R_2 + R_{on}}{63R_2 + 2R_{on}} + V_r. \quad (4)$$

After switching,

$$V_2 = \frac{R_1 + R_{par} \times (R_{on} / (63R_2 + 2R_{on}))}{63R_1 + R_{par}} V_{ref} + V_r. \quad (5)$$

The DAC output variation before and after the switching can be written as

$$\begin{aligned} V_{step1} &= V_2 - V_1 \\ &= \frac{R_1 - (63R_2 R_{par} / (63R_2 + 2R_{on}))}{63R_1 + R_{par}} V_{ref}. \end{aligned} \quad (6)$$

Voltage step size in the second stage before switching is illustrated as

$$V_{step2} = \frac{R_{par} (R_2 / (63R_2 + 2R_{on}))}{63R_1 + R_{par}} V_{ref}. \quad (7)$$

In order to achieve a high accuracy output of the DAC, V_{step1} has to be equal to V_{step2}, which means that

$$R_{on} = \frac{R_2 - R_1}{2}. \quad (8)$$

Due to the fact that MOSFET is working at the linear region under switching on condition, then R_{on} can be rewritten as

$$R_{on} = \frac{1}{\mu_n C_{ox} (W/L) (V_{GS} - V_{TH})}. \quad (9)$$

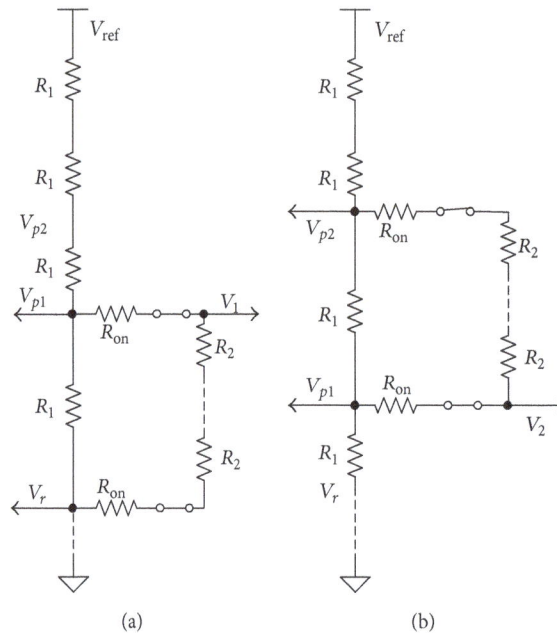

FIGURE 7: The schematic diagram of switching.

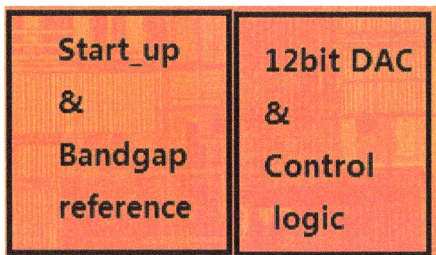

FIGURE 8: Photography of the proposed DAC chip.

Then the width and length ration of the switches in the first stage have to satisfy the following relation:

$$\frac{W}{L} = \frac{1}{R_{\text{on}}\mu_n C_{\text{ox}} (W/L) (V_{\text{GS}} - V_{\text{TH}})}. \qquad (10)$$

Therefore, by tuning the width and length ration of the switches in the first stage, the switching resistance effect on the DAC accuracy is reduced.

6. Test Results and Discussion

The proposed DAC circuit is fabricated by using CSMC 0.5 μm 5 V 1P4M process. Its photography is shown in Figure 8.

At the first instance, the 12-bit digital commands from "000000000000" to "111111111111" are tested at three temperature points, −55°C, 25°C, and 125°C, by using a PXI test system and a probe station. The variation of the output DAC is illustrated in Figure 9. It is almost unchanged under the temperature test which indicates that the proposed DAC is every stable at this point. However, this cannot prove that the

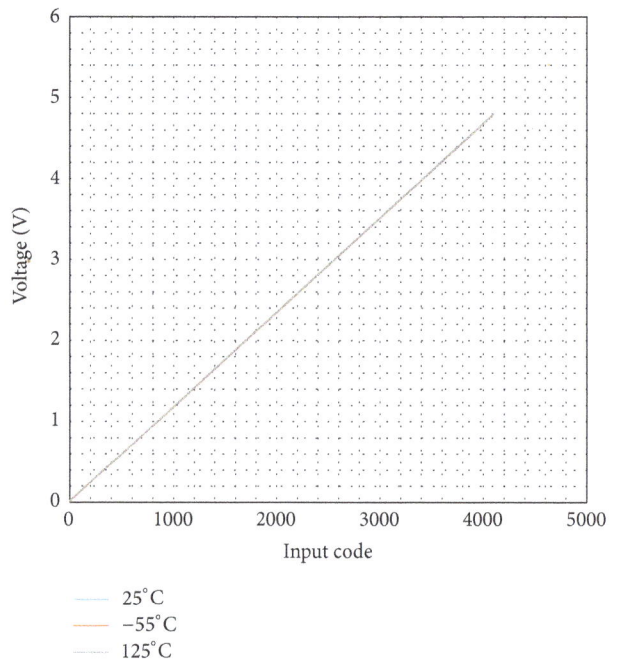

FIGURE 9: DAC output with different input code.

designed DAC are good enough definitely. Therefore, in order to evaluate the DAC performance more deeply, the NDL and INL results are illustrated as follows.

In Figure 10, the DNL verse input command is shown. The overall DNL is less than 0.45 LSB. There are no periodic properties (there is no sudden changes for every 64 commands within total 4096 commands) that have been observed; this indicates that the proposed DAC the step-size accuracy is kept well by the circuit shown in Figure 6.

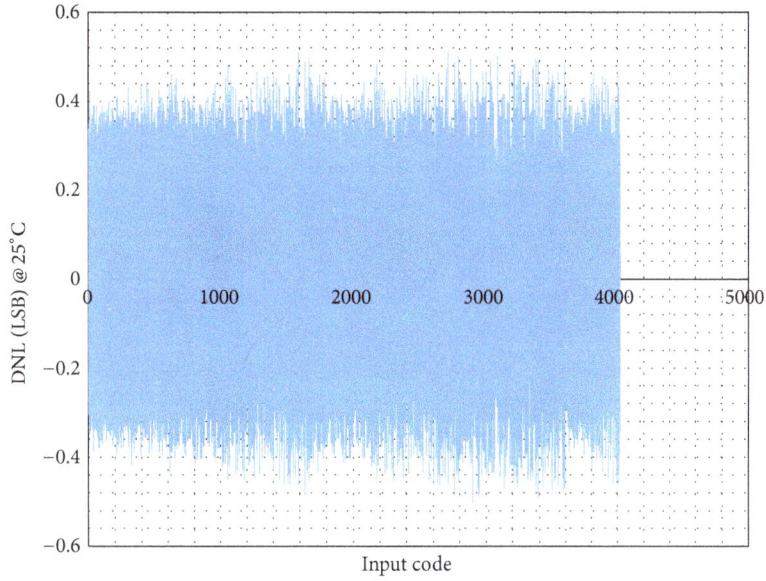

FIGURE 10: DNL at 25°C.

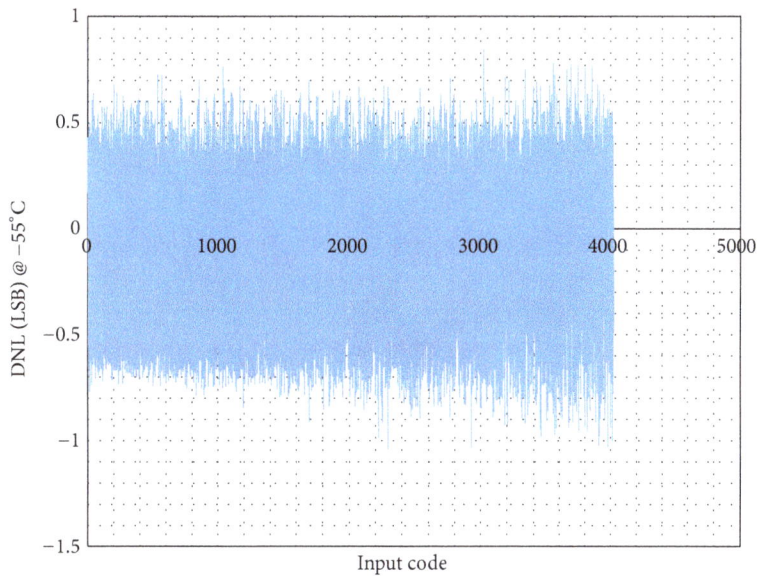

FIGURE 11: DNL at −55°C.

The DNL under −55°C is illustrated in Figure 11. As shown, the DNL is getting worse with a maximum increase of 0.5 LSB. This is mainly due to the fact that the MOSFET has contributed to the power divider circuit in Figure 6. For MOSFET under low temperature condition, the resistance value for each MOSFET highly depends on its bias voltage which varies rapidly. Those variations have effects on the DAC's DNL significantly. In case of high temperature at 125°C, the MOSFET has been less affected by the temperature; therefore, it results in a better DNL variation compared with the low temperature. This conclusion is illustrated in Figure 12. As illustrated, the DNL variation is less than 0.15 LSB with respect DNL value at 25°C.

The variations of DNL at low temperature are very harmful characteristics for DAC design; they could cause a code jumping when DNL is greater than 1. However, they can be reduced by adding more resistor ladders to reduce the effects of the MOSFET switching resistance. The downside is the increasing of chip size. Therefore, tradeoff between the output accuracy and chip size has been considered during DAC design.

The INL versa input commands are shown in Figure 13. The INL value is less than 1.5 LSB. The shape of the INL curve is not monotonous; it has a "sine wave"-like shape, which indicates the INL has no accumulative error for the full command swap.

In Figures 14 and 15, the INL at −55°C and 125°C are shown. The shapes of the INL are preserved, but the INL value has increased up to 2.5 LSB at 125°C. This is because, during the wafer processing, the concentration for producing

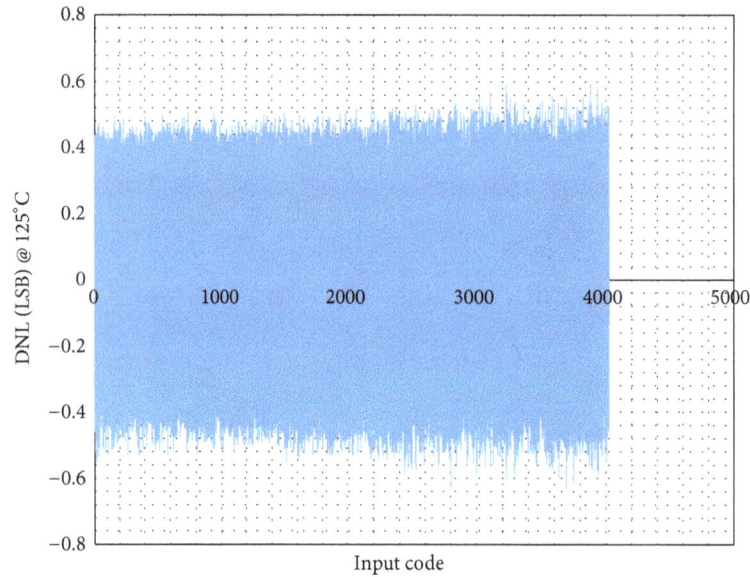

FIGURE 12: DNL at 125°C.

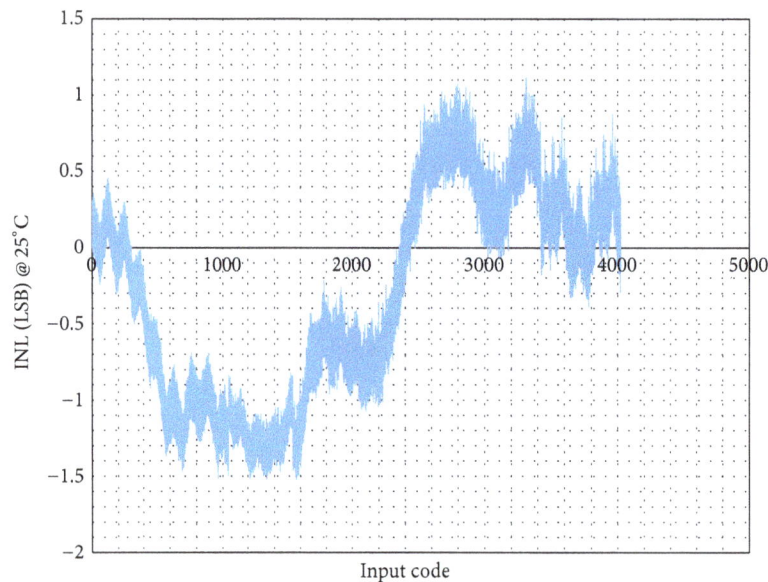

FIGURE 13: INL at 25°C.

the resistor has a huge effect on the quality of the resistors, and the resistors are sensitive to the high temperature as well compared with the low temperature. This results in that the INL has been getting worse along with the increase of temperature. Another source of uncertainty is that the switch on resistance of the MOSFET is highly affected by the temperature which leads to a poor INL performance. Those can be explained by using a resistors mismatch model illustrated in Figure 16.

As shown in Figure 16, during the wafer process, there will be a resistance gradient which results in a resistance change. This can be demonstrated by introducing ΔR to represent the resistance changes with respect to wafer process (in here, the resistance is modeled as a linear incasing of resistance form

the bottom of the resistance ladder to the top). Therefore, the resistance mismatch has created the change of V_{out} under different temperature and leads to the INL and DNL change under temperature.

In case of DNL, it is highly affected by the voltage step size V_{step} as

$$\text{DNL}(k) = \frac{V_{step}(k) - V_{step}(\text{ideal})}{V_{step}(\text{ideal})}. \quad (11)$$

Then the error can be written as follows where $(V_p - V_r)$ is related to the Bandgap reference voltage V_{ref} and ΔR_1:

$$V_{step,error} = (V_p - V_r)\frac{|\Delta R_2|}{63R_2 + 2R_{on}}. \quad (12)$$

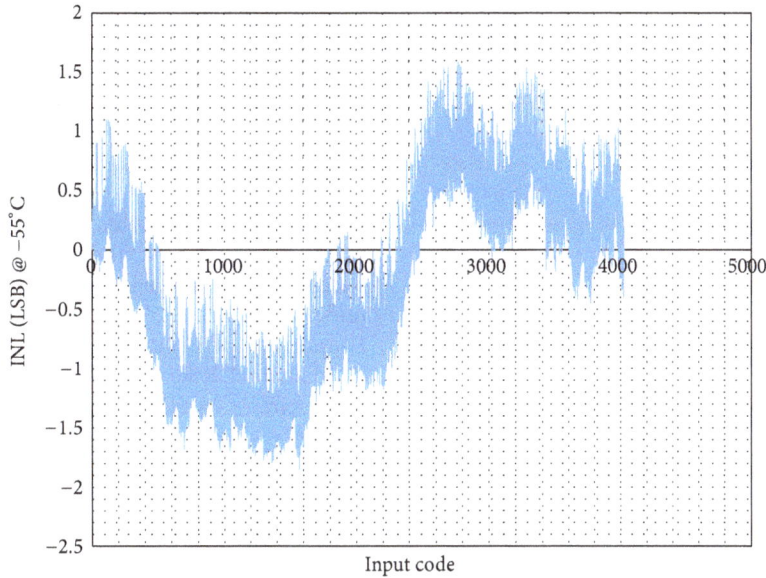

FIGURE 14: INL at −55°C.

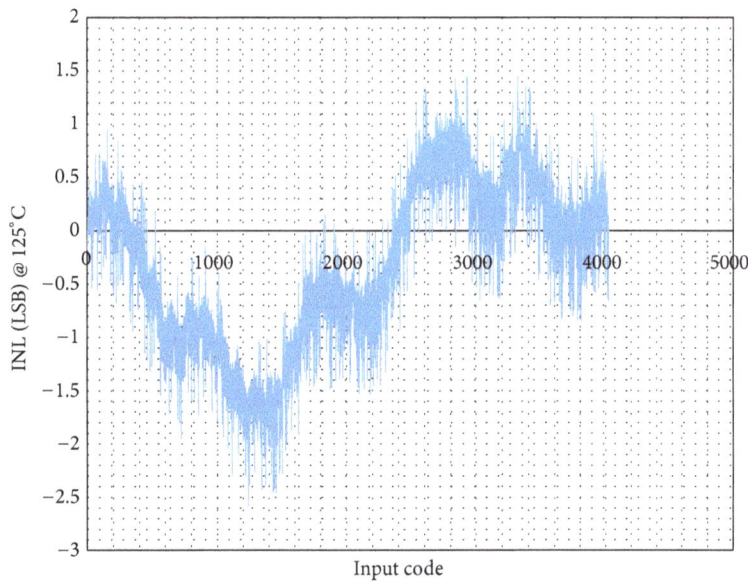

FIGURE 15: INL at 125°C.

Let k_1 and k_2 represent the temperature coefficient of $R_1 (R_2)$ and R_{on}, respectively. The equation of $V_{step,error}$ can be approximated as

$$V_{step,error} = \left(V_p - V_r \right)$$
$$\cdot \frac{\beta}{63 + 2 \left(R_{on} \left(1 + k_2 \Delta T \right) / R_2 \left(1 + k_1 \Delta T \right) \right)}, \quad (13)$$

where β represents the resistance mismatch ratio, $\beta = \Delta R_2 / R_2$, and ΔT represents the change of temperature from room temperature. Then, the step error is ultimately affected by V_{ref}, resistance, and its temperature coefficient. Combined with the design in this paper, we can draw the following

conclusion: when the temperature becomes high, $V_{step,error}$ becomes larger and the DNL performance decreases.

In case of INL, it is the error between the theoretical voltage and actual voltage which can be expressed as

$$\text{INL}(k) = \frac{V(k) - V_{ideal}(k)}{\text{LSB}} = \sum_{n=0}^{k} \text{DNL}(n). \quad (14)$$

As illustrated in Figure 16, when the temperature changes, all the node voltage in the first ladder will change, while the effect of divider resistance will decrease and the absolute value will deviate from the theoretical value, resulting in INL performance decline. On the other hand, as can be seen from the definition of INL, it can be regarded as the accumulation

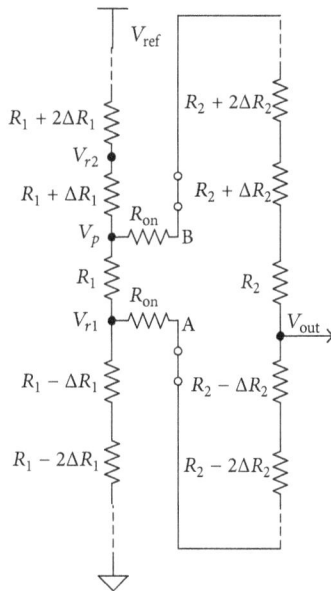

FIGURE 16: The model of resistors mismatch.

of DNL, so the mechanism of temperature influence on DNL is also applicable to the analysis of INL. The same conclusion can be realized that INL performance is decreased under high temperature.

This can be improved by increasing the resistors value at the second ladder and reducing the resistors value at the first ladder at the same time. This method will reduce the effects of the switching on resistance from the MOSFET. However, we have to be alarmed that the power consumption will increase along with the increasing of the resistor value at the second ladder.

7. Conclusion

In conclusion, a new 12-bit digital to analog converter (DAC) circuit based on a low-offset BGR circuit was presented. In order to improve the stability of the BGR circuit, two cascade transistor structures and two self-contained feedback low-offset operational amplifiers are employed to reduce the effects of offset operational amplifier voltage effect on the reference voltage, PMOS current-mirror mismatch, and its channel modulation. A Start-Up circuit with self-bias current architecture and multipoint voltage monitoring was also used to keep the BGR circuit working properly. The DAC-Core circuit with a dual-resistor ladder structure is employed to produce an accuracy output signal. The INL and DNL results of the proposed circuit using CSMC 0.5 μm 5 V 1P4M process at $-55°$C, $25°$C, and $125°$C are presented. The measured DNL of the output voltages is less than 0.45 LSB and INL is less than 1.5 LSB at room temperature, consuming only 3.5 mW from a 5 V supply voltage. The discussion of improving the INL and DNL in the future design is detailed as well to provide a vision of high accuracy, low power consumption, and high stability DAC design.

Competing Interests

The authors declare that there is no conflict of interests regarding the publication of this paper.

Acknowledgments

This work was supported by the National Natural Science Foundation of China under Grant 61401395, the Scientific Research Fund of Zhejiang Provincial Education Department under Grant Y201533913, Zhejiang Provincial Natural Science Foundation of China under Grant LY14F020024, and the Fundamental Research Funds for the Central Universities under Grants 2016QNA4025 and 2016QNA81002.

References

[1] G. M. Rebeiz, K. J. Koh, T. Yu et al., "Highly dense microwave and millimeter-wave phased array T/R modules using CMOS and SiGe RFICs," in *Proceedings of the IEEE 12th Annual Wireless and Microwave Technology Conference (WAMICON '11)*, pp. 1–5, April 2011.

[2] A. Bettidi, D. Carosi, F. Corsaro, L. Marescialli, P. Romanini, and A. Nanni, "MMIC Chipset for wideband multifunction T/R module," in *Proceedings of the IEEE MTT-S International Microwave Symposium (IMS '11)*, Baltimore, Md, USA, June 2011.

[3] F. W. Sexton, "Destructive single-event effects in semiconductor devices and ICs," *IEEE Transactions on Nuclear Science*, vol. 50, no. 3, pp. 603–621, 2003.

[4] F. Faccio, K. Kloukinas, A. Marchioro et al., "Single event effects in static and dynamic registers in a 0.25/spl mu/m CMOS technology," *IEEE Nuclear Science Transactions*, vol. 46, no. 6, pp. 1434–1439, 1999.

[5] A. G. F. Dingwall and V. Zazzu, "An 8-MHz CMOS subranging 8-bit A/D converter," *IEEE Journal of Solid-State Circuits*, vol. 20, no. 6, pp. 1138–1143, 1985.

[6] C.-W. Lu, C.-M. Hsiao, and P.-Y. Yin, "A 10-b two-stage DAC with an area-efficient multiple-output voltage selector and a linearity-enhanced DAC-embedded op-amp for LCD column driver ICs," *IEEE Journal of Solid-State Circuits*, vol. 48, no. 6, pp. 1475–1486, 2013.

[7] C.-W. Lu, P.-Y. Yin, C.-M. Hsiao, M.-C. F. Chang, and Y.-S. Lin, "A 10-bit resistor-floating-resistor-string DAC (RFR-DAC) for high color-depth LCD driver ICs," *IEEE Journal of Solid-State Circuits*, vol. 47, no. 10, pp. 2454–2466, 2012.

[8] J.-S. Kang, J.-H. Kim, S.-Y. Kim et al., "10-bit driver IC using 3-bit DAC embedded operational amplifier for spatial optical modulators (SOMs)," *IEEE Journal of Solid-State Circuits*, vol. 42, no. 12, pp. 2913–2922, 2007.

[9] K. Parthasarathy, T. Kuyel, Y. Zhongjun, C. Degang, and R. Geiger, "A 16-bit resistor string DAC with full-calibration at final test," in *Proceedings of the IEEE International Test Conference (ITC '05)*, Austin, Tex, USA, November 2005.

[10] C. M. Andreou, S. Koudounas, and J. Georgiou, "A novel wide-temperature-range, 3.9 ppm/◦C CMOS bandgap reference circuit," *IEEE Journal of Solid-State Circuits*, vol. 47, no. 2, pp. 574–581, 2012.

4

Development of an Integrated Cooling System Controller for Hybrid Electric Vehicles

Chong Wang,[1] Qun Sun,[1] and Limin Xu[2]

[1]*School of Mechanical and Automotive Engineering, Liaocheng University, Liaocheng, China*
[2]*School of International Education, Liaocheng University, Liaocheng, China*

Correspondence should be addressed to Chong Wang; cwang1@126.com

Academic Editor: Ephraim Suhir

A hybrid electrical bus employs both a turbo diesel engine and an electric motor to drive the vehicle in different speed-torque scenarios. The cooling system for such a vehicle is particularly power costing because it needs to dissipate heat from not only the engine, but also the intercooler and the motor. An electronic control unit (ECU) has been designed with a single chip computer, temperature sensors, DC motor drive circuit, and optimized control algorithm to manage the speeds of several fans for efficient cooling using a nonlinear fan speed adjustment strategy. Experiments suggested that the continuous operating performance of the ECU is robust and capable of saving 15% of the total electricity comparing with ordinary fan speed control method.

1. Introduction

A hybrid electrical vehicle (HEV) employs both a turbo diesel engine and an electric motor to drive the vehicle in different speed-torque scenarios. An effective thermomanagement system is required to dissipate excessive heat from the engine, intercooler, and the motor to avoid part damage during continuous operations [1]. On buses or coaches the power systems are often mounted in the rear of the vehicles and the radiators are often mounted on one side of the vehicles and cooled by lateral wind pulled in by electrical fans, which consume a large amount of electricity.

Studies on HEV thermomanagement system have been extensive but widely spread in different areas, such as whole vehicle thermomanagement [2–5], pumps and thermostats adjustment [6, 7], and battery thermomanagement issues. Within vehicle thermomanagement strategies, cooling fan speed adjustment is a specific topic and often relates to analysis of radiator, flow, and the air. A recent comprehensive study [2] has already addressed several analytical models although it has not led to final controller prototype.

Till now, many vehicles such as Volkswagen Passat B5 still use two-state on-off cooling fan control method, Volkswagen Polo is able to switch cooling fan between high speed and low speed, and Audi A6 adjust fan speed in a linear manner according to water temperature. Some recent studies paid more attention to the nonlinear engine and radiator thermocharacteristics and the corresponding nonlinear PWM control techniques [6–9], which so far has not yielded mass produced equipment.

For large HEV buses or coaches the power consumption of the cooling fans is significant, which requires more advanced cooling control ECU to perform energy saving and vehicle thermomanagement.

2. The Control System Scheme and Cooling Theories

This study is focused on fan drive and control unit for efficient cooling of the engine, intercooler, and the motor. The system concept can be described in Figure 1. Large diameter fans up to 385 mm and 300 w are mounted over the radiators, intercooler, and the motor. Temperature sensors are attached to these parts to measure temperatures and send signals to the controller ECU. The ECU calculates proper control strategies and the duty cycles of PMW outputs and then drives the fans to the proper speeds. Eventually the system controls the water tank outlet temperature below 85°C, the intercooler temperature below 40°C, and the motor temperature below 70°C.

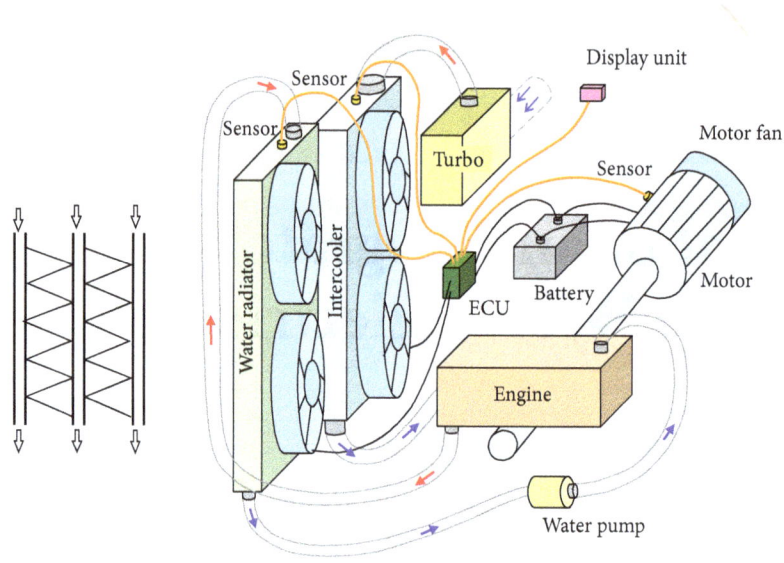

FIGURE 1: The radiator structure and the cooling control system scheme.

According to literatures [10–12], the heat exchanges through a ribbon-tubular radiator can be calculated as the following. The heat transfer area between the coolant and the metal parts of the radiator is

$$F_w = 2 \cdot (L_1 + W_1) \cdot H_1 \cdot N_1, \tag{1}$$

where L_1 and W_1 are the length and width of the pipe cross-section, H_1 is the length of the pipes, and N_1 is the number of pipes. The thermoarea between the metal parts and the air can be expressed as

$$F_A = 2 \cdot \left(2 \cdot \sqrt{HH^2 + \frac{w^2}{4}} + w \right) \cdot H_1 \cdot \frac{LL}{w} \cdot N_2, \tag{2}$$

where HH is wave height and wave distance of the fins, LL is the core thickness, and N_2 is the number of air side heat exchange channels. Another parameter, logarithmic mean temperature deviation, is used to describe the mean temperature difference of two flows during a thermotransformation process of a radiator [13, 14].

$$\Delta t_m = \psi \cdot \frac{(t_{h1} - t_{c2}) - (t_{h2} - t_{c1})}{\ln \left((t_{h1} - t_{c2}) / (t_{h2} - t_{c1}) \right)}, \tag{3}$$

where t_{h1} and t_{h2} are the temperatures of hot flow at the inlet and outlet, t_{c1} and t_{c2} are the temperatures of cold flow at the inlet and outlet, respectively, and ψ is a correction factor ranging from 0.95 to 0.98 according to literature [12]. The amount of thermal exchanges through the radiator can be separately calculated on the first heat dissipation surface around the pipes and the second heat dissipation surface around the fins, as below

$$Q_1 = h_1 \cdot F_1 \cdot \left(t_w - t_f \right)$$
$$Q_2 = h_2 \cdot F_2 \cdot \left(t_m - t_f \right) = h_2 \cdot F_2 \cdot \eta_f \cdot \left(t_w - t_f \right), \tag{4}$$

where Q_1 and Q_2 are the quantities of heat through the first and the second surfaces, h_1 and h_2 are the heat transfer coefficients between the air and the first as well as the second heat dissipation surfaces, F_1 and F_2 are the areas of the first and second heat dissipation surfaces, t_w, t_f, and t_m are the temperatures of the tube walls, the air, and the fin surfaces. The heat distribution along the fins can not be homogeneous; thus a parameter η_f is derived to represent the efficiency of the fins, as below

$$\eta_f = \frac{t_m - t_f}{t_w - t_f}. \tag{5}$$

Based on the above, the total equivalent efficiency can be derived to describe the efficiency using only the pipe temperature t_w and the air temperature t_f, as below

$$\eta_0 = \frac{F_1 + F_2 \cdot \eta_f}{F}. \tag{6}$$

Omitting other derivation details, the amount of heat exchanges in a steady state is given by

$$Q = \frac{1}{1 / (h_c \cdot F_A \cdot \eta_0) + 1 / (h_h \cdot F_W)} \cdot \Delta t_m, \tag{7}$$

where h_c represents the heat transfer coefficient between cold flow and the metal surface and h_h is the heat transfer coefficient between the hot flow and the metal surface. The thermal equilibrium equation can be written as

$$Q = m_c \cdot c_{pc} \cdot (t_{c2} - t_{c1}) \cdot 1000$$
$$= m_h \cdot c_{ph} \cdot (t_{h1} - t_{h2}) \cdot 1000, \tag{8}$$

where m_c and m_h are the quantities of the cold and hot flows and c_{pc} and c_{ph} are constant-pressure specific heat coefficients for the cold and hot flows. Following the above calculations,

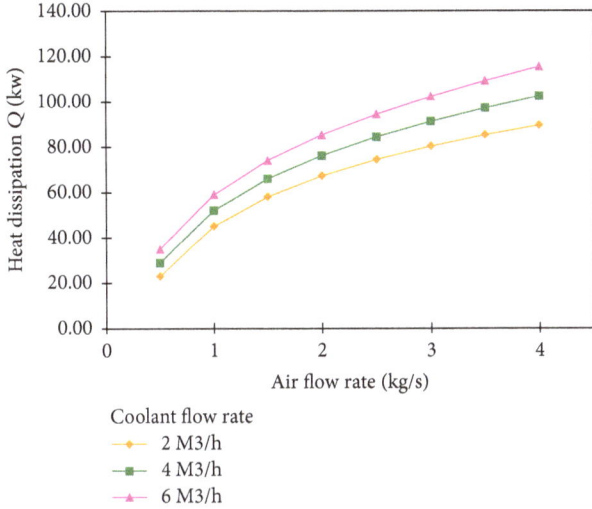

FIGURE 2: Heat dissipation ability of the radiator and fan.

the heat dissipation abilities of an existing radiator can be derived as shown in Figure 2.

Since the air flow quantity is proportional with fan speed [15], the above graph shows the nonlinear heat dissipation abilities that may be fitted by logarithm functions. Since the outlet temperature can not be adjusted again, the controller should use the inlet temperature as an input and then adjust the fan speed to achieve proper heat dissipation in the radiator, so as to get the proper temperature at the outlet from where the coolant goes into the engine. The above analysis indicates a nonlinear fan speed control strategy.

3. Control Strategies

The control strategies are set to use minimal fan electricity to achieve the expected temperature targets, that is, water tank outlet 85°C, intercooler outlet 40°C, and motor 70°C. When a vehicle firstly starts, the engine needs to warm up to the most efficient working temperature around 60°C; until this the water tank fan is not needed. When water tank inlet temperature is above 60°C the fan starts working with 30% PWM duty ratio and rises to 100% duty ratio when inlet temperature reaches 95°C. Between 60°C and 95°C, the fan speeds according to curving fitting with Figure 2 can be summarized into a logarithm function as below:

$$\lambda_1 = \frac{\ln(T-60)}{\ln(95-60)}. \tag{9}$$

The intercooler fans start running when the inlet air temperature is above 40°C and the fan speed can be controlled following a similar function:

$$\lambda_2 = \frac{\ln(T-40)}{\ln(90-40)}. \tag{10}$$

The motor fan speed is set in a linear manner when the temperature is between 50°C and 75°C.

4. Controller Hardware Design

The control unit ECU contains two electronic boards specifically designed in the lab, one is a +5 V low voltage signal processing and control board with a STC89S52 single chip microcomputer (MCU) as the core, and the other is a +24 V high voltage fan drive board. The control board adopts STC89S52 MCU as the core processor for its high reliability, low costs, and well recognized reputation in many fields. As shown in Figure 3, it has a MAX813 watchdog chip to prevent the software from running out or in a dead loop and adopts an 8-bit A-D conversion chip ADC0809 to measure temperature sensors and the voltage signals were filtered in separate RC filters for each channel and then connected to ADC0809.

The outputs of the control board are five-way PWM signals, calculated by the MCU with timers and amplified through five TLP-521 optical couplers before being connected to the external motor drive board. The PWM signals generated by the control board are connected to the motor drive board, amplified separately in five channels, and then drive the fan motor. An example channel is explained in Figure 4.

The PWM signal is firstly connected to the input pin of a MIC4421 MOSFET driver chip, which has independent 15 V power supply that can amplify the 5 V PWM waves to 15 V PWM waves. The amplified PWM signal then goes to two large power P75NF75 MOSFET chips, parallelly connected to reduce the heat as in Figure 4. The motor current goes through the MOSFET before grounding; thus the PWM signal eventually controls the shut-off and switch-on of MOSFET and hence the motor current, which controls fan speed according to PWM theories [16–18].

Two prototypes of the ECU have been made in the lab following comprehensive schematic and PCB designs. The motor board uses LM78L15 chip to obtain the required +15 V voltage for the MOSFET driver MIC4421, whilst the fan motor current directly comes from the vehicle battery and goes through the drive board via thick copper overlay and tin solder that allows up to 15 A current for each motor drive channel. Demonstration of the prototype is given in Figure 5.

The temperature sensors were actual products for Honda Accord Gen 7. When temperature rises, the resistance of the sensor reduces, in a nonlinear manner as shown in Figure 6. The characteristic curve was tested out by positioning the sensor in hot water and measuring resistance values. Thus, the ECU can use the fitted curve function to calculate actual temperature in water tank, and so forth.

5. Controller Software Design

The ECU software mainly performs the following functionalities: measure temperatures, display temperatures, decide fan speed, adjust PWM duty cycles, and output PWM signals. The PWM signals are generated in MCU using timers in a manner as shown in Figure 7.

The timer $T0$ is set to control the PWM substep time interval s_n. The cycle period Tc_n and hence base frequency are obtained by counting the substeps to a fixed number and inverting the signal, for example, at 10 substeps, whilst the

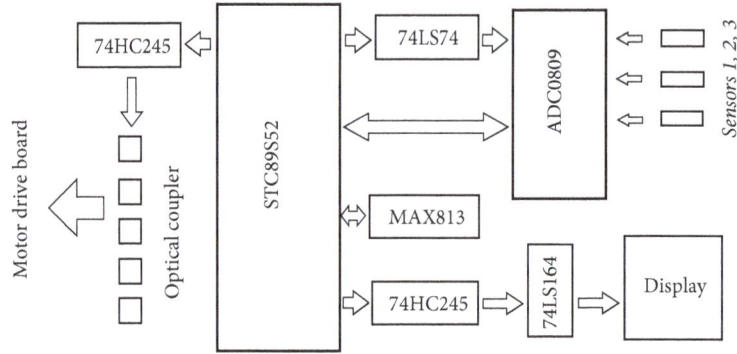

FIGURE 3: The control board sketch diagram.

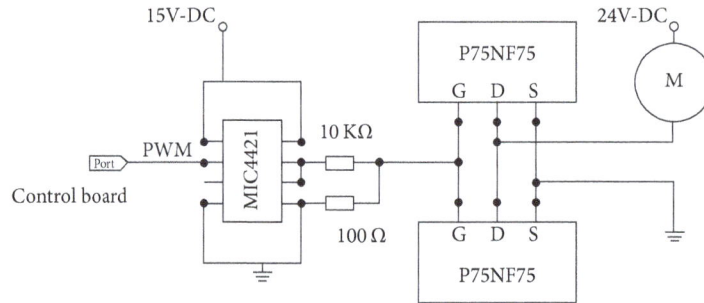

FIGURE 4: The sketch diagram of one channel of the motor drive board.

FIGURE 5: The assembled prototype of the ECU.

duty ratio is obtained by inverting the signal at a counted number within a full cycle. This approach gives more flexible control of the PWM signals comparing with some modern MCUs equipped with PWM port but difficult to amend PWM parameters within a program section.

The overall software flowchart is shown in Figure 8. When the ECU starts with the vehicle, the software firstly initializes the parameters such as timers, interrupt ports, and especially ADC0809 configurations. Since the temperature changes are slow, a large sampling gap up to 1 s is set in ADC0809. Then the MCU uses P1 port to control acquisition of the three signal channels and display the temperature values via serial port. When timer $T0$ is met, the timer interrupt processing sections count the events and invert PWM signals if required. When timer $T1$ is met, the software loops back to reread the temperature data from ADC0809. When neither of the timers is met, the software keeps refreshing the display unit.

A watchdog control signal is sent to the MAX813 chip to ensure the software is in healthy running. When the watchdog chip detects error the software restarts the MCU.

6. Experiments

Experiments were carried out in the lab to verify fan control performance and on a vehicle to verify actual usability and optimal control strategies.

6.1. Bench Tests. The bench tests in the lab as demonstrated in Figure 9 were carried out to evaluate sensor sensitivity, PWM fan drive effectiveness, and durability. A major issue that has arisen was due to the frequency of PWM signals, which according to literatures [16–18] may lead to noises, vibration, or even damage to the motor if the frequency is not properly configured and falls in resonance with the mechanical parts.

The PWM frequency effects can be seen in the test results in Table 1. As literature suggested [17], the proper PWM base frequency should be either as high as 10 K–20 K Hz or as low as 50–200 Hz to minimize noises and vibration, whilst the 1 K-2 K Hz frequency is the worst range. For the above reasons, the final PWM base frequency was chosen to be 100 Hz and small duty ratios less than 30% were avoided, so that the fans could stop directly when the calculated duty ratio reduces to 30%.

6.2. On-Vehicle Tests. On-vehicle tests were carried out on a LCK6118P hybrid electrical bus provided by Zhongtong Bus Holding Co., Ltd., in Liaocheng city, China. The actual configuration of the fans and the working environment is

$$y = 0.0004x^2 - 0.0698x + 3.3092$$

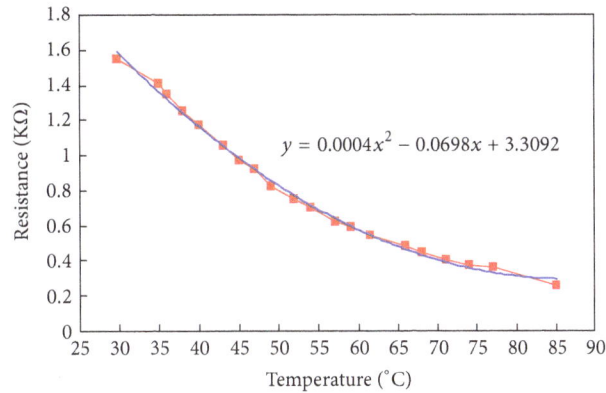

FIGURE 6: The temperature sensor and electrical characteristic.

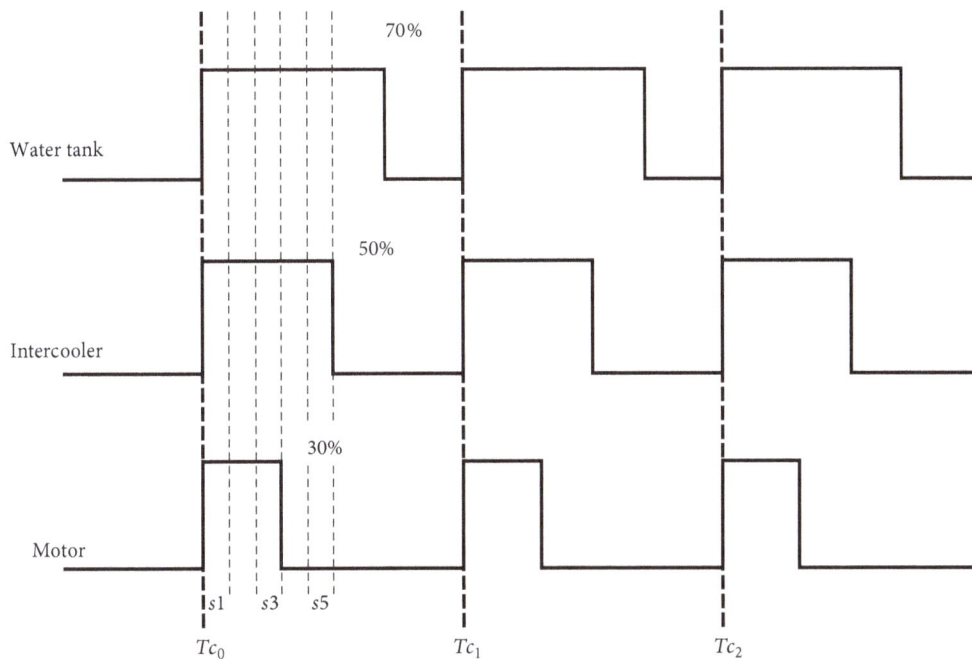

FIGURE 7: The sketch diagram of the generated PWM signals.

shown in Figure 10. Temperature values and fan speeds were read from the ECU and recorded for analysis, and extra temperature sensors were also added at the outlet of the water tank to analyze the changes and effectiveness.

The equipment specifications are provided in Table 2. The tests were carried out in the vehicle test yard within Zhongtong Bus Holding Co., Ltd., using different speed settings for varying durations, with specified numbers of starts and stops.

The tests are still undergoing and more experiments would be carried out along bus routes in the city. So far initial results indicate some sensible relationships between fan speeds and the reduction of coolant temperatures measured at the inlet and outlet of the radiator. From the recorded data in the ECU, temperature and fan speed data pairs were selected and plotted in Figure 11. When the inlet coolant temperatures are 95° and 90°, respectively, the temperature

reduction measured at the outlet changes with fan speed in a nonlinear manner, which is consistent with the theoretically predicted trends in Figure 2.

For the intercooler, the fluid inside the radiator is turbo pressurized air, which is compressed in the turbo and cooled in the intercooler and then fed into the diesel engine to boost the engine power. The temperature reductions with fan speeds are shown in Figure 12. The motor mainly drives the vehicle in high torque stop-start short periods and therefore it is not tested in the current study.

Comparing with old fashion on-off control system that always uses 100% fan speed, the above strategies would significantly reduce power consumption. The cooling fans totally cost 1.5 KW electricity when they are fully running, which counts 62.5% of a typical 100 AH battery. Although a battery can be recharged after starting, the main drive motor consumes the majority of the electricity. According to

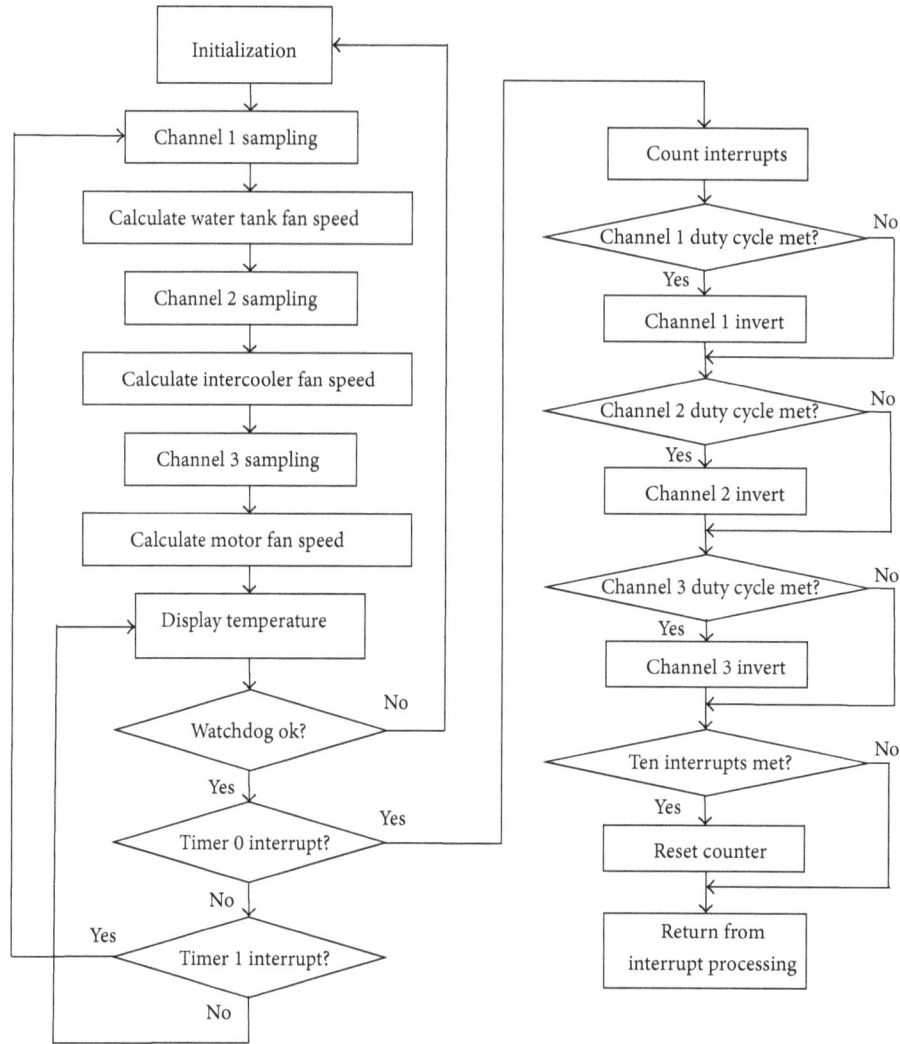

FIGURE 8: The flow chart of the ECU control software.

FIGURE 9: In lab bench test of the ECU control system.

Figure 11 it can be seen that, in order to control radiator outlet temperature to be no more than 85°C, then the fan speed should be set to 100% if the inlet temperature is 95°C, and only use 60% speed if the inlet temperature is 90°C.

According to China's national standard GBT 18386-2005, typical vehicle tests on city road can be simulated by a number of identical test cycles and each cycle is described in Figure 13. Within such cycle, it can be assumed that the acceleration

pedal is released during the speed reduction periods and zero-speed periods. Experiments indicated that the radiator inlet temperature can well drop from 95°C to 90°C 10 seconds after the acceleration pedal is released. Thus, there is together 22 seconds out of the total 200 seconds that only needs 60% fan speed. Taking into account the temperature changing periods, it is reasonable to assume that the cooling system can save at least 15% of the total electrical energy.

FIGURE 10: Actual fan and on-vehicle configurations for driving test.

TABLE 1: PWM frequency and fan speed test results.

Base frequency (Hz)	Duty ratio (%)	Performance	Durability
100	10	Noisy	Not tested
100	20	Slightly noisy	Not tested
100	30	Quiet & smooth	Above 24 hours
100	50	Quiet & smooth	Above 24 hours
100	90	Quiet & smooth	Above 24 hours
1000	30	Very noisy	Not tested
1000	90	Noisy	Not tested
10000	30	Slightly noisy	Not tested
10000	90	Quiet & smooth	24 hours with reboots

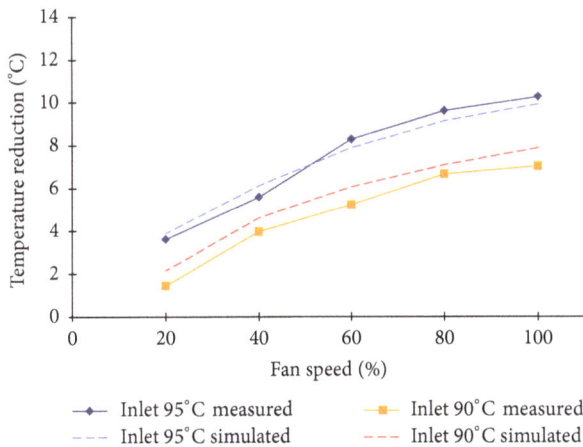

FIGURE 11: Test results for water tank fan speed and temperature reduction.

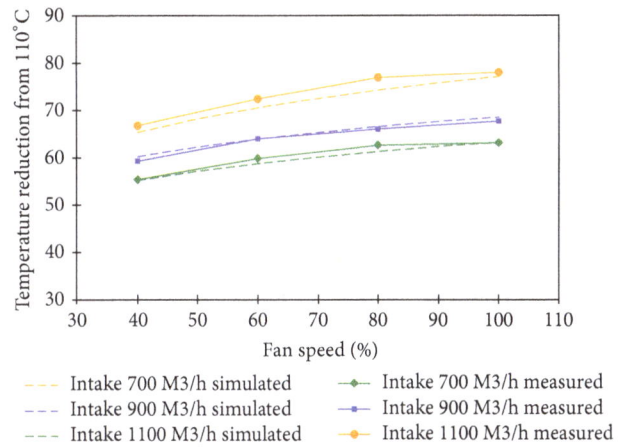

FIGURE 12: Test results for intercooler fan speed and temperature reduction.

7. Conclusion

This study presents a design of an integrated cooling system controller ECU for hybrid electrical vehicle. Temperatures in the water tank, intercooler, and main drive motor are measured by the MCU to control optimal fan speeds. The ECU also consists of a five-way large power fan motor drive board to perform PWM fan speed control. Comprehensive experiments have been carried out to verify the ECU performance and to identify optimal fan speed control strategies. The PWM fan drive frequency has been determined in bench tests and the speed adjustment strategies were firstly analyzed using theories and then compared with experimental data. The ECU design is robust and it can save at least 15% of the

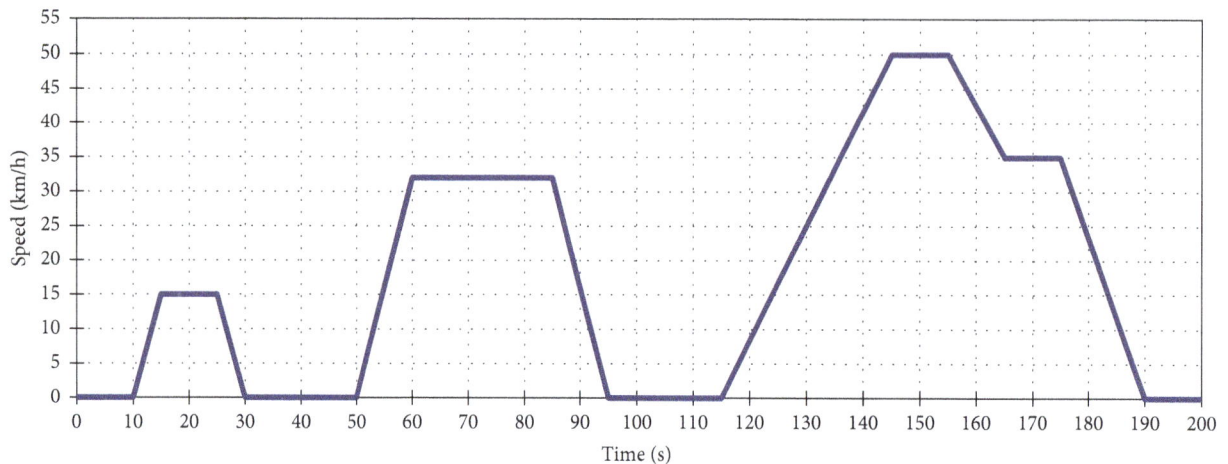

FIGURE 13: A typical city test cycle according to GBT 18386-2005.

TABLE 2: Specifications of the equipment for experiments.

Equipment	Specification
Engine model	YC6L280-42
Engine type	Turbo Diesel
Rated power/speed	206 KW/2200 r/min
Max. torque/speed	1100 N·m ≤ 1500 r/min
Rated coolant temperature	≤95°C
Water pump flow rate	≤420 L/min
Heat with rated power	Q_w = 117 Kw
Heat with rated torque	Q_w = 82.5 Kw
Fan diameter	385 mm
Fan electrical power	300 W
Radiator total area	48.76 m^2
Radiator front area	0.5724 m^2
Radiator dissipation power	150 KW
Radiator wave length	4.2 mm
Radiator wave height	10 mm
Radiator number of pipes	135
Radiator number of ribbons	46
Radiator pipe size	13 × 2 (mm)
Intercooler total area	20.87 m^2
Intercooler front area	0.3465 m^2
Intercooler dissipation power	50 KW
Intercooler wave length	3.5 mm
Intercooler wave height	8 mm
Intercooler number of pipes	74
Intercooler number of ribbons	38
Intercooler pipe size	13 × 2 (mm)

total electricity on normal city routes. Further work should be expanded to take into account extra influential factors such as vehicle speed, environmental temperature and make use of CAN bus data to create a more advanced controller.

Conflicts of Interest

The authors declare that they have no conflicts of interest.

Acknowledgments

This study is supported by Liaocheng University Research Fund 13LD2001.

References

[1] K. Reif, *Automotive Mechatronics: Automotive Networking, Driving Stability Systems, Electronics*, Springer, 2015.

[2] S. Park, *A comprehensive thermal management system model for hybrid electric vehicles [Ph.D. thesis]*, University of Michigan, Ann Arbor, Mich, USA, 2011.

[3] C. Alaoui and Z. M. Salameh, "A novel thermal management for electric and hybrid vehicles," *IEEE Transactions on Vehicular Technology*, vol. 54, no. 2, pp. 468–476, 2005.

[4] Y.-H. Hung, Y.-X. Lin, C.-H. Wu, and S.-Y. Chen, "Mechatronics design and experimental verification of an electric-vehicle-based hybrid thermal management system," *Advances in Mechanical Engineering*, vol. 8, no. 2, pp. 1–9, 2016.

[5] K. J. Kelly, T. Abraham, K. Bennion, D. Bharathan, S. Narumanchi, and M. O'Keefe, "Assessment of thermal control technologies for cooling electric vehicle power electronics," in *Proceedings of the 23rd International Electric Vehicle Symposium*, Anaheim, Calif, USA, 2007.

[6] H. H. Pang and C. J. Brace, "Review of engine cooling technologies for modern engines," *Proceedings of the Institution of Mechanical Engineers Part D: Journal of Automobile Engineering*, vol. 218, no. 11, pp. 1209–1215, 2004.

[7] S. S. Butt, R. Prabel, and H. Aschemann, "Robust input-output linearization with input constraints for an engine cooling system," in *Proceedings of the American Control Conference (ACC '14)*, pp. 4555–4560, IEEE, June 2014.

[8] M. H. Salah, T. H. Mitchell, J. R. Wagner, and D. M. Dawson, "Nonlinear-control strategy for advanced vehicle thermal-management systems," *IEEE Transactions on Vehicular Technology*, vol. 57, no. 1, pp. 127–137, 2008.

[9] P. Setlur, J. R. Wagner, D. M. Dawson, and E. Marotta, "An advanced engine thermal management system: nonlinear control and test," *IEEE/ASME Transactions on Mechatronics*, vol. 10, no. 2, pp. 210–220, 2005.

[10] W. Lothar Seybold, F. Filsinger, B. Gruber, I. Taxis, and A. Lazaridis, "Optimization of an engine coolant radiator for

vehicle thermal management," in *Proceedings of the 16th Internationales Stuttgarter Symposium Part of the series Proceedings*, pp. 1465–1482, 2016.

[11] J.-X. Liu, S.-C. Qin, Z.-Y. Xu, A. Zhang, Y. Xi, and X.-L. Zhang, "Comparative analysis of heat exchange performance of vehicle radiator based on CFD numerical simulation," *Journal of South China University of Technology*, vol. 40, no. 5, pp. 24–29, 2012.

[12] C. Park and A. K. Jaura, "Thermal analysis of cooling system in hybrid electric vehicles," Tech. Rep. SAE 2000-01-0710, 2002.

[13] W. Zhu, Z. Tian, and Y. Zhang, "Prediction model of heat transfer and resistance of pipe belt type automobile radiator," *Modern Business Trade Industry*, vol. 5, pp. 186–188, 2012.

[14] H. Fei, J. Tian, S. Lei, Y. Chen, and W. Yan, "CFD analysis of the thermal management system of the engine room of the passenger car," *Internal Combustion Engine & Power Plant*, vol. 1, pp. 25–28, 2014.

[15] T. Ismael, S. B. Yun, and F. Ulugbek, "Radiator heat dissipation performance," *Journal of Electronics Cooling and Thermal Control*, vol. 6, no. 2, pp. 88–96, 2016.

[16] S. Shrivastava, J. Rawat, and A. Agrawal, "Controlling DC motor using microcontroller (PIC16F72) with PWM," *International Journal of Engineering Research*, vol. 1, no. 2, pp. 45–47, 2012.

[17] J. M. Najm Abad, B. Salami, H. Noori, A. Soleimani, and F. Mehdipour, "A neuro-fuzzy fan speed controller for dynamic thermal management of multi-core processors," in *Proceedings of the 11th ACM Conference on Computing Frontiers*, article no. 29, ACM, New York, NY, USA, 2014.

[18] V. Bhatia and G. Bhatia, "Room temperature based fan speed control system using pulse width modulation technique," *International Journal of Computer Applications*, vol. 81, no. 5, pp. 35–40, 2013.

5

Review and Selection Strategy for High-Accuracy Modeling of PWM Converters in DCM

Yu-Jun Mao,[1] **Chi-Seng Lam ⑩,**[1] **Sai-Weng Sin,**[1,2] **Man-Chung Wong,**[1,2] **and Rui Paulo Martins ⑩**[1,2,3]

[1]*State Key Laboratory of Analog and Mixed-Signal VLSI, University of Macau, Macau 999078, China*
[2]*Department of Electrical and Computer Engineering, Faculty of Science and Technology, University of Macau, Macau 999078, China*
[3]*On Leave from Instituto Superior Técnico, Universidade de Lisboa, Lisbon 1649-004, Portugal*

Correspondence should be addressed to Chi-Seng Lam; cslam@umac.mo

Academic Editor: Luigi Piegari

Among various modeling methods for DC-DC converters introduced in the past two decades, the state-space averaging (SSA) and the circuit averaging (CA) are the most general and popular exhibiting high accuracy. However, their deduction approaches are not entirely equivalent since they incorporate different averaging processes, thus yielding different small signal transfer functions even under identical operating conditions. Some research studies claimed that the improved SSA can obtain the highest accuracy among all the modeling methods, but this paper discovers and clearly verifies that this is not the case. In this paper, we first review and study these two modeling methods for various DC-DC converters operating in the discontinuous conduction mode (DCM). We also streamline the general model-deriving processes for DC-DC converters, and test and compare the accuracy of these two methods under various conditions. Finally, we provide a selection strategy for a high-accuracy modeling method for different DC-DC converters operating in DCM and verified by simulations, which revealed necessary and beneficial for designing a more accurate DCM closed-loop controller for DC-DC converters, thus achieving better stability and transient response.

1. Introduction

A PWM DC-DC converter operates either in the continuous conduction mode (CCM) or in the discontinuous conduction mode (DCM). For small inductances or light loads, DCM operation is occasionally unavoidable in DC-DC converters, their design being intentionally in DCM to reduce both the inductor's size and the switching frequency. Then, it is necessary to develop a modeling analysis for the DC-DC converter operation in DCM to design a closed-loop controller. In previous works [1–22], various small signal modeling methods of PWM DC-DC converters in CCM and DCM are proposed, where different modeling methods provide either an analytical equation or an equivalent circuit and can be categorized as

reduced-order or full-order models, as summarized in Table 1.

Among these small signal modeling methods, the improved state-space averaging (SSA) and the circuit averaging (CA) are the latest to present, with high accuracy, an analytical equation, and an equivalent circuit, respectively. The improved SSA method [6] claims that it has the highest accuracy among all the other modeling methods and can be applied to any circuit composed by inductors and capacitors. But, this conclusion is only proven and verified in a boost converter under specific operating conditions [6]. Besides, this may not hold in other DC-DC converters and under different operating conditions. Moreover, Zeng et al. [13] used the CA method from [5] to deduce the small signal model of a KY converter, but when applying the improved

TABLE 1: Summary of small signal modeling methods.

Method name	Order	Model type
Conventional state-space averaging (SSA) [1]	Reduced order (low accuracy)	Analytical equation
Converter cell [2]	Reduced order (low accuracy)	Equivalent circuit
Full-order SSA [3, 8, 11, 12, 20]	Full order (high accuracy)	Analytical equation
Circuit averaging (CA) [4, 5, 8–10, 13, 14, 17–19, 21, 22]	Full order (high accuracy)	Equivalent circuit
Improved SSA [6, 7, 15, 16]	Full order (high accuracy)	Analytical equation

SSA method to the same case, the CA method yields a better modeling accuracy under some conditions that are discussed in this paper. Contradictions met in CA and SSA methods motivate further analysis and retesting their accuracy in various DC-DC converters under different operating conditions in DCM, in order to derive a selection criterion that allows higher accuracy in the small signal modeling method.

Also, the rather complicated (and not general) original derivation processes of the CA and the improved SSA methods, in which the whole small signal model derivation process should be repeated whenever the DC-DC circuit topology changes, whenever the DC-DC circuit topology changes, leads to unnecessary and time-consuming efforts. Then, another motivation for this paper is to generalize the whole derivation process to attain a general and intuitive model-deriving solution.

The main contributions of this paper are as follows:

(1) To propose and deduct a general and intuitive derivation process of the improved SSA and CA modeling methods for different DC-DC converters, such that their corresponding DCM small signal models can be easily determined.

(2) To study, retest, and compare, through simulations, the accuracy of the improved SSA and CA modeling methods for various DC-DC converters under different operating conditions. Since most of the previous works are either based on the improved SSA or CA, they lack a detailed comparison among them.

(3) To propose a selection strategy of the DCM small signal modeling method in order to obtain high accuracy for different DC-DC converters under different operating conditions.

This paper contributes significantly to the design of a stable and fast transient response DCM closed-loop controller for different DC-DC converters. Section 2 presents the DC analysis of different DC-DC converters in DCM. Section 3 introduces two small signal relationship calculation methods based on the large-signal equation. Section 4 discusses the general DCM large-signal and small signal modeling deduction based on the improved SSA method applied to different DC-DC converters. Then, Section 5 determines the general DCM large-signal and small signal modeling deduction based on the CA method. Section 6 compares the simulation results of the small signal modeling obtained through the improved SSA and the CA methods. Section 7 presents a selection strategy of the DCM small

signal modeling for different DC-DC converters. Finally, the conclusions are drawn in Section 8.

2. DC Analysis of DC-DC Converters in DCM

In the discussion hereafter, if we use the capital letter X to denote the averaged value of a specific variable in one switching frequency (DC value), then the lowercase letter x denotes its large-signal value and $\hat{x}$ denotes its small signal value. The relationship between these three variables is [10]

$$x = X + \hat{x}. \tag{1}$$

2.1. DCM Operation. The DCM operation of the DC-DC converters consists of three intervals. Here, we use D_1, D_2, and D_3 to denote the duty ratio of each interval, respectively, and T_s denotes the switching period. Figure 1 shows the inductor current, i_L, waveform of the DC-DC converters operating in DCM [10].

For a typical DC-DC converter, i_L starts to rise during the charging of the inductor L and begins to drop while the L is discharging. In this paper, we use V_{on} (charging) and V_{off} (discharging) to denote the voltage across the L in the first and the second intervals, respectively. From Figure 1, we can obtain

$$I_{pk} = \frac{V_{on}}{L}D_1 T_s, \tag{2}$$

where I_{pk} denotes the peak value of inductor current i_L, D_1 denotes the duty ratio of charging interval, and T_s denotes the switching period. From Figure 1, the relationship between the peak value (I_{pk}) and average value (I_L) of the inductor current can be expressed as

$$\frac{1}{2}I_{pk}(D_1 + D_2)T_s = I_L T_s, \tag{3}$$

where I_L denotes the average value of the inductor current i_L while T_s and D_2 denotes the duty ratio of the discharging interval. From Figure 1,

$$\frac{V_{on}}{L}D_1 T_s = \frac{-V_{off}}{L}D_2 T_s = I_{pk}. \tag{4}$$

Equation (4) yields the following

$$D_2 = -\frac{V_{on}}{V_{off}}D_1, \tag{5}$$

where V_{on} (charging) and V_{off} (discharging) denote the voltage across the L in the first and the second intervals,

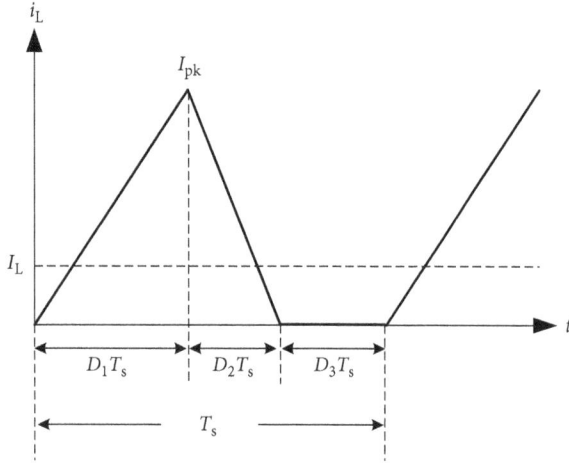

FIGURE 1: Inductor current of DC-DC converters operating in DCM.

respectively. Considering (2), (3), and (5), I_L can be expressed as

$$I_L = \frac{D_1^2 T_s V_{on} (V_{off} - V_{on})}{2 L V_{off}}. \tag{6}$$

2.2. DC Parameters Calculation for Different DC-DC Converters in DCM. Before we derive the small signal model, some DC parameters of the circuit operating in DCM need to be calculated beforehand. Figure 2 shows the four DC-DC converters that will be studied throughout this paper; they are the buck converter, the boost converter, the buck-boost converter, and the KY converter [23], where v_i/i_i and v_o/i_o represent the input and output voltage/current of the converter, respectively, and L represents the inductor, D the diode, S the switch, C the output capacitor, R the loading, C_f the flying capacitor, i_L the inductor current, i_D the diode current, and i_S the switch current. As the equivalent series resistance (ESR) for the passive components are usually small due to high Q design, which are neglected in this paper for simplification. Thus, the left-hand plane zero caused by the ESR of the output capacitor C will not present in the small signal transfer functions for different DC-DC converters.

For calculating the necessary DC parameters of different DC-DC converters, we can just input different V_{on} and V_{off} into (5) and (6). For the buck converter, $V_{on} = V_i - V_o$ and $V_{off} = -V_o$; for the boost converter, $V_{on} = V_i$ and $V_{off} = V_i - V_o$; for the buck-boost converter, $V_{on} = V_i$ and $V_{off} = -V_o$, and for the KY converter, $V_{on} = 2V_i - V_o$ and $V_{off} = V_i - V_o$. On the other hand, using $M = V_o/V_i$ to denote the voltage gain, then D_2 and I_L can be calculated via (5) and (6), as summarized in Table 2.

For the DC relationship between D_1 and M, according to [4], the DC relationships among the inductor current i_L, the switch current i_S, and the diode current i_D are

$$I_S = \frac{D_1}{D_1 + D_2} I_L, \tag{7}$$

$$I_D = \frac{D_2}{D_1 + D_2} I_L. \tag{8}$$

For different DC-DC converters' configurations as in Figure 2, the output current $I_o = I_L$ for the buck and the KY converters and $I_o = I_D$ for the boost and the buck-boost converters. With the D_2 and I_L in Table 2 and the help of (7) and (8), the relationship between D_1 (expressed as $D_1^2 T_s R/2L$) and M can be calculated as presented in Table 2. These parameters are essential for the small-signal transfer function calculation as shown in Sections 4 and 5.

3. Calculation of the Small-Signal Relationship with the Proposed Differentiation Method

If f is a large-signal function of some variables x, y, and z, to attain the small-signal model of these variables, we can express them as the sums of DC and small signal components [10],

$$f = F + \hat{f}, \tag{9}$$

$$x = X + \hat{x}, \tag{10}$$

$$y = Y + \hat{y}, \tag{11}$$

$$z = Z + \hat{z}. \tag{12}$$

By neglecting the DC terms and the high-order small signal terms, we can realize the linear approximation. During the small signal calculation, the following approximation can be used:

$$\frac{1}{1 + \hat{x}} \approx 1 - \hat{x}. \tag{13}$$

For example, the large-signal inductor current of the buck converter given by Kazimierczuk [10] is shown in (14), which can also be given by referring to the expression of I_L in Table 2 and replacing all DC quantities with large-signal quantities. The basis for this replacement is that large-signal analysis is based on DC relationship of circuit variables [10, 15, 19, 21]:

$$i_L = \frac{d_1^2 T_s v_i (1 - m)}{2 L m}, \tag{14}$$

where m denotes the large-signal voltage gain and v_i denotes the small signal input voltage. Expressing the variables as the sums of DC and AC components as (9)–(12) do yield (15), after cancelling the DC components, the small signal expression of the inductor current $\hat{i}_L$ will become (16):

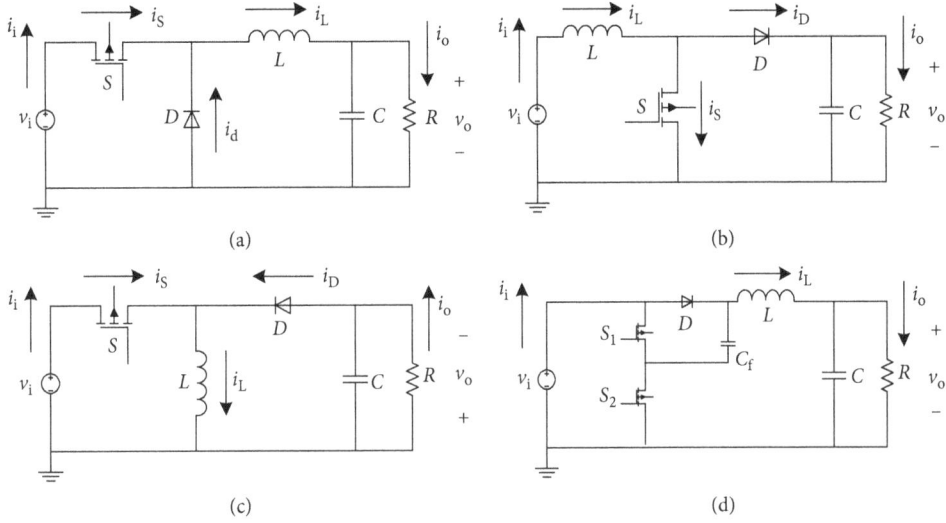

FIGURE 2: Circuit configuration of (a) buck, (b) boost, (c) buck-boost, and (d) KY converters.

TABLE 2: DC parameters for different DC-DC converters.

DC-DC converter	D_2	I_L	$(D_1^2 T_s R)/(2L)$
Buck	$((1-M)/M)D_1$	$(D_1^2 T_s V_i(1-M))/(2LM)$	$(M^2/(1-M))$
Boost	$(1/(M-1))D_1$	$(D_1^2 T_s V_i M)/(2L(M-1))$	$M^2 - M$
Buck-boost	$(1/M)D_1$	$(D_1^2 T_s V_i(1+M))/(2LM)$	M^2
KY	$((2-M)/(M-1))D_1$	$D_1^2 T_s V_i(2-M)/2L(M-1)$	$(M(M-1))/(2-M)$

$$I_L + \hat{i}_L = \frac{(D_1 + \hat{d}_1)^2 T_s (V_i + \hat{v}_i)(1 - M - \hat{m})}{2L(M + \hat{m})} \tag{15}$$

$$\approx \frac{T_s (D_1^2 + 2D_1 \hat{d}_1)(V_i + \hat{v}_i)(1 - M - \hat{m})(1 - (\hat{m}/M))}{2LM},$$

$$\hat{i}_L = \frac{T_s [2D_1 V_i(1-M)\hat{d}_1 + D_1^2(1-M)\hat{v}_i - D_1^2 V_i \hat{m} - D_1^2 V_i(1-M)(\hat{m}/M)]}{2LM} \tag{16}$$

$$= \frac{D_1 T_s V_i(1-M)}{LM} \hat{d}_1 + \frac{D_1^2 T_s(1-M)}{2LM} \hat{v}_i - \frac{D_1^2 T_s V_i}{2LM^2} \hat{m}.$$

However, the above deduction process is quite complicated and time-consuming. To simplify the analysis, we can utilize the calculation of the small signal perturbation as shown in Figure 3.

With a small variation of $\hat{x}$, the corresponding variation of $\hat{f}$ is also very small such that the value $\hat{f}/\hat{x}$ is equal to the derivative of f to x, thus yielding

$$\hat{f} = f'_x \hat{x} + f'_y \hat{y} + f'_z \hat{z}. \tag{17}$$

Following (17) and taking the derivative of (14) with respect to each variable, we can easily get

$$\hat{i}_L = \frac{D_1 T_s V_i(1-M)}{LM} \hat{d}_1 + \frac{D_1^2 T_s(1-M)}{2LM} \hat{v}_i - \frac{D_1^2 T_s V_i}{2LM^2} \hat{m}, \tag{18}$$

where the results in (16) and (18) are equivalent. In the following section, we will use the proposed differentiation method (17) to calculate the small signal relationship.

4. Improved State-Space Averaging Method for Large-Signal and Small Signal Modeling Deduction

In this section, the approach for deducing the DCM small signal models for different DC-DC converters by using the improved SSA method will be discussed.

4.1. General Large-Signal and Corresponding Small Signal Modeling Using the SSA Method. According to the conclusion from [10], the large-signal value relationship and the DC value relationship are identical. Then, (2) and (3) lead to the following large-signal duty ratio d_2:

$$d_2 = \frac{2Li_L}{v_{on} d_1 T_s} - d_1. \tag{19}$$

By applying the SSA to the inductor voltage and using (19), we get

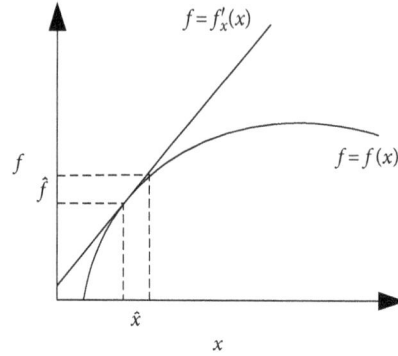

FIGURE 3: Small signal perturbation.

$$L\frac{di_L}{dt} = v_{on}d_1 + v_{off}d_2 = (v_{on} - v_{off})d_1 + \frac{2v_{off}Li_L}{v_{on}d_1T_s}. \quad (20)$$

Taking the derivative of (20) with respect to v_{on}, v_{off}, d_1, and i_L, which are variables in (20), and using (6), we will have the corresponding small signal model given by

$$L\frac{\widehat{di_L}}{dt} = \left(\frac{2V_{on} - V_{off}}{V_{on}}D_1\right)\widehat{v}_{on} + \left(-\frac{V_{on}}{V_{off}}D_1\right)\widehat{v}_{off}$$

$$+ 2(V_{on} - V_{off})\widehat{d}_1 + \left(\frac{V_{off} - V_{on}}{I_L}D_1\right)\widehat{i}_L. \quad (21)$$

Then, using the SSA to the capacitor current yields

$$C\frac{du_C}{dt} = i_o - \frac{u_C}{R}. \quad (22)$$

If $i_o = i_L$, and taking the derivative of (22), the corresponding small signal model becomes

$$C\frac{\widehat{du}_C}{dt} = \widehat{i}_L - \frac{\widehat{u}_C}{R}. \quad (23)$$

If $i_o = i_D$, then (8), (19), and (22) lead to

$$C\frac{du_C}{dt} = i_L - \frac{d_1^2 T_s v_{on}}{2L} - \frac{u_C}{R}. \quad (24)$$

Finally, taking the derivative of (24), we obtain the corresponding small signal model as follows:

$$C\frac{\widehat{du}_C}{dt} = \widehat{i}_L - \frac{D_1 T_s V_{on}}{L}\widehat{d}_1 - \frac{D_1^2 T_s}{2L}\widehat{v}_{on} - \frac{\widehat{u}_C}{R}, \quad (25)$$

where (21), (23), and (25) are the three general equations for deducing the DCM small signal models of different DC-DC converters. Then, after input, the values of the inductor

voltage drops during the charging (V_{on}) and discharging (V_{off}) cycles, as well as the DC parameters of each DC-DC converter from Table 2, into (21), (23), and (25), and the corresponding DCM small signal models can be deduced easily, as detailed next.

4.2. Buck Converter Small Signal Transfer Function. For the buck converter, $V_{on} = V_i - V_o$, $V_{off} = -V_o$, and $i_o = i_L$. Besides, Table 2 gives $I_L = (D_1^2 T_s V_i(1 - M))/(2LM)$ and $(D_1^2 T_s R)/(2L) = (M^2)/(1 - M)$, and substituting them into (21) and (23), gives

$$L\frac{\widehat{di_L}}{dt} = \left(\frac{2V_i - V_o}{V_i - V_o}D_1\right)(\widehat{v}_i - \widehat{v}_o)$$

$$- \left(\frac{V_i - V_o}{V_o}D_1\right)\widehat{v}_o + 2V_i\widehat{d}_1 - \frac{V_i}{I_L}D_1\widehat{i}_L, \quad (26)$$

$$C\frac{\widehat{du}_C}{dt} = \widehat{i}_L - \frac{\widehat{u}_C}{R}. \quad (27)$$

Let $\widehat{v}_i = 0$, then (26) becomes

$$\left(sL + \frac{2LM}{D_1 T_s(1 - M)}\right)\widehat{i}_L = \frac{1}{M^2 - M}D_1\widehat{v}_o + 2V_i\widehat{d}_1. \quad (28)$$

Further, if $u_c = v_o$, then (27) will be

$$\left(sC + \frac{1}{R}\right)\widehat{v}_o = \widehat{i}_L. \quad (29)$$

Considering (28) and (29) simultaneously, the transfer function (V_o over d_1) of the buck converter can be obtained as

$$\frac{\widehat{v}_o(s)}{\widehat{d}_1(s)} = \frac{2V_i}{s^2 LC + s((L/R) + ((2LCM)/(D_1 T_s(1 - M)))) + (((2 - M)D_1)/((1 - M)M))}. \quad (30)$$

4.3. Boost Converter Small Signal Transfer Function.
For the boost converter, $V_{on} = V_i$, $V_{off} = V_i - V_o$, and $i_o = i_D$, and Table 2 gives $I_L = (D_1^2 T_s V_i M)/(2L(M-1))$ and $(D_1^2 T_s R)/(2L) = M^2 - M$; by substituting them into (21) and (25), it leads to

$$L\frac{d\hat{i}_L}{dt} = \left(\frac{V_i + V_o}{V_i}D_1\right)\hat{v}_i - \left(\frac{V_i}{V_i - V_o}D_1\right)(\hat{v}_i - \hat{v}_o)$$
$$+ 2V_o\hat{d}_1 - \frac{V_o}{I_L}D_1\hat{i}_L, \tag{31}$$

$$C\frac{d\hat{u}_C}{dt} = \hat{i}_L - \frac{D_1 T_s V_i}{L}\hat{d}_1 - \frac{D_1^2 T_s}{2L}\hat{v}_i - \frac{\hat{u}_C}{R}. \tag{32}$$

Let $\hat{v}_i = 0$, then (31) will be

$$\left(sL + \frac{2L(M-1)}{D_1 T_s}\right)\hat{i}_L = \frac{1}{1-M}D_1\hat{v}_o + 2V_o\hat{d}_1, \tag{33}$$

and if $u_c = v_o$ and $\hat{v}_i = 0$, then we can obtain the following equation from (32):

$$\left(sC + \frac{1}{R}\right)\hat{v}_o + \frac{D_1 T_s V_i}{L}\hat{d}_1 = \hat{i}_L. \tag{34}$$

Also considering (33) and (34) simultaneously, in this case, the transfer function of the boost converter will become

$$\frac{\hat{v}_o(s)}{\hat{d}_1(s)} = \frac{D_1 T_s V_i((2)/(D_1 T_s)) - s)}{s^2 LC + s((L/R) + ((2LC(M-1))/(D_1 T_s))) + (((2M-1)D_1)/(M(M-1)))}. \tag{35}$$

4.4. Buck-Boost Converter Small Signal Transfer Function.
For the boost converter, $V_{on} = V_i$, $V_{off} = -V_o$, and $i_o = i_D$, and from Table 2, $I_L = (D_1^2 T_s V_i(1+M))/(2LM)$ and $(D_1^2 T_s R)/(2L) = M^2$. Then, by substituting them into (21) and (25), we have

$$L\frac{d\hat{i}_L}{dt} = \left(\frac{2V_i + V_o}{V_i}D_1\right)\hat{v}_i - \frac{V_i}{V_o}D_1\hat{v}_o$$
$$+ 2(V_i + V_o)\hat{d}_1 - \frac{(V_i + V_o)}{I_L}D_1\hat{i}_L, \tag{36}$$

$$C\frac{d\hat{u}_C}{dt} = \hat{i}_L - \frac{D_1 T_s V_i}{L}\hat{d}_1 - \frac{D_1^2 T_s}{2L}\hat{v}_i - \frac{\hat{u}_C}{R}. \tag{37}$$

With $\hat{v}_i = 0$, then (36) will lead to

$$\left(sL + \frac{2LM}{D_1 T_s}\right)\hat{i}_L = -\frac{1}{M}D_1\hat{v}_o + 2(V_i + V_o)\hat{d}_1, \tag{38}$$

and again if $u_c = v_o$ and $\hat{v}_i = 0$, (37) yields

$$\left(sC + \frac{1}{R}\right)\hat{v}_o + \frac{D_1 T_s V_i}{L}\hat{d}_1 = \hat{i}_L. \tag{39}$$

Finally, considering (38) and (39) simultaneously, again the transfer function of the buck-boost converter will be

$$\frac{\hat{v}_o(s)}{\hat{d}_1(s)} = \frac{D_1 T_s V_i((2/(D_1 T_s)) - s)}{s^2 LC + s((L/R) + ((2LCM)/(D_1 T_s))) + ((2D_1)/M)}. \tag{40}$$

4.5. KY Converter Small Signal Transfer Function.
For the KY converter, $V_{on} = 2V_i - V_o$, $V_{off} = V_i - V_o$, and $i_o = i_L$, and from Table 2, $I_L = (D_1^2 T_s V_i(2-M))/(2L(M-1))$ and $(D_1^2 T_s R)/(2L) = (M(M-1))/(2-M)$. Then, substituting them into (21) and (23) will lead to

$$L\frac{d\hat{i}_L}{dt} = \frac{3V_i - V_o}{2V_i - V_o}D_1(2\hat{v}_i - \hat{v}_o)$$
$$- \frac{2V_i - V_o}{V_i - V_o}D_1(\hat{v}_i - \hat{v}_o) + 2V_i\hat{d}_1 - \frac{V_i}{I_L}D_1\hat{i}_L, \tag{41}$$

$$C\frac{d\hat{u}_C}{dt} = \hat{i}_L - \frac{\hat{u}_C}{R}. \tag{42}$$

Let $\hat{v}_i = 0$, then (41) becomes

$$\left(sL + \frac{2L(M-1)}{D_1 T_s(2-M)}\right)\hat{i}_L = \frac{1}{(2-M)(1-M)}D_1\hat{v}_o + 2V_i\hat{d}_1. \tag{43}$$

Then with $u_c = v_o$, (42) gives

$$\left(sC + \frac{1}{R}\right)\hat{v}_o = \hat{i}_L. \tag{44}$$

Again with (43) and (44) considered simultaneously, the transfer function of the KY converter can be obtained as

$$\frac{\hat{v}_o(s)}{\hat{d}_1(s)} = \frac{2V_i}{s^2 LC + s((L/R) + ((2LC(M-1))/(D_1 T_s(2-M)))) + (((M^2 - 4M + 2)D_1)/((2-M)(1-M)M))}. \tag{45}$$

5. Circuit Averaging Method for Large-Signal and Small Signal Modeling Deduction

In this section, the approach for deducing the DCM small signal models for different DC-DC converters by using the CA method will be discussed.

5.1. General Large-Signal and Corresponding Small Signal Modeling Using the CA Method. With the help of Reference [10], according to (7) and (8), the switching network of the DC-DC converter can be transformed into the circuit as shown in Figure 4.

When the circuit reaches the steady-state, the average voltage across the inductor is zero, then Figure 4 imposes,

$$V_{LM} = 0. \tag{46}$$

From Figure 4, we can obtain

$$V_{LS} = V_{LM} + V_{MS} = V_{MS}, \tag{47}$$

$$V_{LD} = V_{LM} + V_{MD} = V_{MD}. \tag{48}$$

By using (5)–(8), they yield the following large-signal equations (considering identical the large-signal value and the DC value relationships):

$$i_S = \frac{d_1^2 T_s v_{on}}{2L}, \tag{49}$$

$$i_D = -\frac{d_1^2 T_s v_{on}^2}{2L v_{off}}. \tag{50}$$

Taking the derivative of (49) and (50), the corresponding small signal equations will emerge as follows:

$$\begin{aligned}
\widehat{i}_S &= \frac{D_1 T_s V_{on}}{L}\widehat{d}_1 + \frac{D_1^2 T_s}{2L}\widehat{v}_{on}, \\
\widehat{i}_D &= -\frac{D_1 T_s V_{on}^2}{L V_{off}}\widehat{d}_1 - \frac{D_1^2 T_s V_{on}}{L V_{off}}\widehat{v}_{on} + \frac{D_1^2 T_s V_{on}^2}{2L V_{off}^2}\widehat{v}_{off},
\end{aligned} \tag{51}$$

and from (47) and (48), $V_{on} = V_{LS} = V_{MS}$ and $V_{off} = V_{LD} = V_{MD}$. Then, the small signal circuit of the switching network in Figure 4 can be transformed into Figure 5, where

$$k_S = \frac{D_1 T_s V_{on}}{L},$$

$$g_S = \frac{D_1^2 T_s}{2L},$$

$$k_D = -\frac{D_1 T_s V_{on}^2}{L V_{off}}, \tag{52}$$

$$g_D = -\frac{D_1^2 T_s V_{on}}{L V_{off}},$$

$$g_M = \frac{D_1^2 T_s V_{on}^2}{2L V_{off}^2}.$$

From Figure 5, we can write

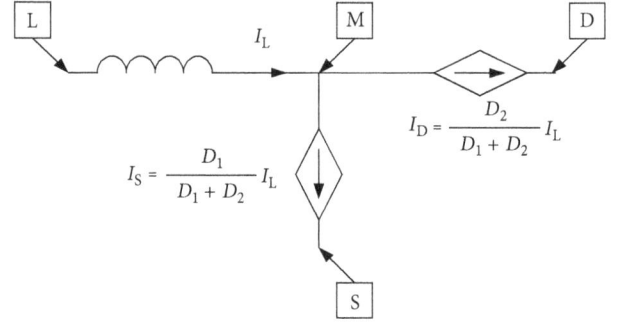

FIGURE 4: Equivalent circuit of the switching network (current source).

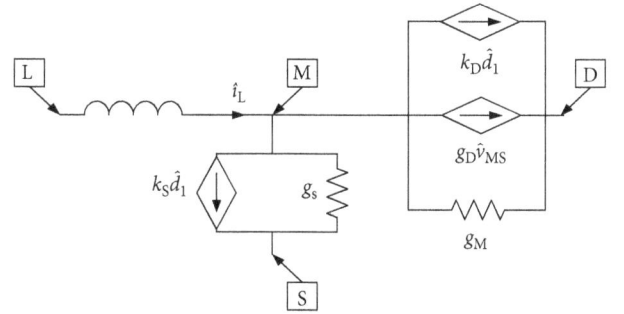

FIGURE 5: Small signal equivalent circuit of the switching network.

$$\begin{aligned}
\widehat{i}_L &= (k_S + k_D)\widehat{d}_1 + (g_S + g_D)\widehat{v}_{MS} + g_M\widehat{v}_{MD}, \\
&= (k_S + k_D)\widehat{d}_1 + (g_S + g_D + g_M)\widehat{v}_{MS} + g_M\widehat{v}_{SD}.
\end{aligned} \tag{53}$$

Then, with $k_{d1} = k_S + k_D$ and $g_{MS} = g_S + g_D + g_M$, we can obtain the factors $k_S, k_D, k_{d1}, g_S, g_D, g_M$, and g_{MS} of the four DC-DC converters (buck, boost, buck-boost, and KY) as shown in Table 3.

With these factors, we can substantially simplify the calculation process, as shown next. Further, we can deduce that

$$\widehat{v}_{LM} = sL\widehat{i}_L, \tag{54}$$

$$\widehat{v}_{MS} = \widehat{v}_{LS} - \widehat{v}_{LM}. \tag{55}$$

Then (53) can be rewritten as

$$\left(sL + \frac{1}{g_{MS}}\right)\widehat{i}_L = \frac{k_{d_1}}{g_{MS}}\widehat{d}_1 + \widehat{v}_{LS} + \frac{g_M}{g_{MS}}\widehat{v}_{SD}, \tag{56}$$

and considering the capacitor, its current becomes

$$C\frac{d\widehat{u}_C}{dt} = \widehat{i}_o - \frac{\widehat{u}_C}{R}. \tag{57}$$

If $i_o = i_L$, (57) will become

$$C\frac{d\widehat{u}_C}{dt} = \widehat{i}_L - \frac{\widehat{u}_C}{R}. \tag{58}$$

Then, with $u_c = v_o$, we have

TABLE 3: Relevant factors of the circuit averaging (CA) method.

DC-DC converter	Buck	Boost	Buck-boost	KY
k_S	$(2M^2V_i)/(D_1R)$	$(2M(M-1)V_i)/(D_1R)$	$(2M^2V_i)/(D_1R)$	$(2M(M-1)V_i)/(D_1R)$
k_D	$((2M(1-M)V_i)/(D_1R))$	$(2MV_i)/(D_1R)$	$(2MV_i)/(D_1R)$	$(2M(2-M)V_i)/(D_1R)$
k_{d1}	$(2MV_i)/(D_1R)$	$(2M^2V_i)/(D_1R)$	$(2M(M+1)V_i)/(D_1R)$	$(2MV_i)/(D_1R)$
g_S	$(M^2)/((1-M)R)$	$(M^2-M)/(R)$	M^2/R	$(M(M-1))/((2-M)R)$
g_D	$(2M)/R$	$(2M)/R$	$(2M)/R$	$(2M)/R$
g_M	$(1-M)/R$	$M/((M-1)R)$	$1/R$	$(M(2-M))/((M-1)R)$
g_{MS}	$1/((1-M)R)$	$M^3/((M-1)R)$	$((M+1)^2)/R$	$M/((2-M)(M-1))$

$$\hat{i}_L = \left(SC + \frac{1}{R} \right)\hat{v}_o, \tag{59}$$

and with $i_o = i_D$, (57) yields

$$C\frac{d\hat{u}_C}{dt} = k_D\hat{d}_1 + g_D\hat{v}_{MS} + g_M\hat{v}_{MD} - \frac{\hat{u}_C}{R}. \tag{60}$$

Finally, with (54)–(56), $u_c = v_o$, (60) can be rewritten as

$$\hat{i}_L = \frac{g_{MS}}{g_D + g_M}\left(SC + \frac{1}{R} \right)\hat{v}_o$$

$$+ \left(k_{d1} - \frac{k_D \cdot g_{MS}}{g_D + g_M} \right)\hat{d}_1 + \left(g_M - \frac{g_M \cdot g_{MS}}{g_D + g_M} \right)\hat{v}_{SD}. \tag{61}$$

Here, (56), (59), and (61) are the 3 general equations to deduce the DCM small signal models for the different DC-DC converters. By just input, the voltage drops V_{LS} and V_{SD} from Figure 5, plus the factors (Table 3) and the DC parameters (Table 2) into (56), (59) and (61), and the corresponding DCM small signal transfer functions can be calculated easily in the following.

5.2. Buck Converter Small Signal Transfer Function.

For the buck converter, from Figure 5, $V_{LS} = V_i - V_o$, $V_{SD} = -V_i$, and $i_o = i_L$, and from Table 2, $I_L = (D_1^2T_sV_i(1-M))/(2LM)$ and $(D^2T_sR)/(2L) = M^2/(1-M)$. Then, with $\hat{v}_i = 0$, (56) and (59) lead to

$$(sL + (1-M)R)\hat{i}_L = \frac{2M(1-M)V_i}{D_1}\hat{d}_1 - \hat{v}_o, \tag{62}$$

$$\hat{i}_L = \left(sC + \frac{1}{R} \right)\hat{v}_o. \tag{63}$$

And from (62) and (63) simultaneously, the transfer function of the buck converter can be obtained as

$$\frac{\hat{v}_o(s)}{\hat{d}_1(s)} = \frac{M(1-M)((2V_i)/D_1)}{s^2LC + s((L/R) + RC(1-M)) + 2 - M}. \tag{64}$$

5.3. Boost Converter Small Signal Transfer Function.

For the boost converter, from Figure 5, $V_{LS} = V_i$, $V_{SD} = -V_o$, $i_o = i_D$, and from Table 2, $I_L = (D_1^2T_sV_iM)/(2L(M-1))$ and $(D_1^2T_sR)/(2L) = M^2 - M$. Then, with $\hat{v}_i = 0$, (56) and (61) imply,

$$\left(sL + \frac{(M-1)R}{M^3} \right)\hat{i}_L = \frac{2(M-1)V_i}{D_1M}\hat{d}_1 - \frac{1}{M^2}\hat{v}_o, \tag{65}$$

$$\hat{i}_L = \frac{M^2}{2M-1}\left(sC + \frac{2M-1}{MR} \right)\hat{v}_o + \frac{2(M-1)M^2V_i}{(2M-1)D_1R}\hat{d}_1. \tag{66}$$

Again, from (65) and (66) simultaneously, the transfer function of the boost converter will become,

$$\frac{\hat{v}_o(s)}{\hat{d}_1(s)} = \frac{((D_1T_sV_i)/M)(((2(M-1))/(D_1^2T_sM)) - s)}{s^2LC + s(((L(2M-1))/(RM)) + ((RC(M-1))/(M^3))) + ((2M-1)/M^3)}. \tag{67}$$

5.4. Buck-Boost Converter Small Signal Transfer Function.

For the buck-boost converter, from Figure 5, $V_{LS} = V_i$, $V_{SD} = -(V_i + V_o)$, and $i_o = i_D$, and from Table 2, $I_L = (D_1^2T_sV_i(1+M))/(2LM)$ and $(D_1^2T_sR)/(2L) = M^2$. Then, with $\hat{v}_i = 0$, (56) and (61) impose

$$\left(sL + \frac{R}{(M+1)^2} \right)\hat{i}_L = \frac{2MV_i}{D_1(M+1)}\hat{d}_1 - \frac{1}{(M+1)^2}\hat{v}_o, \tag{68}$$

$$\hat{i}_L = \frac{(M+1)^2}{2M+1}\left(SC + \frac{2M^2 + 2M + 1}{(M+1)^2R} \right)\hat{v}_o + \frac{2(M^3 + M^2)V_i}{(2M+1)D_1R}\hat{d}_1. \tag{69}$$

Similarly, with (68) and (69) considered simultaneously, the transfer function of the buck-boost converter will be

$$\frac{\widehat{v}_o(s)}{\widehat{d}_1(s)} = \frac{\big((D_1 T_s V_i)/(M+1)\big)\big(\big((2M)/(D_1^2 T_s (M+1))\big) - s\big)}{s^2 LC + s\big(\big((L(2M^2 + 2M + 1))/(R(M+1)^2)\big) + \big((RC)/\big((M+1)^2\big)\big)\big) + \big(2/((M+1)^2)\big)}. \tag{70}$$

6.5. KY Converter Small Signal Transfer Function. For the KY converter, from Figure 5, $V_{LS} = 2V_i - V_o$, $V_{SD} = -V_i$, and $i_o = i_L$, and from Table 2, $I_L = (D_1^2 T_s V_i (2 - M))/(2L(M - 1))$ and $(D_1^2 T_s R)/(2L) = (M(M-1))/((2-M))$. Then, with $\widehat{v}_i = 0$, (56) and (59) entail

$$\left(sL + \frac{(2-M)(M-1)R}{M}\right)\widehat{i}_L = \frac{2(2-M)(M-1)V_i}{D_1}\widehat{d}_1 - \widehat{v}_o, \tag{71}$$

$$\widehat{i}_L = \left(sC + \frac{1}{R}\right)\widehat{v}_o. \tag{72}$$

Finally, from (71) and (72) simultaneously, the transfer function of the KY converter will become

$$\frac{\widehat{v}_o(s)}{\widehat{d}_1(s)} = \frac{(2-M)(M-1)\big((2V_i)/D_1\big)}{s^2 LC + s\big((L/R) + ((RC(2-M)(M-1))/M)\big) + \big(((2-M)(M-1) + M)/M\big)}. \tag{73}$$

6. Simulation Results

We simulated the four DC-DC converters within a Cadence environment of a 65 nm CMOS process and also MATLAB to verify and compare the accuracy of the small signal transfer functions (30), (35), (40), (45), (64), (67), (70), and (73) deduced by the improved SSA and CA methods.

6.1. Buck Converter. The relevant parameters of the buck converter (Figure 2) are $V_i = 1.2$ V, $f_s = 100$ MHz, $C = 10$ nF, $L = 36$ nH, and $R = 40\,\Omega$. The Bode plot comparison between the transfer functions with the SSA and the CA methods ((30) and (64)), and the simulated circuit power stage (PS), is shown in Figure 6, for $d_1 = 0.3$, 0.5, and 0.7.

From Figure 6, we can conclude that in the buck converter, the CA method provides higher accuracy than the SSA method in any duty ratio range d_1, with clearer emphasis at larger d_1 values.

6.2. Boost Converter. The relevant parameters of the boost converter (Figure 2) are $V_i = 1.2$ V, $f_s = 100$ MHz, $C = 10$ nF, $L = 13.5$ nH, and $R = 60\,\Omega$. The Bode plot comparison between the transfer functions with the SSA and the CA methods ((35) and (67)), and the simulated circuit PS, is shown in Figure 7, for $d_1 = 0.3$, 0.5, and 0.7.

From Figure 7, we can conclude that in the boost converter with a small duty ratio value ($d_1 = 0.3$), the SSA method contains slightly better accuracy than the CA method. But, as the d_1 increases, the accuracy of the CA method will improve over the SSA method, again being more evident for larger d_1 values.

6.3. Buck-Boost Converter. The relevant parameters of the buck-boost converter (Figure 2) are $V_i = 1.2$ V, $f_s = 100$ MHz, $C = 40$ nF, $L = 15$ nH, and $R = 150\,\Omega$. The Bode plot comparison between the transfer functions with the SSA and

the CA methods ((40) and (70)), and the simulated circuit PS, is shown in Figure 8, for $d_1 = 0.3$, 0.5, and 0.7.

From Figure 8, the analysis of the simulation results of the boost converter is similar to those of the buck-boost converter, leading exactly to the same conclusions.

6.4. KY Converter. Finally, for the KY converter (Figure 2), $V_i = 1.2$ V, $f_s = 100$ MHz, $C = 10$ nF, $C_f = 10$ nF, $L = 3.6$ nH, and $R = 60\,\Omega$, and the Bode plot comparison between the transfer functions with the SSA and the CA methods ((45) and (73)), and the simulated circuit PS, is shown in Figure 9, for $d_1 = 0.3$, 0.5, and 0.7.

From Figure 9, the analysis of the simulation results of the KY converter are similar to those of the buck converter, leading exactly to the same conclusions.

7. High-Accuracy Modeling Method for Different DC-DC Converters in DCM—Selection Strategy

7.1. Derivation. By using the methods presented in [3] and [6], the approximate poles and zeros for different DC-DC converters in DCM with the SSA and CA methods can be calculated and summarized in Table 4. The previous Bode plot simulation results clearly demonstrate that the phase-frequency responses of the DC-DC converters power stages generally show a larger phase lag than the small signal models given by both the SSA and CA methods. Based on this, if the modeling method presents a smaller value of the second pole or zero (leading to a larger phase lag), it will exhibit a better accuracy of the system phase-frequency response. From Table 4, we propose a selection strategy of high-accuracy small signal modeling method for the DC-DC converters as in Table 5.

7.2. Verification. For verification of the proposed selection strategy, we simulated a buck converter and a boost

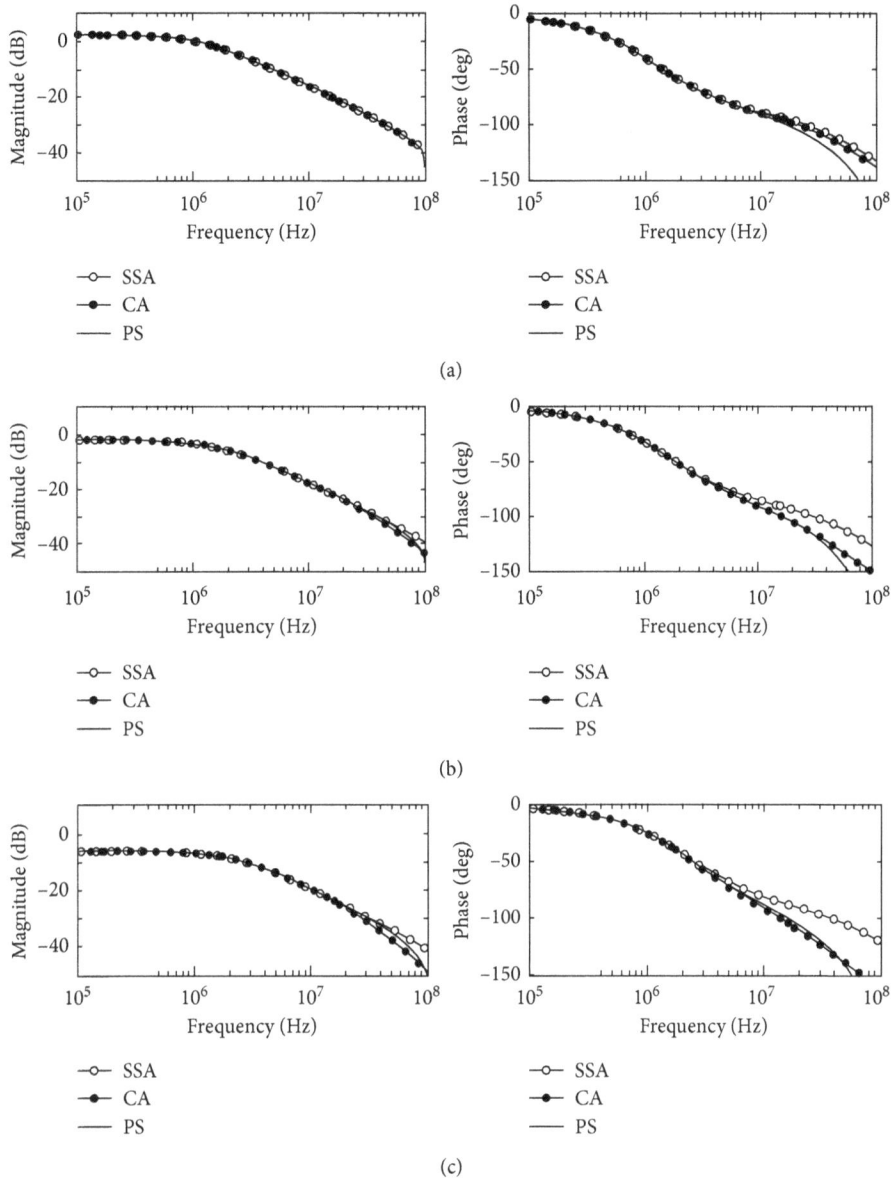

FIGURE 6: Small signal models comparison of the buck converter with SSA and CA methods: (a) $d_1 = 0.3$; (b) $d_1 = 0.5$; and (c) $d_1 = 0.7$.

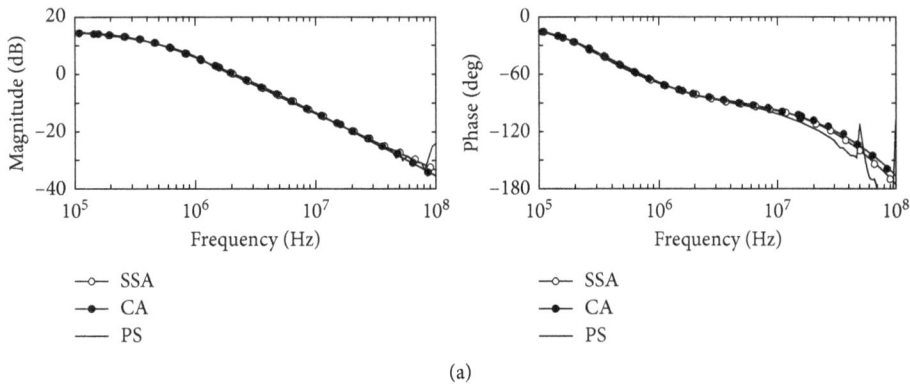

(a)

FIGURE 7: Continued.

(b)

(c)

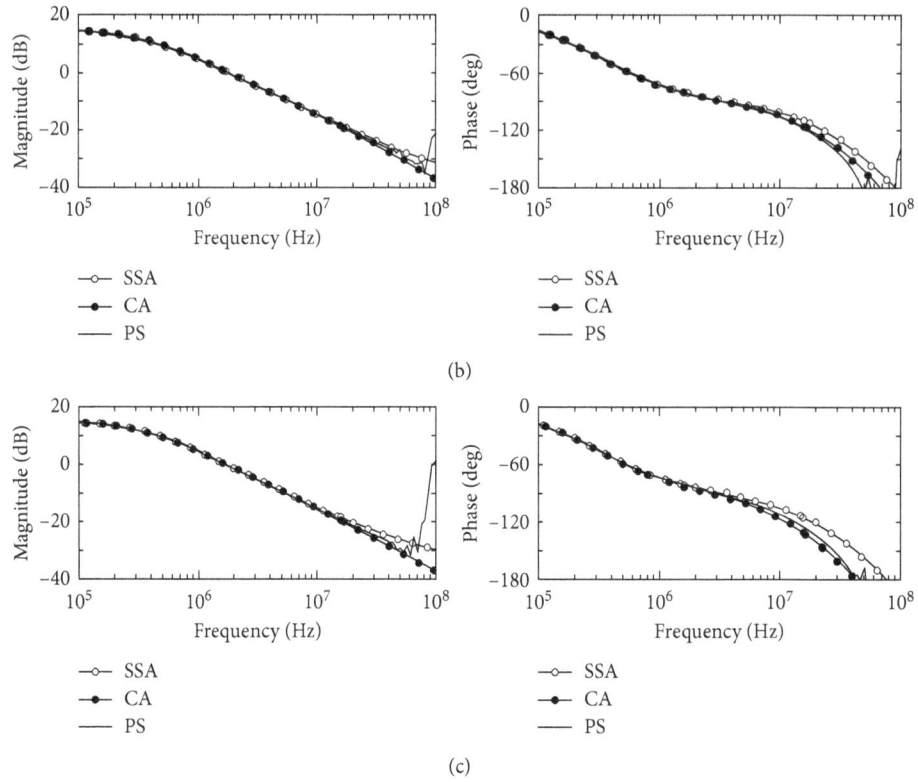

FIGURE 7: Small signal models comparison of the boost converter with SSA and CA methods: (a) $d_1 = 0.3$; (b) $d_1 = 0.5$; and (c) $d_1 = 0.7$.

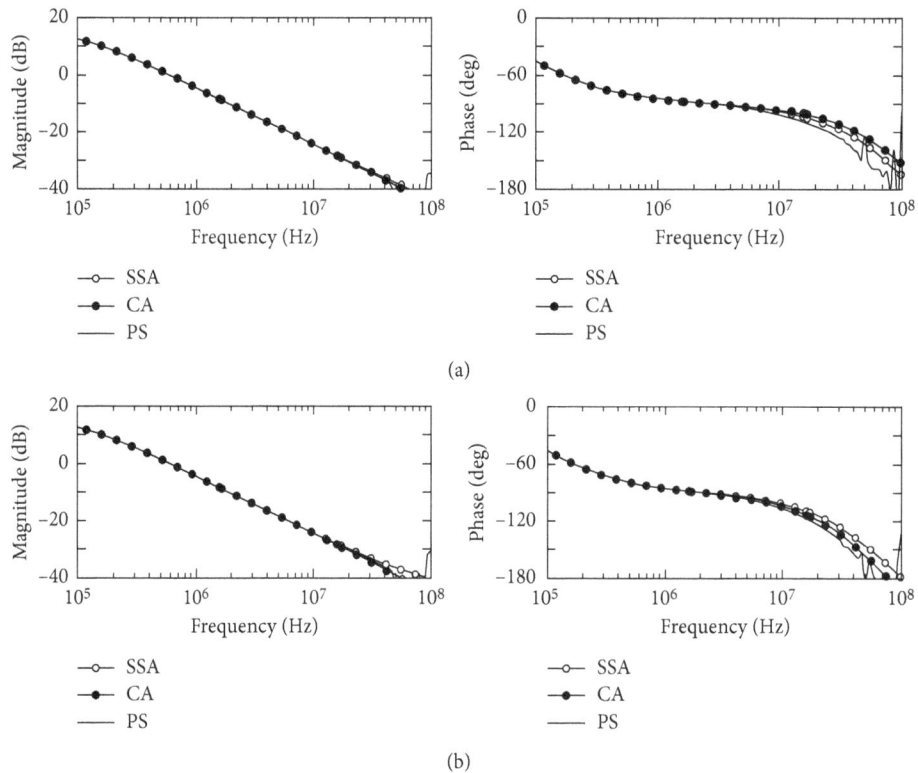

(a)

(b)

FIGURE 8: Continued.

(c)

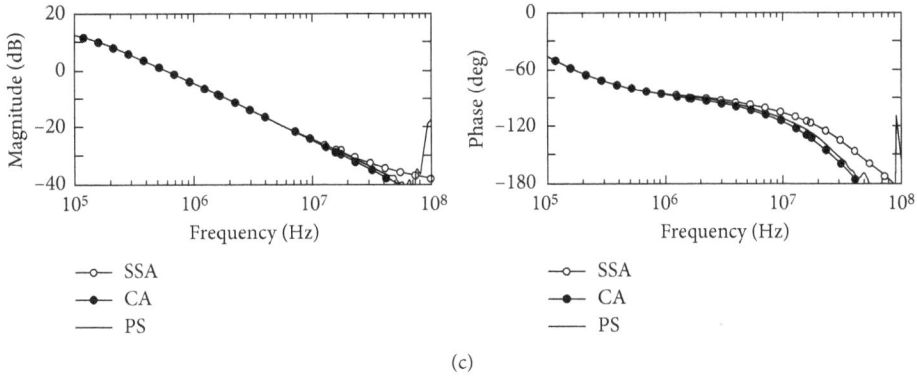

FIGURE 8: Small signal models comparison of the buck-boost converter with SSA and CA methods: (a) $d_1 = 0.3$; (b) $d_1 = 0.5$; and (c) $d_1 = 0.7$.

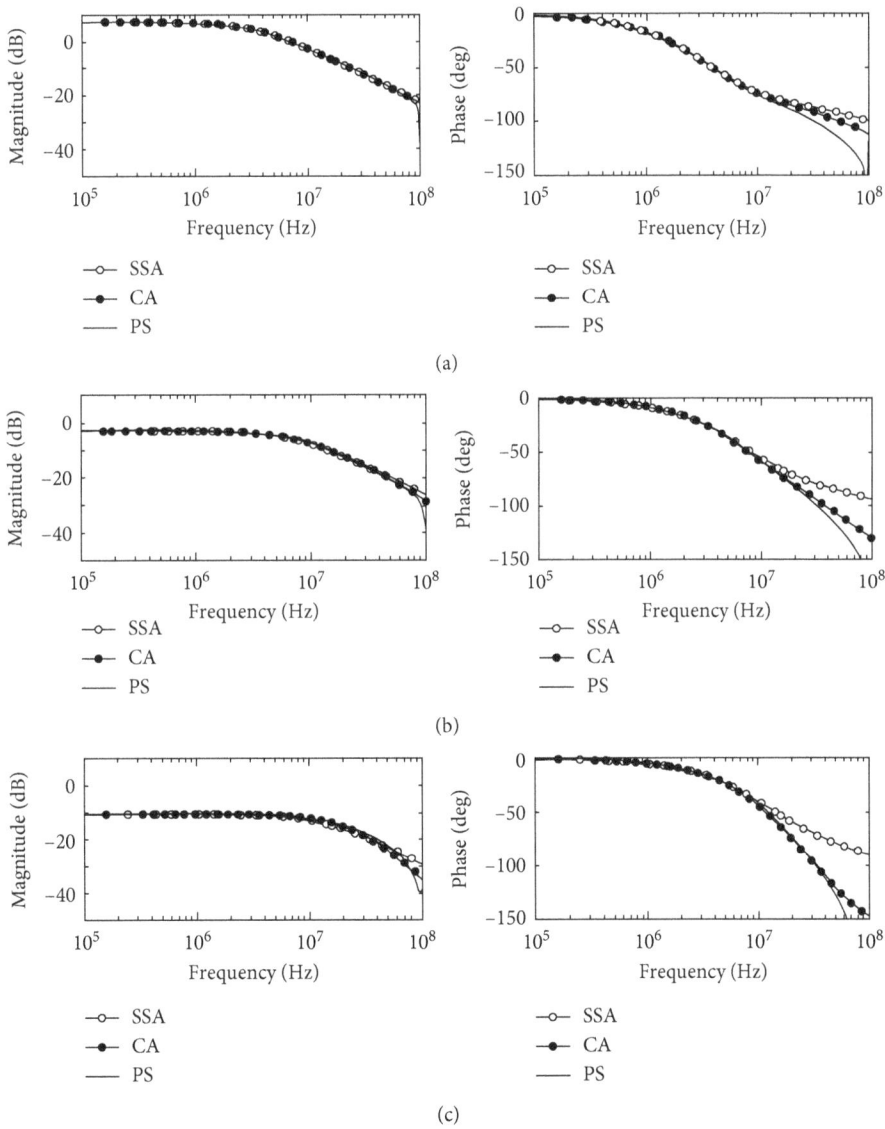

(a)

(b)

(c)

FIGURE 9: Small signal models comparison of the KY converter with SSA and CA methods: (a) $d_1 = 0.3$; (b) $d_1 = 0.5$; and (c) $d_1 = 0.7$.

converter with the Cadence Spectre simulator to demonstrate that the converter operates with higher stability with a more accurate small signal modeling method used in the compensator design. The simulated results of the system phase-frequency response and the closed-loop controlled converters' load transient response are presented in

TABLE 4: Approximate poles and zeros for various DC-DC converters in DCM.

Name	Method	First pole	Second pole	Zero
Buck	SSA	$(2-M)/(RC(1-M))$	$(2M)/(d_1T_s(1-M))$	
	CA	$(2-M)/(RC(1-M))$	$(2M^2)/(d_1^2T_s)$	
Boost	SSA	$(2M-1)/(RC(M-1))$	$(2(M-1))/(d_1T_s)$	$2/(d_1T_s)$
	CA	$(2M-1)/(RC(M-1))$	$(2((M-1)/M)^2)/(d_1^2T_s)$	$(2((M-1)/M))/(d_1^2T_s)$
Buck-boost	SSA	$2/(RC)$	$(2M)/(d_1T_s)$	$2/(d_1T_s)$
	CA	$2/(RC)$	$2(M/(M+1))^2/d_1^2T_s$	$2(M/(M+1))^2/d_1^2T_s$
KY	SSA	$(M^2-4M+2)/(RC(M^2-3M+2))$	$(2(M-1))/(d_1T_s(2-M))$	
	CA	$(M^2-4M+2)/(RC(M^2-3M+2))$	$(2(M-1)^2)/(d_1^2T_s)$	

TABLE 5: Selection strategy between SSA and CA for high-accuracy small signal modeling.

Name	Criterion	Modeling selection
Buck	$(2M/(d_1T_s(1-M))) < ((2M^2)/(d_1^2T_s))$	SSA
	$(2M/(d_1T_s(1-M))) > ((2M^2)/(d_1^2T_s))$	CA
Boost	$\min(((2(M-1))/(d_1T_s)), (2/d_1T_s)) < (2((M-1)/M)^2/(d_1^2T_s))$	SSA
	$\min(((2(M-1))/(d_1T_s)), (2/d_1T_s)) > (2((M-1)/M)^2/(d_1^2T_s))$	CA
Buck-boost	$\min(((2M)/(d_1T_s)), (2/d_1T_s)) < (2(M/(M+1))^2/(d_1^2T_s))$	SSA
	$\min(((2M)/(d_1T_s)), (2/d_1T_s)) > (2(M/(M+1))^2/(d_1^2T_s))$	CA
KY	$(2(M-1))/(d_1T_s(2-M)) < ((2(M-1)^2)/(d_1^2T_s))$	SSA
	$(2(M-1))/(d_1T_s(2-M)) > ((2(M-1)^2)/(d_1^2T_s))$	CA

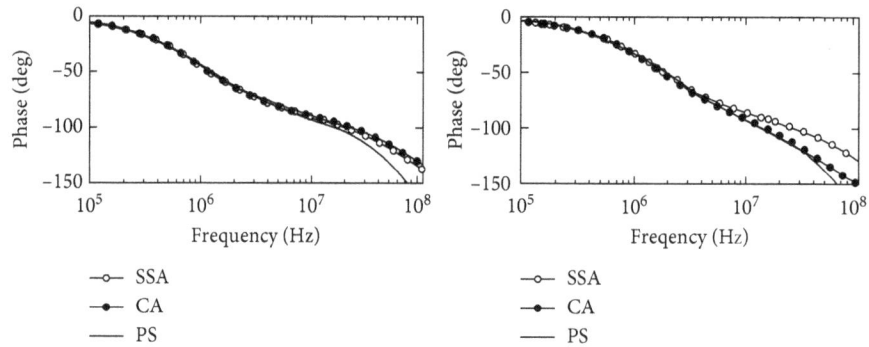

FIGURE 10: Simulation verification with the buck converter ($V_i = 1.2$ V, $f_s = 100$ MHz, $C = 10$ nF, $L = 40$ nH, and $R = 40\,\Omega$): (a) $d_1 = 0.2$, $(2M/(d_1T_s(1-M))) < ((2M^2)/(d_1^2T_s))$; and (b) $d_1 = 0.5$, $(2M/(d_1T_s(1-M))) > ((2M^2)/(d_1^2T_s))$.

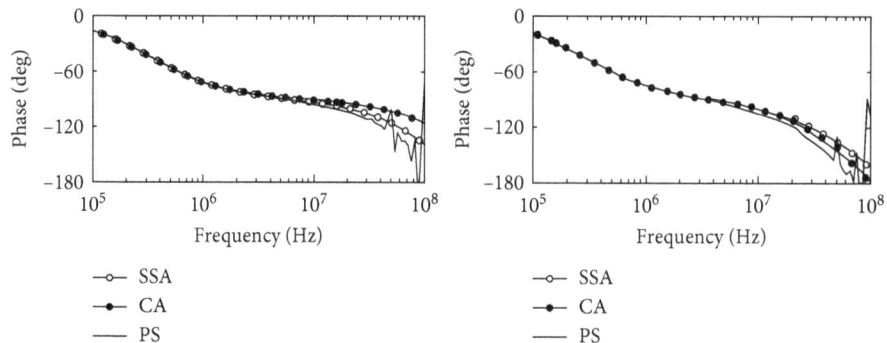

FIGURE 11: Simulation verification with the boost converter ($V_i = 1.2$ V, $f_s = 100$ MHz, $C = 20$ nF, $L = 2$ nH, and $R = 15\,\Omega$): (a) $d_1 = 0.2$, $\min(((2(M-1))/(d_1T_s)), (2/d_1T_s)) < (2((M-1)/M)^2/(d_1^2T_s))$; and (b) $d_1 = 0.5$, $\min(((2(M-1))/(d_1T_s)), (2/d_1T_s)) > (2((M-1)/M)^2/(d_1^2T_s))$.

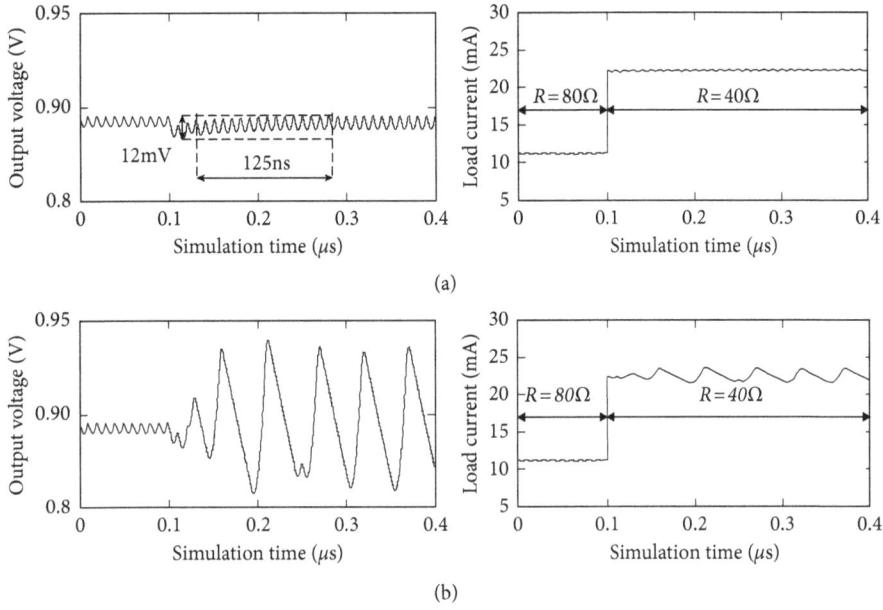

FIGURE 12: V_o and i_o of the designed buck converter with $((2M)/(d_1 T_s (1-M))) > (2M^2/d_1^2 T_s)$ ($V_i = 1.2$ V, $f_s = 100$ MHz, $C = 10$ nF, $L = 40$ nH, and $R = 80/40\,\Omega$) and PM = 45°. (a) Compensator based on CA. (b) Compensator based on SSA.

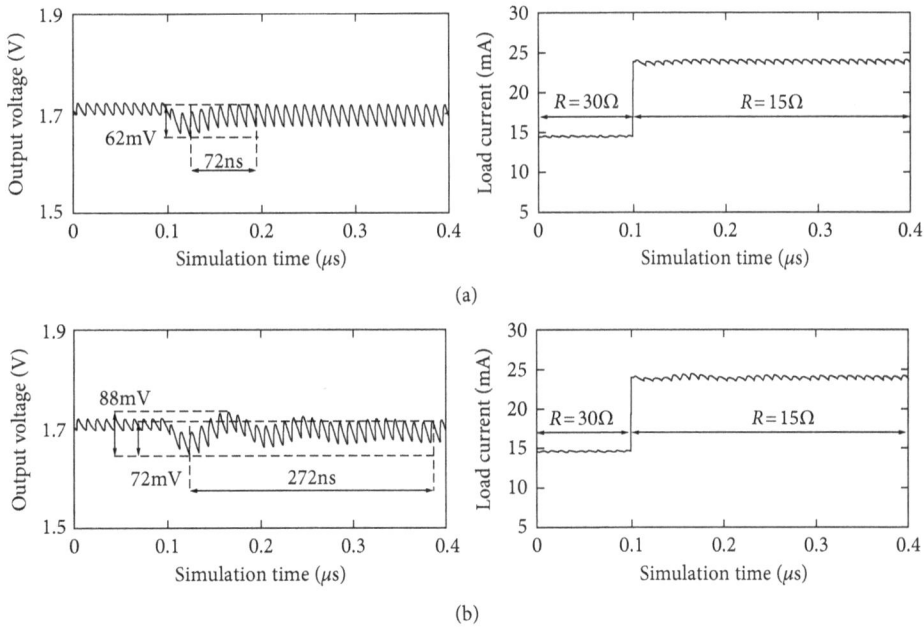

FIGURE 13: V_o and i_o of the designed boost converter with $\min(((2(M-1))/(d_1 T_s)), (2/d_1 T_s)) < (2((M-1)/M)^2/(d_1^2 T_s))$ ($V_i = 1.2$ V, $f_s = 100$ MHz, $C = 20$ nF, $L = 2$ nH, and $R = 30/15\,\Omega$) and PM = 45°. (a) Compensator based on SSA. (b) Compensator based on CA.

Figures 10–13. These figures confirm the correctness of the proposed selection strategy in Table 5.

Figure 12 presents the simulated output voltage V_o and load current i_o of the designed closed-loop controlled buck converter during a load transient, applying a Type II compensator. As indicated in Table 5, the CA method is more accurate with the condition $((2M)/(d_1 T_s (1-M))) > ((2M^2)/(d_1^2 T_s))$, and the simulated result (Figure 12) also confirms that the closed-loop controller designed with

the CA method exhibits better stability and transient response than the SSA method, even though both cases have the same phase margin (PM) of 45°. On the other hand, Figure 13 shows the simulated V_o and the i_o of the closed-loop controlled boost converter during a load transient, applying also a Type II compensator. In this case, as indicated in Table 5, the SSA modeling method is more accurate with the condition $(\min((2(M-1))/(d_1 T_s)), (2/d_1 T_s)) < ((2((M-1)/M)^2)/(d_1^2 T_s))$ and the simulated result

(Figure 13) also confirms that the closed-loop controller designed with SSA method obtain a better stability and transient response. When there is a sudden change of the i_o, the boost converter with the SSA method responds faster than that with the CA method.

Unlike the conclusion made in [6], this paper shows that, in some cases, the CA method exhibits better accuracy than the SSA method. Figures 12 and 13 confirm that an accurate modeling method is critical to design the appropriate closed-loop controller of the DC-DC converter, demonstrating that the selection strategy given in Table 5 is essential and necessary in the design. The general and streamlined small signal deduction process for both modeling methods can be further applied conveniently to similar DC-DC converter topologies.

8. Conclusions

This paper presented the review, study, DCM small signal modeling deduction and simulation verification by using the improved SSA and CA methods for four DC-DC converters. This paper first proposed a general and intuitive deriving process for the improved SSA and CA modeling methods, such that the corresponding DCM small signal models for DC-DC converters can be easily determined. Then, this paper discovers that the CA can obtain higher accuracy than the improved SSA at some operating conditions, as some research studies claimed that the improved SSA can obtain the highest accuracy among all the modeling methods. Finally, this paper provided a selection strategy for a high-accuracy modeling method for various DC-DC converters operating in DCM, verified by simulations, which is necessary and beneficial in the design of a more accurate DCM closed-loop controller for DC-DC converters, achieving better stability and transient response.

Conflicts of Interest

The authors declare that there are no conflicts of interest regarding the publication of this paper.

Acknowledgments

This work was supported in part by the Science and Technology Development Fund, Macao SAR (FDCT) (120/2016/A3) and in part by the Research Committee of the University of Macau (MYRG2015-00030-AMSV and MYRG2017-00090-AMSV).

References

[1] R. D. Middlebrook and S. Cuk, "A general unified approach to modelling switching-converter power stages," in *Proceedings of IEEE Annual Power Electronics Specialists Conference (PESC)*, pp. 18–34, Cleveland, OH, USA, 1976.

[2] R. Tymerski and V. Vorperian, "Generation, classification and analysis of switched-mode DC-to-DC converters by the use of converter cells," in *Proceedings of International Telecommunications Energy Conference*, pp. 181–195, Toronto, Canada, 1986.

[3] D. Maksimovic and S. Cuk, "A unified analysis of PWM converters in discontinuous modes," *IEEE Transactions on Power Electronics*, vol. 6, no. 3, pp. 476–490, 1991.

[4] E. Mamarelis, G. Petrone, and G. Spagnuolo, "An hybrid digital-analog sliding mode controller for photovoltaic applications," *IEEE Transactions on Industrial Informatics*, vol. 9, no. 2, pp. 1094–1103, 2013.

[5] V. Vorperian, "Simplified analysis of PWM converters using model of PWM switch. II. Discontinuous conduction mode," *IEEE Transactions on Aerospace and Electronic Systems*, vol. 26, no. 3, pp. 497–505, 1990.

[6] J. Sun, D. M. Mitchell, M. F. Greuel, P. T. Krein, and R. M. Bass, "Averaged modeling of PWM converters operating in discontinuous conduction mode," *IEEE Transactions on Power Electronics*, vol. 16, no. 4, pp. 482–492, 2001.

[7] K. Mandal, S. Banerjee, and C. Chakraborty, "A new algorithm for small-signal analysis of DC–DC converters," *IEEE Transactions on Industrial Informatics*, vol. 10, no. 1, pp. 628–636, 2014.

[8] R. H. G. Tan and M. Y. W. Teow, "A comprehensive modeling, simulation and computational implementation of buck converter using MATLAB/Simulink," in *Proceedings of IEEE Conference on Energy Conversion*, pp. 37–42, Johor Bahru, Malaysia, 2014.

[9] J. P. Torreglosa, P. García, L. M. Fernández, and F. Jurado, "Predictive control for the energy management of a fuel-cell–battery–supercapacitor tramway," *IEEE Transactions on Industrial Informatics*, vol. 10, no. 1, pp. 276–285, 2014.

[10] M. K. Kazimierczuk, *Pulse-Width Modulated DC-DC Power Converters*, Wiley, West Sussex, UK, 2008.

[11] J. Sun, D. M. Mitchell, M. F. Greuel, P. T. Krein, and R. M. Bass, "Modeling of PWM converters in discontinuous conduction mode. A reexamination," in *Proceedings of IEEE Annual Power Electronics Specialists Conference (PESC)*, pp. 615–622, Fukuoka, Japan, 1998.

[12] W. R. Liou, W. B. Lacorte, A. B. Caberos et al., "A programmable controller IC for DC/DC converter and power factor correction applications," *IEEE Transactions on Industrial Informatics*, vol. 9, no. 4, pp. 2105–2113, 2013.

[13] W. L. Zeng, C.-S. Lam, W.-M. Zheng et al., "DCM operation analysis of ky converter," *Electronics Letters*, vol. 51, no. 24, pp. 2037–2039, 2015.

[14] Y. Qiu, X. Y. Chen, C. Q. Zhong, and C. Qi, "Uniform models of PWM DC–DC converters for discontinuous conduction mode considering parasitics," *IEEE Transactions on Industrial Electronics*, vol. 61, no. 11, pp. 6071–6080, 2014.

[15] A. Davoudi, J. Jatskevich, and T. De Rybel, "Numerical state-space average-value modeling of PWM DC-DC converters operating in DCM and CCM," *IEEE Transactions on Power Electronics*, vol. 21, no. 4, pp. 1003–1012, 2006.

[16] M. U. Iftikhar, P. Lefranc, D. Sadarnac, and C. Karimi, "Theoretical and experimental investigation of averaged modeling of non-ideal PWM DC-DC converters operating in DCM," in *Proceedings of IEEE Annual Power Electronics Specialists Conference (PESC)*, pp. 2257–2263, Rhodes, Greece, 2008.

[17] R. Trinchero, I. S. Stievano, and F. G. Canavero, "Steady-state analysis of switching power converters via augmented time-invariant equivalents," *IEEE Transactions on Power Electronics*, vol. 29, no. 11, pp. 5657–5661, 2014.

[18] T. Pavlovic, T. Bjazic, and Z. Ban, "Simplified averaged models of DC-DC power converters suitable for controller design and microgrid simulation," *IEEE Transactions on Power Electronics*, vol. 28, no. 7, pp. 3266–3276, 2013.

[19] E. Van Dijk, J. N. Spruijt, D. M. O'Sullivan, and J. B. Klaassens, "PWM-switch modeling of DC-DC converters," *IEEE Transactions on Power Electronics*, vol. 10, no. 6, pp. 659–665, 1995.

[20] C. P. Basso, *Switch-Mode Power Supplies*, McGraw-Hill, New York, NY, USA, 2008.

[21] R. W. Erickson and D. Maksimovic, *Fundamentals of Power Electronics*, Kluwer Academic Publisher, Norwell, MA, USA, 2001.

[22] G. Nirgude, R. Tirumala, and N. Mohan, "A new, large-signal average model for single-switch DC-DC converters operating in both CCM and DCM," in *Proceedings of IEEE Annual Power Electronics Specialists Conference (PESC)*, vol. 3, pp. 1736–1741, Vancouver, BC, Canada, 2001.

[23] K. I. Hwu and Y. T. Yau, "KY converter and its derivatives," *IEEE Transactions on Power Electronics*, vol. 24, no. 1, pp. 128–137, 2009.

A 3.9 μs Settling-Time Fractional Spread-Spectrum Clock Generator using a Dual-Charge-Pump Control Technique for Serial-ATA Applications

Takashi Kawamoto,[1] **Masato Suzuki,**[2] **and Takayuki Noto**[2]

[1]*Hitachi Central Research Laboratory, 1-280 Higashi-Koigakubo, Kokubunji-shi, Tokyo 185-8601, Japan*
[2]*Renesas Electronics Corporation, Tokyo 185-8601, Japan*

Correspondence should be addressed to Takashi Kawamoto; takashi.kawamoto.hv@hitachi.com

Academic Editor: John N. Sahalos

A low-jitter fractional spread-spectrum clock generator (SSCG) utilizing a fast-settling dual-charge-pump (CP) technique is developed for serial-advanced technology attachment (SATA) applications. The dual-CP architecture reduces a design area to 60% by shrinking an effective capacitance of a loop filter. Moreover, the settling-time is reduced by 4 μs to charge a current to the capacitor by only main-CP in initial period in settling-time. The SSCG is fabricated in a 0.13 μm CMOS and achieves settling time of 3.91 μs faster than 8.11 μs of a conventional SSCG. The random jitter and total jitter at 250 cycles at 1.5 GHz are less than 3.2 and 10.7 psrms, respectively. The triangular modulation signal frequency is 31.5 kHz and the modulation deviation is from −5000 ppm to 0 ppm at 1.5 GHz. The EMI reduction is 10.0 dB. The design area and power consumption are 300 × 700 μm and 18 mW, respectively.

1. Introduction

Serial Advanced Technology Attachment (SATA) is widely used as a low-cost, high-speed interface for external storage devices like HDDs and optical disc drives (ODDs) such as blu-ray discs, DVDs, and CDs. However, electromagnetic interference (EMI) is a particular problem with SATA devices [1]. One approach to reducing the EMI is to apply a spread-spectrum clock generator (SSCG) to a SATA-PHY.

Figure 1 shows a common block diagram of the SATA-PHY. It consists of a parallel-to-serial converter (P/S), a spread-spectrum clock generator (SSCG), a driver (DRV), a receiver (RCV), a clock and data recovery (CDR) circuit, and a serial-to-parallel converter (S/P). The SSCG generates a transmission clock signal (F_{VCO}). The P/S converts a transmission parallel data (TD) into a transmission serial data by using the F_{VCO}. This transmission serial data is transmitted by the DRV. The F_{VCO} frequency should be modulated to reduce the EMI in accordance with the SATA specification [1]. A received serial data is inputted to the CDR via the RCV. The CDR generates the recovery data (DATA) and recovery clock (CLK) from the received data. The serial-to-parallel converter (S/P) converts from the DATA to the received parallel data (RD) by using the CLK. In this SATA-PHY, the SSCG is applied a fractional SSCG because of a large EMI reduction [2–18]. The fractional SSCG should be narrow loop bandwidth because the quantized noise originated from a $\Sigma\Delta$ modulator is removed. Therefore, the fractional SSCG essentially has large design area and long settling-time. There were some approaches to reduce design area in previous works [15, 17, 18]. However, those could not consist with shrinking design area, shorting settling-time, and reducing EMI and jitter. Therefore, we introduced a capacitance multiplication technique to the fractional SSCG and then we proposed fast-settling technique by controlling the CP.

Figure 2 depicts the states defined by the SATA specification and the SATA-PHY power consumption [1]. In a sync state, a communication is successfully established between a host and a device. In the sync state, the SATA-PHY operates the SSCG. The slumber state is a standby state. The SATA-PHY can stop the SSCG because allowed wake-up period

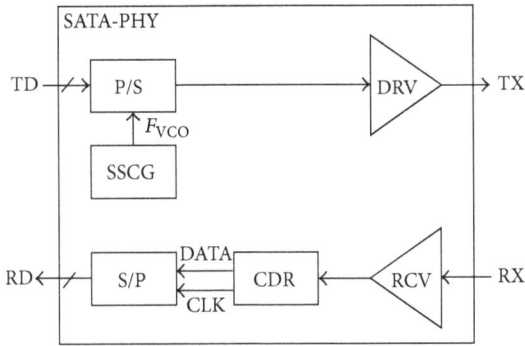

FIGURE 1: Block diagram of SATA-PHY.

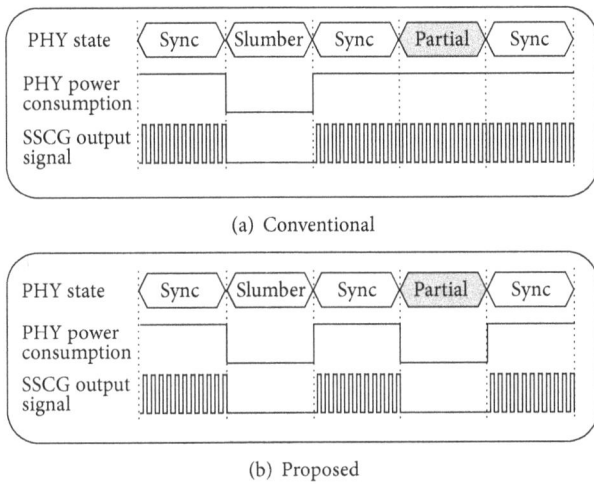

(a) Conventional

(b) Proposed

FIGURE 2: Proposed low power SATA-PHY operation.

FIGURE 3: Conventional SSCG block diagram.

The F_{VCO} frequency is thus modulated by the triangular wave. This SSCG can reduce EMI more substantially than other SSCGs because the linearity of the modulation can be obtained accurately by utilizing the logic circuits [2–7].

This SSCG has two main jitter sources. One is a VCO jitter originating from the thermal and flicker noise of the MOS transistors. The other is a $\Sigma\Delta$ modulator jitter originating from the quantization noise of the $\Sigma\Delta$ modulator. The SSCG output jitter is the sum of these two jitters. To remove the quantization noise that is high-pass characteristics, the loop bandwidth should be designed narrow. Thus, the settling-time is necessarily longer and the design area is large.

Several approaches have been presented for reducing the design area. The high-resolution fractional divider technique shifts the modulator quantization noise to the higher frequency side and so it achieves wider bandwidth [5]. However, it is difficult to reduce the EMI by much because spurious jitter originating from the high-resolution fractional divider remains in the modulation bandwidth. All-digital SSCGs have been presented as a means of substantially reducing design area [17, 18]. However, their output jitter is still large because their digitally controlled ring oscillators generate large jitter and it is difficult to operate them accurately if there are PVT variations. The capacitance multiplication technique has been presented to reduce the design area as an approach in which the operation is based on that of a conventional SSCG [2, 7, 20]. However, the settling time is necessarily long because the loop bandwidth is set to be narrow.

To achieve a low-cost SATA-PHY suitable for portable applications, the design area, settling-time, power consumption, jitter, and EMI must all be reduced at the same time. To consummate these aggressive demands, we have proposed the dual-CP SSCG architecture with fast-settling CP control technique.

The rest of the paper is organized as follows. Section 2 describes the overall dual-CP SSCG architecture in detail. Section 3 presents the fast-settling CP control technique we have developed to achieve short settling time. Section 4 describes a CP circuit design to achieve a dual-CP architecture that is robust against PVT variations. Section 5 describes the VCO with high-frequency limiter. Section 6 presents

from slumber to sync is a long period. The partial state is also a standby state. However, allowed wake-up period from partial to sync is a short period of less than 10 μs. Thus, it cannot stop the SSCG because the settling time of the SSCG is longer than 10 μs as shown in Figure 2(a). The SATA-PHY is set to the partial state many times in HDD and ODD applications. If the SSCG can achieve a settling time less than 10 μs, the SATA-PHY can stop the SSCG in partial state as shown in Figure 2(b). This is attractive for portable applications because of saving power. To achieve this operation, the SSCG has to have a settling time of less than about 4 μs, taking the wake-up time of the other blocks into consideration. This stringent settling time requirement is far shorter than that of a conventional SSCG.

Figure 3 shows a block diagram of a conventional SSCG based on a fractional PLL [3, 4, 6, 10–20]. It consists of a phase frequency detector (PFD), a charge pump (CP), a 3rd order loop filter (LF), a voltage controlled oscillator (VCO), a multimodulus divider (MMD), a programmable counter (PGC), a $\Sigma\Delta$ modulator ($\Sigma\Delta$), and a wave generator (WG). The WG is a logic circuit and generates a triangular wave as a spread-spectrum modulation. The $\Sigma\Delta$ modulates the triangular signal and then generates divide ratio (N) that the average of the N is modulated by the triangular wave.

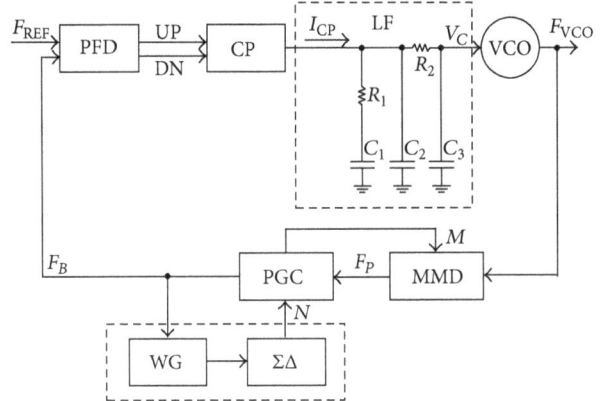

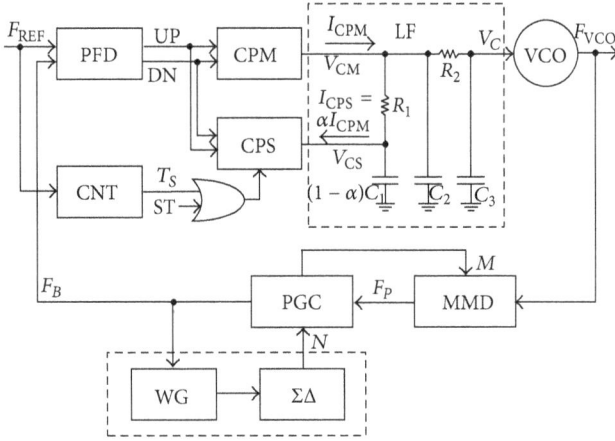

FIGURE 4: Block diagram of the proposed dual-CP SSCG with fast-settling CP control technique.

measurement results for evaluation purposes, and Section 7 concludes with a short summary of the key points.

2. Overall Dual-CP Architecture

Figure 4 shows our dual-CP SSCG architecture with fast-settling CP control technique to reduce the design area, settling time, power consumption, jitter, and EMI all at the same time. It includes a conventional SSCG, an additional CP (CPS), and a counter (CNT). A third-order $\Sigma\Delta$ modulator and a third-order low-pass loop filter are applied to reduce the $\Sigma\Delta$ modulator jitter. The PFD compares a phase of the reference clock signal (F_{REF}) with that of the feedback clock signal (F_B) and then generates up and down signals (UP, DN). Two CPs, a main CP (CPM) and an auxiliary CP (CPS), are applied to fulfill the need for high capacitance via the use of a capacitance multiplier. When the UP is generated by the PFD, the CPM charges the C_1 by the current (I_{CPM}), and the CPS discharges the C_1 by the current (I_{CPS}). In this architecture, the open loop transfer function of the main CP (CPM) path ($F_{\mathrm{CPM}}(s)$) is given by

$$F_{\mathrm{CPM}}(s) = \left(I_{\mathrm{CPM}} \left(C_1 R_1 \right) s + 1 \right)$$
$$\cdot \left(s^3 + \left(\frac{1}{C_3 R_2} + \frac{1}{C_1 R_1} + \frac{1}{C_2 R_1} + \frac{1}{C_2 R_2} \right) s^2 \right.$$
$$\left. + \left(\frac{C_1 + C_2 + C_3}{C_1 C_2 C_3 R_1 R_2} \right) s \right)^{-1} . \quad (1)$$

The open loop transfer function of the auxiliary CP (CPS) path ($F_{\mathrm{CPS}}(s)$) is given by

$$F_{\mathrm{CPS}}(s) = I_{\mathrm{CPS}}$$
$$\cdot \left(s^3 + \left(\frac{1}{C_3 R_2} + \frac{1}{C_1 R_1} + \frac{1}{C_2 R_1} + \frac{1}{C_2 R_2} \right) s^2 \right.$$
$$\left. + \left(\frac{C_1 + C_2 + C_3}{C_1 C_2 C_3 R_1 R_2} \right) s \right)^{-1} . \quad (2)$$

If the $I_{\mathrm{CPM}} = \alpha * I_{\mathrm{CPS}}$ and α is less than 1, the open loop transfer function from the PFD to the VCO control voltage (V_C) is given by

$$F(s) = \left(I_{\mathrm{CPM}} \left(C_1 R_1 \right) s + I_{\mathrm{CPM}} \left(1 - \alpha \right) \right)$$
$$\cdot \left(s^3 + \left(\frac{1}{C_3 R_2} + \frac{1}{C_1 R_1} + \frac{1}{C_2 R_1} + \frac{1}{C_2 R_2} \right) s^2 \right.$$
$$\left. + \left(\frac{C_1 + C_2 + C_3}{C_1 C_2 C_3 R_1 R_2} \right) s \right)^{-1} . \quad (3)$$

The zero of the open loop transfer function is given by

$$Fu = \frac{(1 - \alpha)}{C_1 R_1} . \quad (4)$$

On the other hand, that of the conventional one is given by

$$Fu = \frac{1}{C_1 R_1} . \quad (5)$$

Our dual-CP technique, therefore, results in C_1 being ($1 - \alpha$) times smaller than that of the conventional one.

There is a key design point in this dual-CP architecture. Figure 5 shows the difference in the CP and VCO characteristics for a conventional SSCG and proposed dual-CP SSCG. The locking range, which means the V_C range at which the charge current (I_{CPP}) is almost the same as the discharge current (I_{CPN}), is wide because the SSCG loop has a tolerance for the current difference "$I_{\mathrm{CPP}} - I_{\mathrm{CPN}}$" in the conventional SSCG, as shown in Figure 5(a). The SSCG does not have to lock out of the lock range, which means that the CP current difference between the I_{CPP} and the I_{CPN} is large. However, in this case, the jitter originating from CP current difference becomes large. Thus, the SSCG output jitter becomes larger. Therefore, to meet the SATA jitter specification, the SSCG should lock into the lock range. In a conventional SSCG, the locking-point can thus be designed at a higher voltage area to reduce the VCO jitter because the low VCO sensitivity (K_V) brings about in the low VCO jitter.

In proposed dual-CP SSCG, the jitter originating from CP is more sensitive than that of the conventional one. Thus, the lock range becomes narrower as shown in Figure 5(b). This is because the SSCG loop is affected by the current differences "$I_{\mathrm{CPMP}} - I_{\mathrm{CPSN}}$" and "$I_{\mathrm{CPMN}} - I_{\mathrm{CPSP}}$". This means that it is important for the current difference to have a sufficient tolerance for PVT variations. The narrow lock range makes it possible to design the VCO sensitivity (K_V) to be higher than that of the conventional one.

Figure 6 shows a typical example of the open loop transfer function. As aspect of the EMI, the loop bandwidth should be designed wider because harmonic elements of the modulation triangle signal that fundamental frequency is 31.5 kHz can pass through the loop bandwidth. In our previous work, the 15th harmonics of the triangular signal should be passed in order to obtain the large EMI reduction [3]. However, as aspect of the jitter, as the loop bandwidth is wider, the jitter originated from $\Sigma\Delta$ modulator quantized

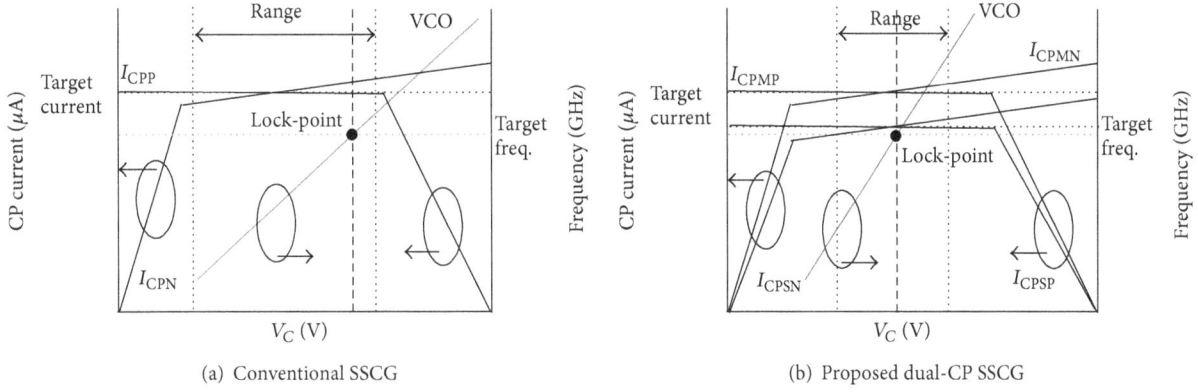

(a) Conventional SSCG

(b) Proposed dual-CP SSCG

FIGURE 5: The explanation of CP current characteristics and VCO frequency current characteristics of the conventional SSCG (a) and the proposed dual-CP SSCG (b).

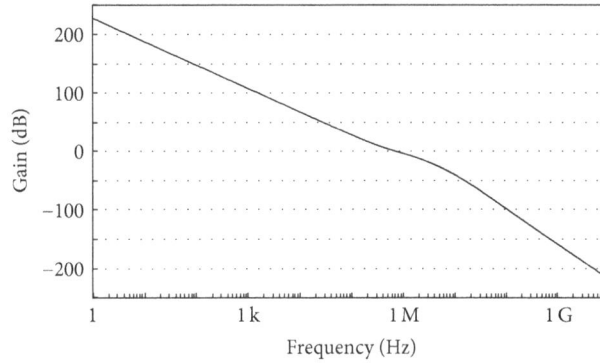

FIGURE 6: SSCG open loop frequency characteristics.

noise is larger. And the jitter originated from the VCO phase noise is larger as the loop bandwidth is narrower. Therefore, the loop bandwidth is designed at about 650 kHz. This is wide enough to meet the jitter specification and reduce the EMI, but this structure cannot achieve a settling-time of less than 4 μs. Such a settling-time is achieved by utilizing the CP control technique we describe in Section 3. It is important for the SSCG to have a sufficient tolerance for CP current variation due to PVT variation.

Figure 7 shows the effects of the CP current variation on the loop design. The current difference "$I_{CPM} - I_{CPS}$" affects the phase margin very little as shown in Figure 7(a). Even if the current difference varies −50%, the effect on the phase margin is less than 5% as shown in Figure 7(a). On the other hand, the current "I_{CPM}" or "I_{CPS}" has a huge influence impact on the loop bandwidth. Even if the different current varies −50%, the phase margin is affected at less than 50% as shown in Figures 7(b) and 7(c). The CP should be designed such that the variation of the current difference should remain less than about ±40%, taking the jitter specification into consideration. The main CP current (I_{CPM}) and auxiliary CP current (I_{CPS}) have a huge impact on the loop bandwidth and phase margin. The variation of the I_{CPM} and I_{CPS} should be designed at less than ±20%, taking the jitter specification and loop stability into consideration. A CP design that is robust against PVT variation is presented in Section 4.

3. Fast-Settling CP Control Technique

We have developed a fast-setting CP control technique. Figure 8 shows the concept of proposed fast-settling CP control technique. In the conventional SSCG, the settling-time is long because the large C_1 is charged by the small CP current as shown in Figure 8(a) [7]. As shown in Figure 8(a), in this SSCG, a charging speed (Δ_{CONV}) that means a slope in the settling period is given by

$$\Delta_{conv} = \frac{(1 - \alpha) I_{CPM}}{C_1}. \tag{6}$$

In our dual-CP SSCG architecture, the C_1 is smaller than that of the conventional one. Thus, in the dual-CP SSCG, if a charged current is same, the charging speed is faster than that of the conventional one. In the dual-CP SSCG, the differential charge current is small; however, the I_{CPM} is larger than the charge current of the conventional SSCG. Thus, the charging speed can be faster if the only CPM charges the C_1. As shown in Figure 8(b), in this case, the charging speed (Δ_{PROP}) is given by

$$\Delta_{PROP} = \frac{I_{CPM}}{C_1}. \tag{7}$$

The charging speed (Δ_{PROP}) achieved with our technique is $1/(1 - \alpha)$ times faster than the conventional one. In this

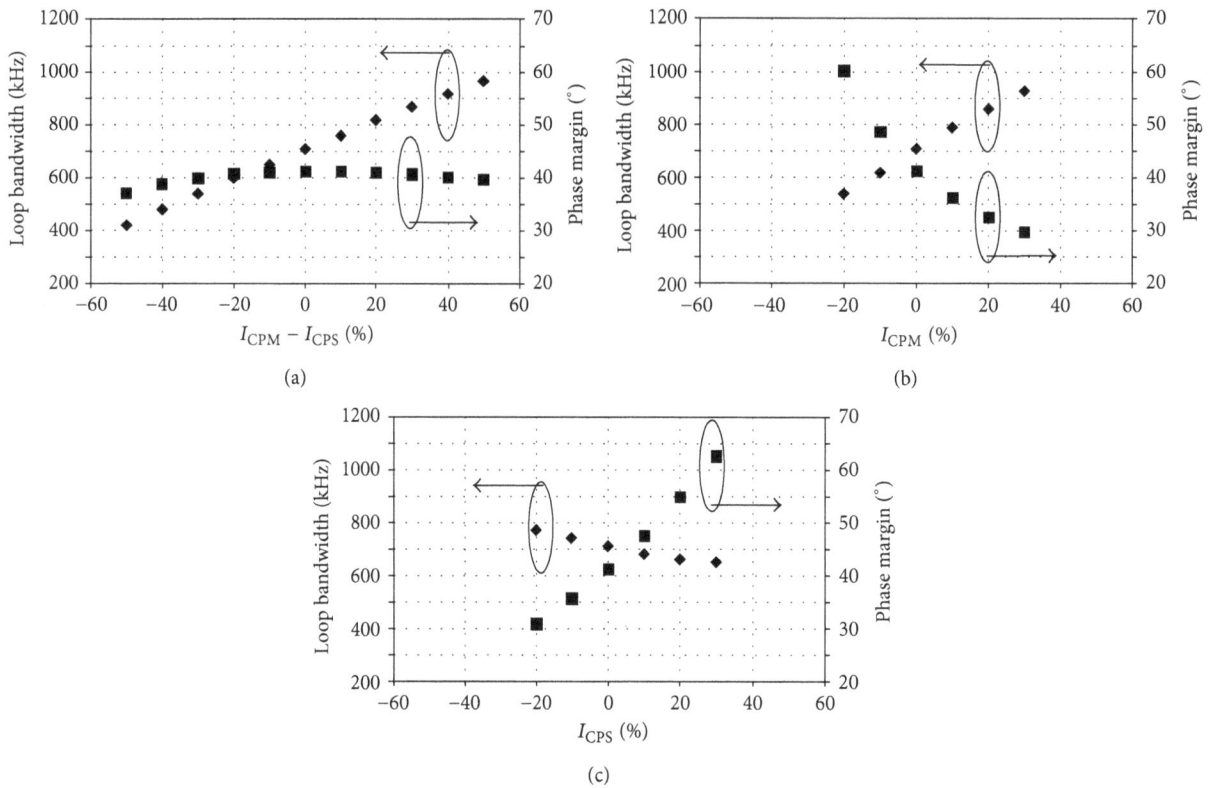

FIGURE 7: The effect of the loop bandwidth and phase margin due to CP parameters. (a) The effect by different current ($I_{CPM} - I_{CPS}$). (b) The effect by CPM (I_{CPM}). (c) The effect by CPS (I_{CPS}).

case, the CPS activates in the middle of the settling period. If the CPS activates at early in the settling period, the effect of the boosting charge by the CPM is weak. On the other hand, if the CPS activates at late in the settling period, a large overshoot may occur because the operation is unstable and settling period is prolonged rather than shortened. Moreover, the MMD cannot operate due to the overshoot and then the SSCG falls into a malfunction as shown in Figure 8(b). To overcome this trade-off, the VCO with high frequency limiter is applied to the SSCG as shown in Figure 8(c) [3]. When the CPS is activated, the overshoot occurs. However, the F_{VCO} frequency cannot be more than 2.2 GHz, which is the MMD maximum operating frequency by the high frequency limiter. Therefore, even if the overshoot occurs, the SSCG can be locked. To reduce the settling period, the CPS activation time is as long as possible. However, the settling period becomes longer due to large dumping if the CPS is activated after cross-over 1.5 GHz of the SSCG output signal. Thus, the CPS should be activated before cross-over 1.5 GHz of the SSCG output signal. Moreover, even if the CPS is activated right before cross-over 1.5 GHz, the large dumping might occur in the case that the phase difference between the reference clock and the feedback clock becomes large. It is difficult to control the phase difference at the CPS activation point. Thus, the CPS should activate relative less than cross over 1.5 GHz to have a margin of the lock period of the phase difference. In our proposed SSCG, the CPS activation time is set to 1 μs to reduce settling period when the SSCG achieves the settling

period of less than 4 μs. As shown in Figure 4, the CPS is controlled by the CNT. The CNT is the counter that makes the CPS activation time by counting the F_{REF}. As shown in Figure 8, the SSCG is activated when the standby signal (T_S) is set to low. At this time, the CPS is not activated because the T_S is set to high. After a certain period that is made by the CNT, the T_S is set to low and the CPS activates.

Figure 9 shows the behavior simulation results for the settling time. This simulation is not designed for a settling time of less than 4 μs but designed to verify the fast settling period by using the CPS control. The conventional dual-CP SSCG achieves a settling time of less than about 22 μs in this simulation. On the other hand, when only the CPM operates, the overshoot occurs and the operation is unstable. When we apply our CP control technique to this SSCG so that the CPS is activated at 3 μs, the settling behavior is the same as that when only the CPM is activated before 3 μs. After the CPS is activated at 3 μs, the settling behavior deviates from that when only the CPM is activated and then directed to the target frequency slowly. After small overshoot occurs, the SSCG becomes locked at 18 μs. In this case, our technique enables the settling time to be shortened to about 4 μs, which is almost the same as the period during which the CPS is stopped. As the CPS is stopped for as long a time as possible, the settling time can be shortened. However, this technique has little effect when the CPS is activated after the overshoot occurs. Moreover, large overshoot may occur due to a small damping factor and the settling time may be longer than that

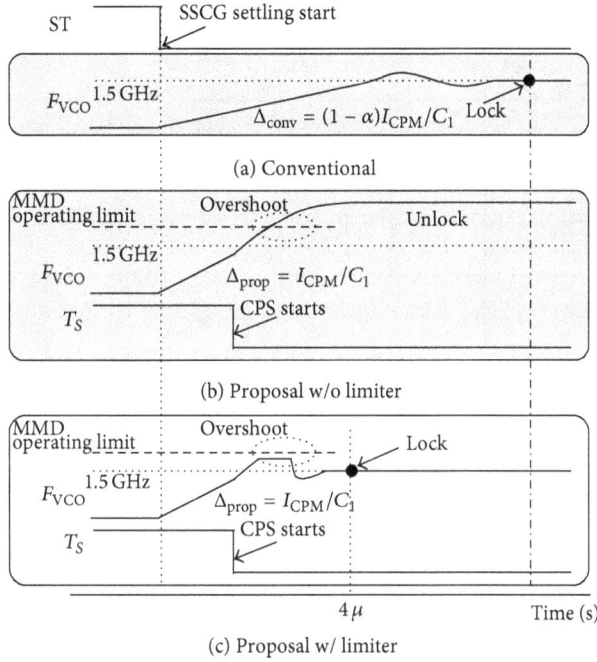

FIGURE 8: Explanation of proposed fast-lock dual-CP SSCG settling operation of the conventional SSCG (a) and the proposed dual-CP controlled technique without VCO high frequency limiter (b) and with limiter (c).

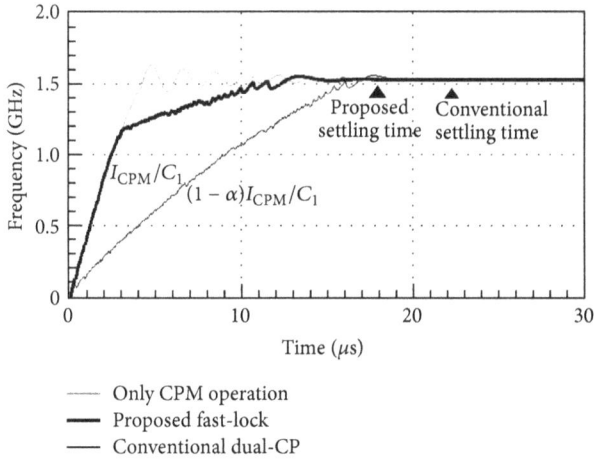

FIGURE 9: Simulation results of the conventional and proposed settling time.

without our technique's function when the CPS is activated just before the overshoot appears. Therefore, the timing is set such that the CPS is activated just before the overshoot occurs. This timing is made by counter that counts F_{REF}. In our design, the CPS is activated at about 1.0 μs, taking the CP current and filter capacitance into consideration.

4. Dual-CP Circuit Design

Figure 10 shows a circuit diagram of the CPM and the CPS. The PFD output signals (UP, UPB, DN, and DNB) are

connected to the gate of the switch MOSs (M8, M21, M9, M10, M16, M15, M17, and M18). The VB is connected to the band-gap reference (BGR) and the reference current (I_{CP}) is generated as M1 drain current. The main CP current (I_{CPMP} and I_{CPMN}) and the auxiliary CP current (I_{CPSP} and I_{CPSN}) are generated from the I_{CP} by utilizing the current mirror and are given by

$$\frac{I_{CPMP}}{I_{CP}} = \frac{W_{M5}}{W_{M3}}, \qquad \frac{I_{CPMN}}{I_{CP}} = \frac{W_{M7}}{W_{M6}},$$

$$\frac{I_{CPSP}}{I_{CP}} = \frac{W_{M13}}{W_{M3}}, \qquad \frac{I_{CPSN}}{I_{CP}} = \frac{W_{<14}}{W_{M6}}.$$

(8)

The transistor width is described as W_{MN}, where N is the transistor number. Thus, the CP current ratio (α) is given by

$$\alpha = \frac{I_{CPSP}}{I_{CPMP}} = \frac{W_{M13}}{W_{M5}} = \frac{I_{CPSN}}{I_{CPMN}} = \frac{W_{M14}}{W_{M7}}. \qquad (9)$$

Figure 11 shows the simulation results for the CPM and CPS charge/discharge characteristics. The simulation conditions are that the process, voltage, and temperature are typical, 1.35 V, and −40°C, respectively. The horizontal axis is the control voltage (V_C) and the vertical axis is the CP current. PMOS currents (I_{CPMP} and I_{CPSP}) appear as absolute values in the Figure. In our work, the CPM current (I_{CPMP} and I_{CPMN}) and CPS current (I_{CPSP} and I_{CPSN}) are designed at 44 μA and 32 μA, respectively. Therefore, the designed current difference value is 12 μA and α is 76%. In general, a PMOS transistor has accurate saturation characteristics and a narrow saturation region and an NMOS transistor has inaccurate saturation characteristics and a wide saturation region. In this work,

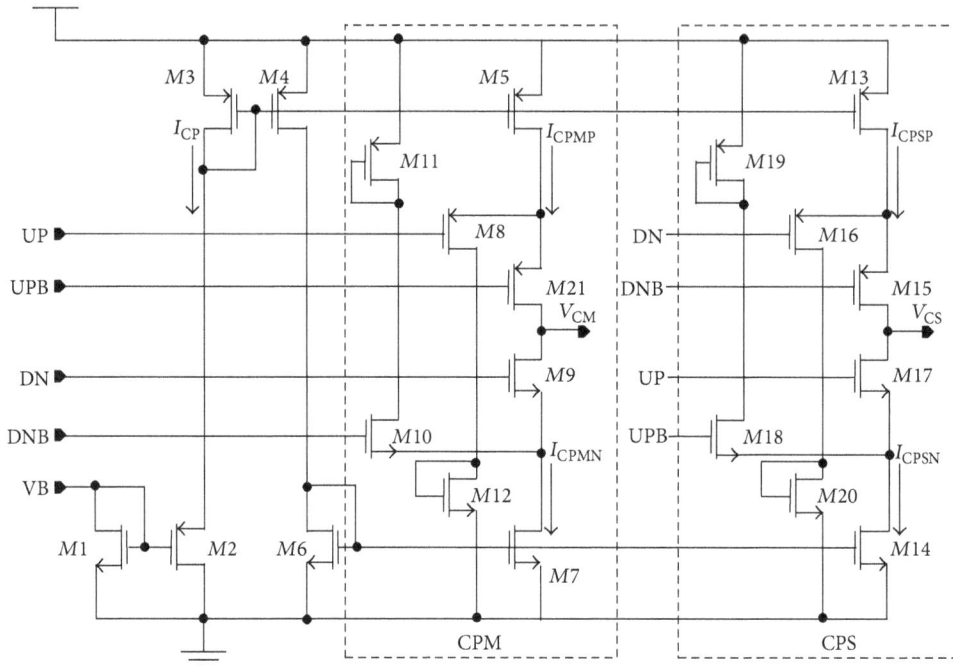

FIGURE 10: Circuit diagram of CP. The CP consists of the CPM and the CPS.

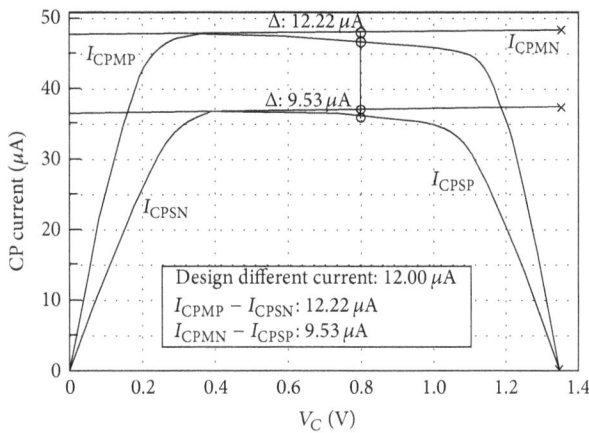

FIGURE 11: Circuit diagram of CP. The CP consists of the CPM and the CPS.

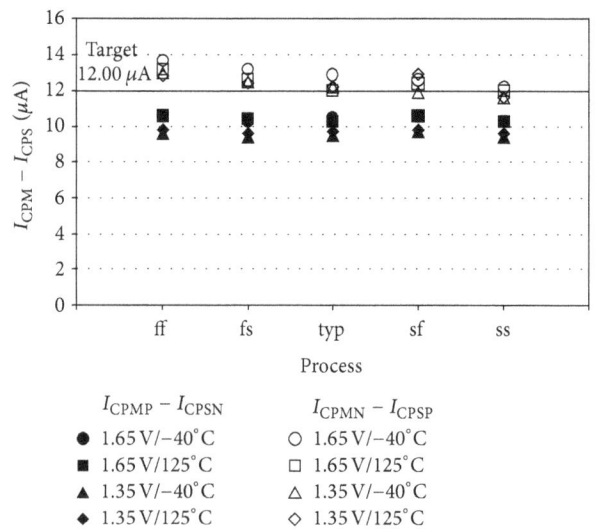

FIGURE 12: Simulation result of different CP current ($I_{CPMP} - I_{CPSN}$, $I_{CPMN} - I_{CPSP}$). Design target current is that I_{CPM} (I_{CPMP} and I_{CPMN}) and I_{CPS} (I_{CPSP} and I_{CPSN}) are 44 mA and 32 mA, respectively.

these problems are resolved merely by using a simple circuit design because other solutions, such as using an OpAmp or other additional circuits, increase design area and power consumption. The main reason for the difference between the design targets and simulation results is the channel length modulation. The NMOS length is designed to be long to decrease the effects of channel length modulation. Figure 12 shows the simulation results for the current difference between the main CP and the auxiliary CP. Since the main CP current and the auxiliary CP current are designed at 44 μA and 32 μA, respectively, the design target for the current difference is 12 μA. As shown in Figure 12, the variation of the current difference is designed at less than ±20%.

And then, this CP has offset between charge current and discharge current. This offset causes jitter. There are some techniques to overcome the offset. However, these techniques cause large power. In our SSCG, the main jitter sources are VCO and $\Sigma\Delta$ modulator and jitter is designed sufficiently even if the CP has offset. Therefore, the CP circuit as shown in Figure 10 is applied to prefer the power to the jitter.

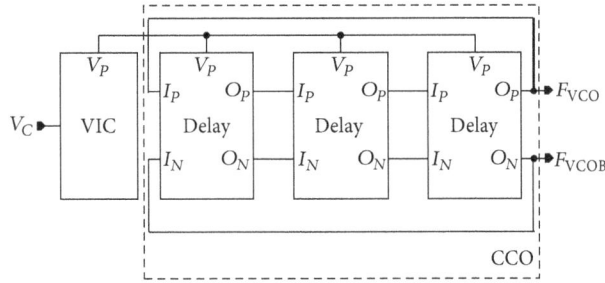

FIGURE 13: VCO block diagram [3].

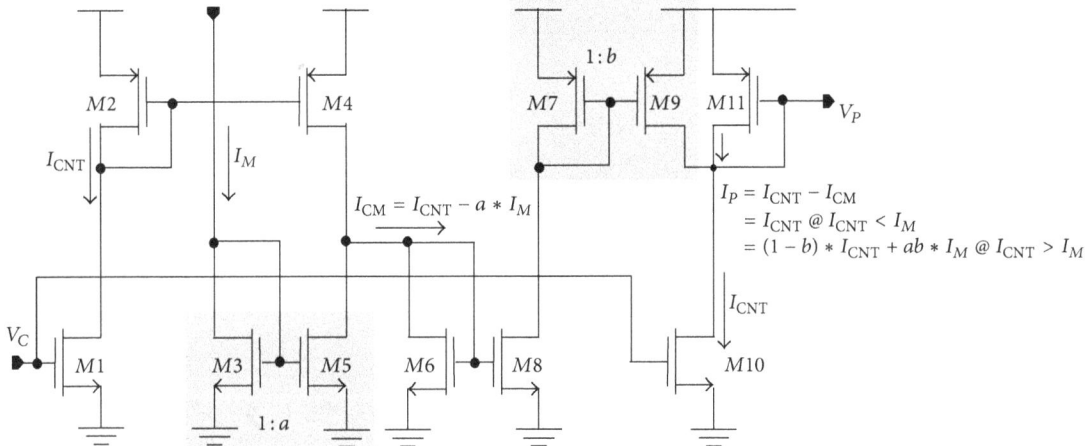

FIGURE 14: VIC circuit diagram [3].

5. VCO with High Frequency Limiter

The MMD can operate at less than 2.2 GHz under the worst condition. If the SSCG output signal frequency exceeds 2.2 GHz in the settling period due to the dual-CP control technique, the SSCG falls into an unlocked state. To prevent this malfunction, a VCO with a high frequency limiter is applied as shown in Figure 13 [3]. This VCO consists of a voltage-current converter (VIC) and a current-controlled oscillator (CCO) as shown in Figures 14 and 15. Figure 16 shows the explanation of the VCO with the high frequency limiter. The VIC converts a control voltage (V_C) to a control current (I_P). The VIC performs a high current limiter. The CCO generates output clock signals (F_{VCO} and F_{VCOB}) where the frequency is controlled by the I_P. Therefore, this VCO can perform the high frequency limiter. In the VIC, $M1$ converts the V_C to an I_{CNT}. An I_{CM} that is calculated as $I_{CNT} - I_M$ is generated at $M4$ drain node. The I_P that is a $M11$ drain current is calculated as $I_{CNT} - I_{CM}$. When the I_{CNT} smaller than the I_M, the I_P is the I_{CNT} because the I_{CM} is zero. On the other hand, when the I_{CNT} larger than the I_M, the I_P is calculated as $(1 - b) * I_{CNT} + ab * I_M$ that is nearly $ab * I_M$. The a and b are current mirror ratios of $M3 : M5$ and $M7 : M9$, respectively. The I_P is expected constant current against the V_C. However, if the I_{CNT} that is the $M1$ drain current is different from the I_{CNT} that is $M10$ drain current, the I_P may not be constant. The I_{CNT} that is $M10$ drain current is likely to be smaller than one of the $M1$

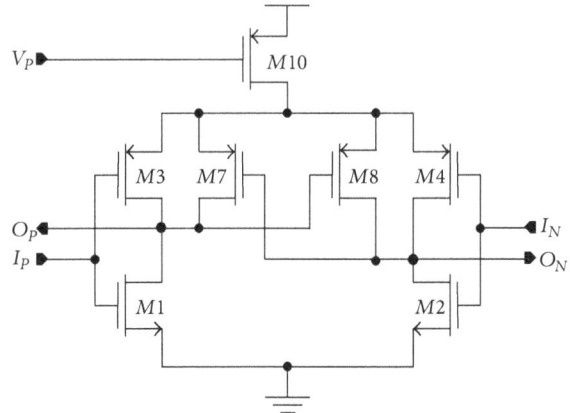

FIGURE 15: Delay cell circuit diagram [3].

because the $M10$ has heavier loads that are the $M9$ and $M11$ than the $M1$. In this case, the I_P characteristics have negative slope against the V_C. These negative characteristics cause that SSCG falls into the unlock state because the SSCG loop may be positive feedback. To prevent from this malfunction, the current mirror ratio between the $M7$ and $M9$ is $1 : b$ in order that the I_P characteristics have positive slope against the V_C when the I_{CNT} is larger than I_M.

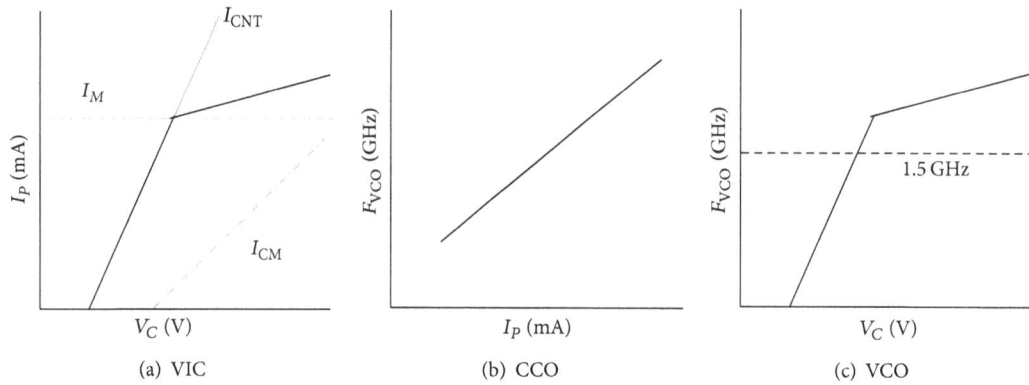

FIGURE 16: Explanation of the VCO with high frequency limiter [3].

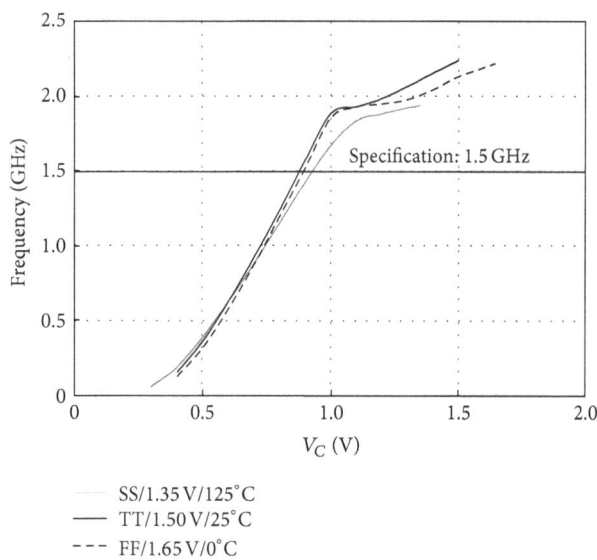

FIGURE 17: Postlayout simulation result of VCO frequency-voltage characteristics.

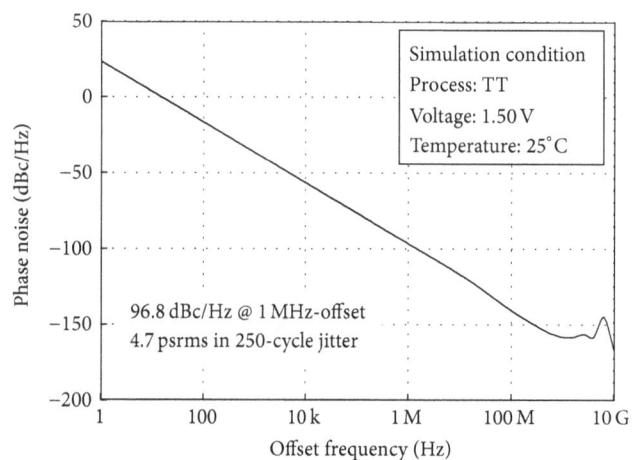

FIGURE 18: Postlayout simulation result of VCO phase noise characteristics.

Figure 17 shows the postlayout simulation results for the VCO frequency-voltage characteristics. As shown in Figure 11, the locking-point should be set at the range from 0.5 V to 0.8 V because the current differences ($I_{CPMP} - I_{CPSN}$ and $I_{CPMN} - I_{CPSP}$) are nearly equal to the design targets. The maximum frequency of the VCO output signal should be set at less than 2.2 GHz under the worst condition. As shown in Figure 17, the VCO oscillates at 1.5 GHz at about 0.8 V and the maximum frequency is less than 1.9 GHz under the worst condition.

Figure 18 shows the VCO phase noise characteristics in TT condition. The phase noise at 1 MHz offset frequency is 96.8 dBc/Hz. The jitter in 250-cycle is 4.7 psrms. Main jitter sources are thermal noise of the $M10$ and $M11$ in the VIC in Figure 14.

6. Measurement Result

We fabricated our SSCG using a 0.13 μm CMOS process. Figure 19 shows the settling-time results with and without

the fast-settling dual-CP control technique. The sample is SS. In Figure 19, there are four results. Figures 19(a) and 19(b) show the fast settling-time setup of the SSCG without and with the proposed control. Figure 19(b) shows 4 μs settling-time by using proposed control technique. On the other hand, Figures 19(c) and 19(d) show the same setup as Figure 9 without and with proposed control to demonstrate the silicon results of the settling-time same as the simulation results as shown in Figure 9. The measurement condition is 1.35 V/125°C. In Figures 19(a) and 19(b), without proposed control technique, the settling-time was 8.11 μs as shown in Figure 19(a). With it, the settling-time was 3.91 μs, as shown in Figure 19(b), which is less than 4 μs. The CPS began operating at about 1.0 μs. Soon after, an overshoot appeared and the SSCG output signal frequency became nearly 2.2 GHz. However, the VCO with its high-frequency limiter prevented it from exceeding 2.2 GHz and leading to malfunctions. In Figures 19(c) and 19(d) that are shown to compare between simulation results in Figure 9 and silicon results in Figure 19. Without proposed control technique, measurement and simulation results are 24.8 μs and 22.5 μs, respectively. With control technique, measurement and simulation results are 19.4 μs and 18.2 μs, respectively. In the

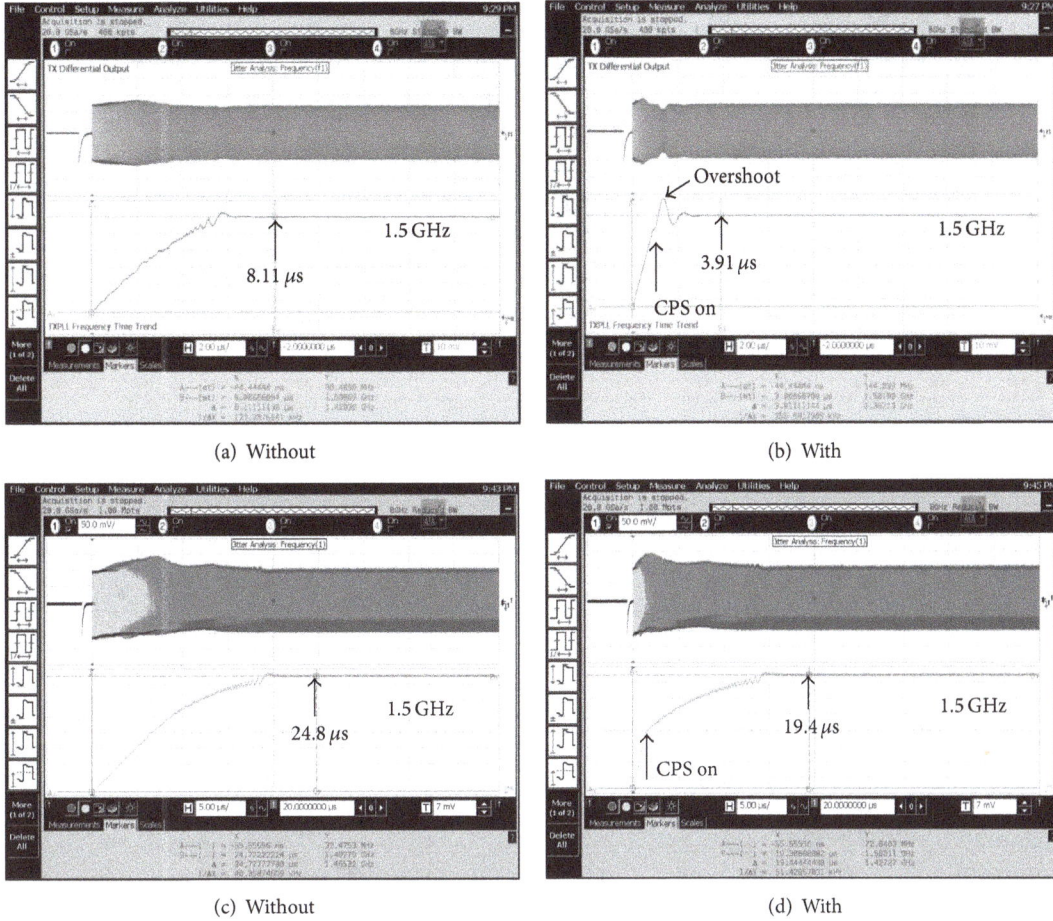

(a) Without

(b) With

(c) Without

(d) With

FIGURE 19: Measurement result of the proposed SSCG settling-period. The sample is SS. Measurement condition is 1.35 V/125°C. Without proposed fast-lock dual-CP function (a), settling-period could be less than 8.11 ms. With function (b), settling-period could be less than 3.91 µs. (c) and (d) are same PLL parameter as Figure 9. Without function (c), measurement and simulation settling period are 24.8 µs and 22.5 µs, respectively. With function (d), measurement and simulation settling period are 19.4 µs and 18.2 µs, respectively.

settling-time, measurement results are similar to simulation results.

Figure 20 shows the measurement results for the SSCG output signal frequency. The signal was modulated by a triangular wave whose frequency was 31.5 kHz. The modulation deviation of the 1.5 GHz output signal was from +50 ppm to −4259 ppm, which met the SATA specification of from +350 ppm to −5000 ppm.

Figure 21 shows the measurement results of SSCG output signal spectrum. The EMI reduction was 10.0 dB with the SSC.

Figure 22 shows the measurement results for RJ and TJ under various conditions; the results met the SATA specification for all PVT variations. The RJ was less than 3.2 psrms. The domain jitter source was the VCO. The CP jitter due to the current mismatch of the dual-CP was far smaller than the VCO jitter.

Figure 23 shows the measurement result of the VCO frequency-voltage characteristics in worst condition. The 1.5 GHz locking frequency was achieved at 1.0 V. The maximum frequency is less than 2.2 GHz.

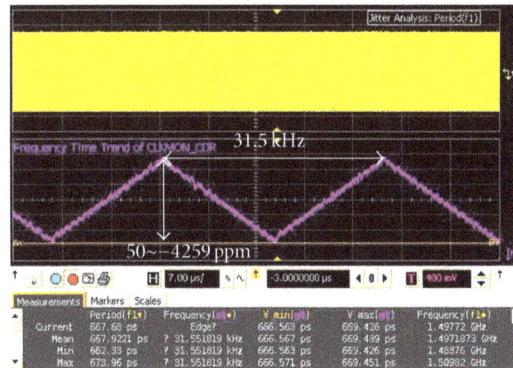

FIGURE 20: Measurement result of the output signal frequency modulated by triangular signal. Modulation frequency is 31.5 kHz and modulation deviation is from +50∼−4259 ppm at 1.5 GHz.

Figure 24 shows the measurement results for the CP current. The CP currents were measured by using an output pin between the CP and the LF. When the I_{CPMP} was measured, the CPM was enabled and the CPS was disabled,

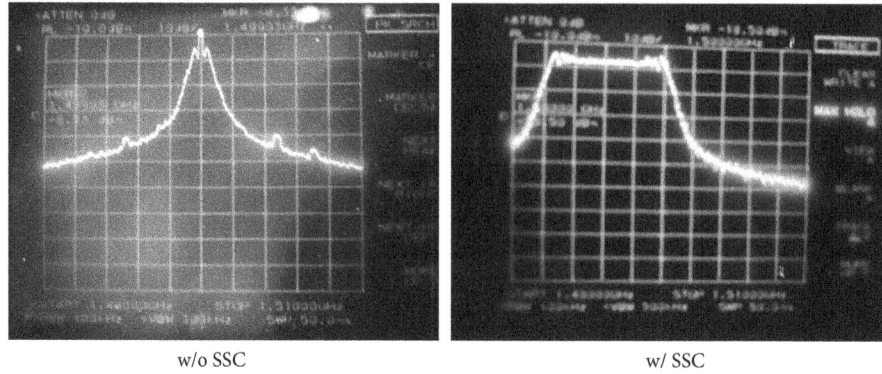

w/o SSC   w/ SSC

FIGURE 21: Measurement results of the SSCG output signal spectrum with and without SSC.

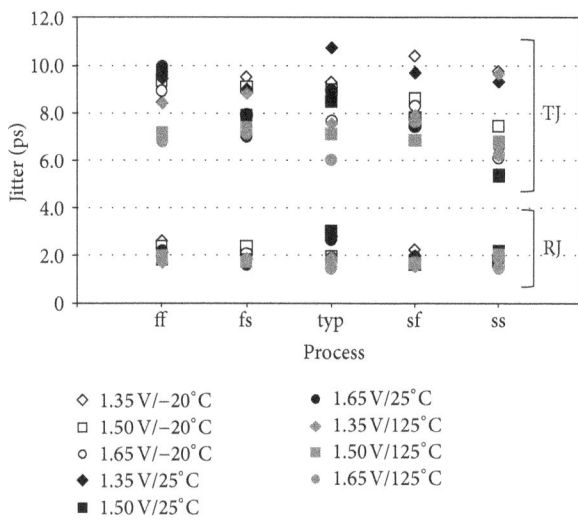

FIGURE 22: Measurement result of output signal jitter of RJ and TJ.

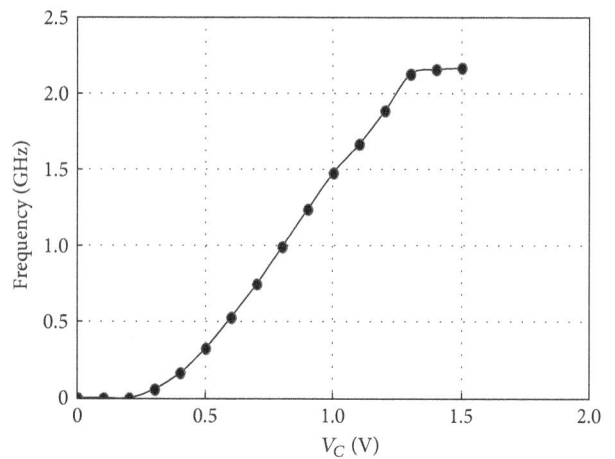

FIGURE 23: Measurement result of VCO frequency-voltage characteristics in worst condition (SS/1.35 V/125°C).

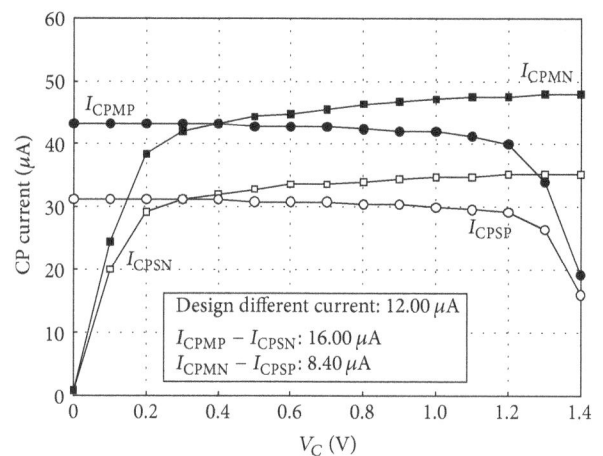

FIGURE 24: Measurement result of CP current.

and then the UP and the DN were set to high and low, respectively. The CPM currents (I_{CPMP} and I_{CPMN}) and CPS current (I_{CPSP} and I_{CPSN}) were designed at 44 μA and 32 μA, respectively. The NMOS currents (I_{CPMN} and I_{CPSN}) did not have accurate saturation characteristics due to channel length modulation.

Figure 25 shows the measurement results for the current difference between the CPM and CPS under various conditions. The design target for the current difference was 12.0 μA. The variation of the current difference was less than ±30%. Under the "ff" and "fs" conditions, the variation was larger than under the other conditions. This was because the channel length modulation caused the ratio of the current mirror consisting of PMOSs to deviate from the ideal ratio.

Our SSCG generated an output signal with a frequency of 1.5 GHz, which meets the SATA specification. As summarized in Table 1, its EMI was reduced by 10.0 dB, its power consumption was 18 mW, and its settling-time was less than 4 μs; the latter had been unachievable with previous SSCGs that applied to the SATA specifications [2–6].

Figure 26 shows a chip microphotograph. The design area was 300 × 700 μm. The 3.91 μs locking-time was faster than the other previous works. The proposed SSCG can make the SATA-PHY reduce the power in the partial state because the SSCG can be disabled. For portable devices, a battery lifetime

TABLE 1: Comparison table.

Item	Unit	[3]	[5]	[2]	[4]	This work
Output frequency	GHz	1.5	1.5	1.5	1.5	1.5
Random jitter @ 250 cycles	Ps rms	<3.3	—	5	3.2	<3.2
Total jitter @ 250 cycles	ps rms	<3.6	—	—	16.8	<10.7
EMI reduction	dB	10.0	8.2	—	9.8	10.0
Locking-time	μs	30	4.2	50	6.3	3.91
Technology	μm	0.13	0.13	0.18	0.18	0.13
Power	mW	30	12	27	77	18
Area	mm^2	0.3570	0.1120	0.2112	0.3100	0.2100

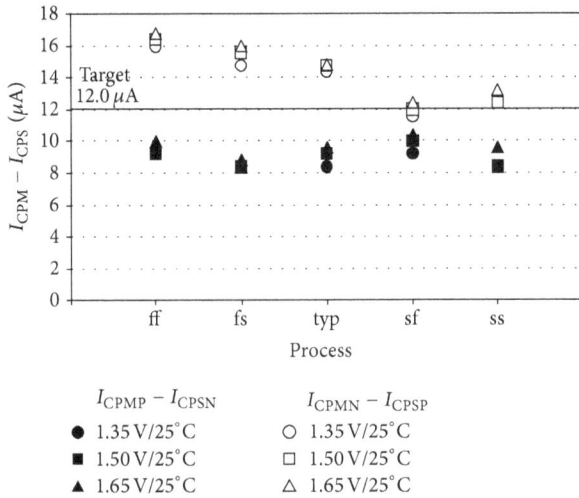

FIGURE 25: Measurement result of different CP current.

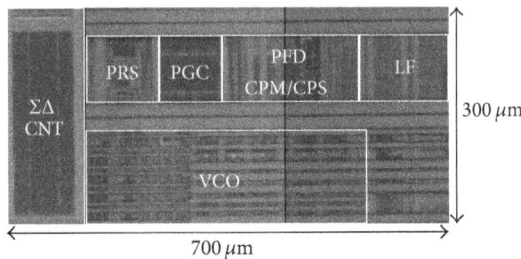

FIGURE 26: Test chip microphotograph.

is critical issue. Our proposed low power SATA-PHY can be one of the solutions to overcome the battery issue.

7. Conclusion

A fast-settling spread-spectrum clock generator (SSCG) for Serial Advanced Technology Attachment (SATA) application has been developed. The SSCG's settling time is shortened through the use of a charge-pump (CP) control technique. A prototype of our SSCG achieved 3.91 μs settling-time,

300×700 μm design area, 18 mW power consumption, 3.2 psrms random jitter, and 10.0 dB EMI reduction. A SATA-PHY with our SSCG consumes less power in the partial state in SATA applications because it can stop the SSCG. This makes it well suited for portable applications.

Conflict of Interests

The authors declare that there is no conflict of interests regarding the publication of this paper.

References

[1] Serial ATA International Organization, "Serial ATA Revision 2.6 Specification".

[2] Y.-B. Hsieh and Y.-H. Kao, "A new spread spectrum clock generator for SATA using double modulation schemes," in *Proceedings of the IEEE Custom Integrated Circuits Conference (CICC '07)*, pp. 297–300, September 2007.

[3] T. Kawamoto, T. Takahashi, H. Inada, and T. Noto, "Low-jitter and large-EMI-reduction spread-spectrum clock generator with auto-calibration for serial-ATA applications," in *Proceedings of the IEEE Custom Integrated Circuits Conference (CICC '07)*, pp. 345–348, September 2007.

[4] H.-R. Lee, O. Kim, G. Ahn, and D.-K. Jeong, "A low-jitter 5000ppm spread spectrum clock generator for multi-channel SATA transceiver in 0.18 μm CMOS," in *Proceedings of the IEEE International Solid-State Circuits Conference (ISSCC '05)*, Digest of Technical Papers, pp. 162–163, February 2005.

[5] P.-Y. Wang and S.-P. Chen, "Spread spectrum clock generator," in *Proceedings of the IEEE Asian Solid-State Circuits Conference (ASSCC '07)*, pp. 304–307, November 2007.

[6] J.-S. Pan, T.-H. Hsu, H.-C. Chen et al., "Fully integrated CMOS SoC for 56/18/16 CD/DVD-dual/RAM applications with on-chip 4-LVDS channel WSG and 1.5 Gb/s SATA PHY," in *Proceedings of the IEEE ISSCC, Digest of Technical Papers*, pp. 266–267, February 2006.

[7] Y. Moon, G. Ahn, H. Choi, N. Kim, and D. Shim, "A quad 6Gb/s multirate CMOS transceiver with TX rise/fall-time control," in *Proceedings of the Digest of Technical Papers IEEE International Conference on Solid-State (ISSCC '06)*, pp. 233–242, IEEE, San Francisco, Calif, USA, February 2006.

[8] S. Pellerano, S. Levantino, C. Samori, and A. L. Lacaita, "A 13.5-mW 5-GHz frequency synthesizer with dynamic-logic frequency divider," *IEEE Journal of Solid-State Circuits*, vol. 39, no. 2, pp. 378–383, 2004.

[9] H.-S. Li, Y.-C. Cheng, and D. Puar, "Dual-loop spread-spectrum clock generator," in *Proceedings of the IEEE International Solid-State Circuits Conference (ISSCC '99)*, Digest of Technical Papers, pp. 184–185, February 1999.

[10] J.-Y. Michel and C. Neron, "A frequency modulated PLL for EMI reduction in embedded application," in *Proceedings of the 12th Annual IEEE International ASIC/SOC Conference*, pp. 362–365, Washington, DC, USA, 1999.

[11] M. Kokubo, T. Kawamoto, T. Oshima et al., "Spread-spectrum clock generator for serial ATA using fractional PLL controlled by ΔΣ modulator with level shifter," in *Proceedings of the IEEE International Solid-State Circuits Conference (ISSCC '05)*, vol. 1 of *Digest of Technical Papers*, pp. 160–590, February 2005.

[12] J. Shin, I. Seo, J. Kim et al., "A low-jitter added SSCG with seamless phase selection and fast AFC for 3rd generation serial-ATA," in *Proceedings of the IEEE Custom Integrated Circuits Conference (CICC '06)*, pp. 409–412, San Jose, Calif, USA, September 2006.

[13] H.-H. Chang, I.-H. Hua, and S.-I. Liu, "A spread-spectrum clock generator with triangular modulation," *IEEE Journal of Solid-State Circuits*, vol. 38, no. 4, pp. 673–676, 2003.

[14] C. D. LeBlanc, B. T. Voegeli, and T. Xia, "Dual-loop direct VCO modulation for spread spectrum clock generation," in *Proceedings of the IEEE Custom Integrated Circuits Conference (CICC '09)*, pp. 479–482, September 2009.

[15] Y.-H. Kao and Y.-B. Hsieh, "A low-power and high-precision spread spectrum clock generator for serial advanced technology attachment applications using two-point modulation," *IEEE Transactions on Electromagnetic Compatibility*, vol. 51, no. 2, pp. 245–254, 2009.

[16] T. Kawamoto, T. Takahashi, S. Suzuki, T. Noto, and K. Asahina, "Low-jitter fractional spread-spectrum clock generator using fast-settling dual charge-pump technique for serial-ATA application," in *Proceedings of the 35th European Solid-State Circuits Conference (ESSCIRC '09)*, pp. 380–383, September 2009.

[17] S.-Y. Lin and S.-I. Liu, "A 1.5 GHz all-digital spread-spectrum clock generator," *IEEE Journal of Solid-State Circuits*, vol. 44, no. 11, pp. 3111–3119, 2009.

[18] D. de Caro, C. A. Romani, N. Petra, A. G. M. Strollo, and C. Parrella, "A 1.27 GHz, all-digital spread spectrum clock generator/synthesizer in 65 nm CMOS," *IEEE Journal of Solid-State Circuits*, vol. 45, no. 5, pp. 1048–1060, 2010.

[19] S. Levantino, L. Romanò, S. Pellerano, C. Samori, and A. L. Lacaita, "Phase noise in digital frequency dividers," *IEEE Journal of Solid-State Circuits*, vol. 39, no. 5, pp. 775–784, 2004.

[20] K. Shu, E. Sánchez-Sinencio, J. Silva-Martínez, and S. H. K. Embabi, "A 2.4-GHz monolithic fractional-N frequency synthesizer with robust phase-switching prescaler and loop capacitance multiplier," *IEEE Journal of Solid-State Circuits*, vol. 38, no. 6, pp. 866–874, 2003.

7

On Improving the Performance of Dynamic DCVSL Circuits

Pratibha Bajpai,[1] Neeta Pandey,[1] Kirti Gupta,[2] Shrey Bagga,[2] and Jeebananda Panda[1]

[1]*Department of Electronics and Communication Engineering, Delhi Technological University, Delhi, India*
[2]*Department of Electronics and Communication Engineering, Bharati Vidyapeeth's College of Engineering, Delhi, India*

Correspondence should be addressed to Neeta Pandey; n66pandey@rediffmail.com

Academic Editor: Ephraim Suhir

This contribution aims at improving the performance of Dynamic Differential Cascode Voltage Switch Logic (Dy-DCVSL) and Enhanced Dynamic Differential Cascode Voltage Switch Logic (EDCVSL) and suggests three architectures for the same. The first architecture uses transmission gates (TG) to reduce the logic tree depth and width, which results in speed improvement. As leakage is a dominant issue in lower technology nodes, the second architecture is proposed by adapting the leakage control technique (LECTOR) in Dy-DCVSL and EDCVSL. The third proposed architecture combines features of both the first and the second architectures. The operation of the proposed architectures has been verified through extensive simulations with different CMOS submicron technology nodes (90 nm, 65 nm, and 45 nm). The delay of the gates based on the first architecture remains almost the same for different functionalities. It is also observed that Dy-DCVSL gates are 1.6 to 1.4 times faster than their conventional counterpart. The gates based on the second architecture show a maximum of 74.3% leakage power reduction. Also, it is observed that the percentage of reduction in leakage power increases with technology scaling. Lastly, the gates based on the third architecture achieve similar leakage power reduction values to the second one but are not able to exhibit the same speed advantage as achieved with the first architecture.

1. Introduction

Digital design space is filled with a variety of logic styles suitable for different applications [1–6]. Conventionally static CMOS has predominance over remaining styles due to the low static power consumption. Differential Cascode Voltage Switch Logic (DCVSL) [6] is a static style which is beneficial from circuit delay, layout density, logic flexibility, and power consumption. The DCVSL has been employed to develop various circuits for fault tolerance [7], ternary logic [8], micro pipelining [9], delay cell [10], ring oscillator [11], capacitor neutralization [12], and so forth. It is well known that static logic styles suffer from high power consumption when output switches its logic state, a situation which worsens with increasing clock frequencies. The performance can be improved by using the dynamic version of DCVSL which is based on precharge-evaluation logic. Many clocked versions of DCVSL style named dynamic DCVSL (Dy-DCVSL) and enhanced DCVSL (EDCVSL) [13] are presented in the literature. As the speed of the dynamic circuit depends on logic tree depth and width [13], this paper proposes an architecture to reduce logic tree depth by employing transmission gates in logic function realization. Apart from the speed issue, leakage currents are yet another concern that shows predominance in submicron technologies. Leakage loss occurs when the output is stable (i.e., low output or high output). A new architecture incorporating the leakage control technique [14] in dynamic DCVSL circuits is put forward, which reduces leakage current. The features of the former architectures are combined to present a third architecture.

The paper is arranged in five sections including the present one. Section 2 briefly presents existing dynamic DCVSL and EDCVSL circuits. Section 3 describes the proposed dynamic DCVSL circuits. The functional verification and performance of the proposal are placed in Section 4 and conclusions are drawn in Section 5.

2. Dynamic Differential Cascode Voltage Switch Logic Family

Differential Cascode Voltage Switch Logic (DCVSL) is a differential style derived from conventional CMOS logic and

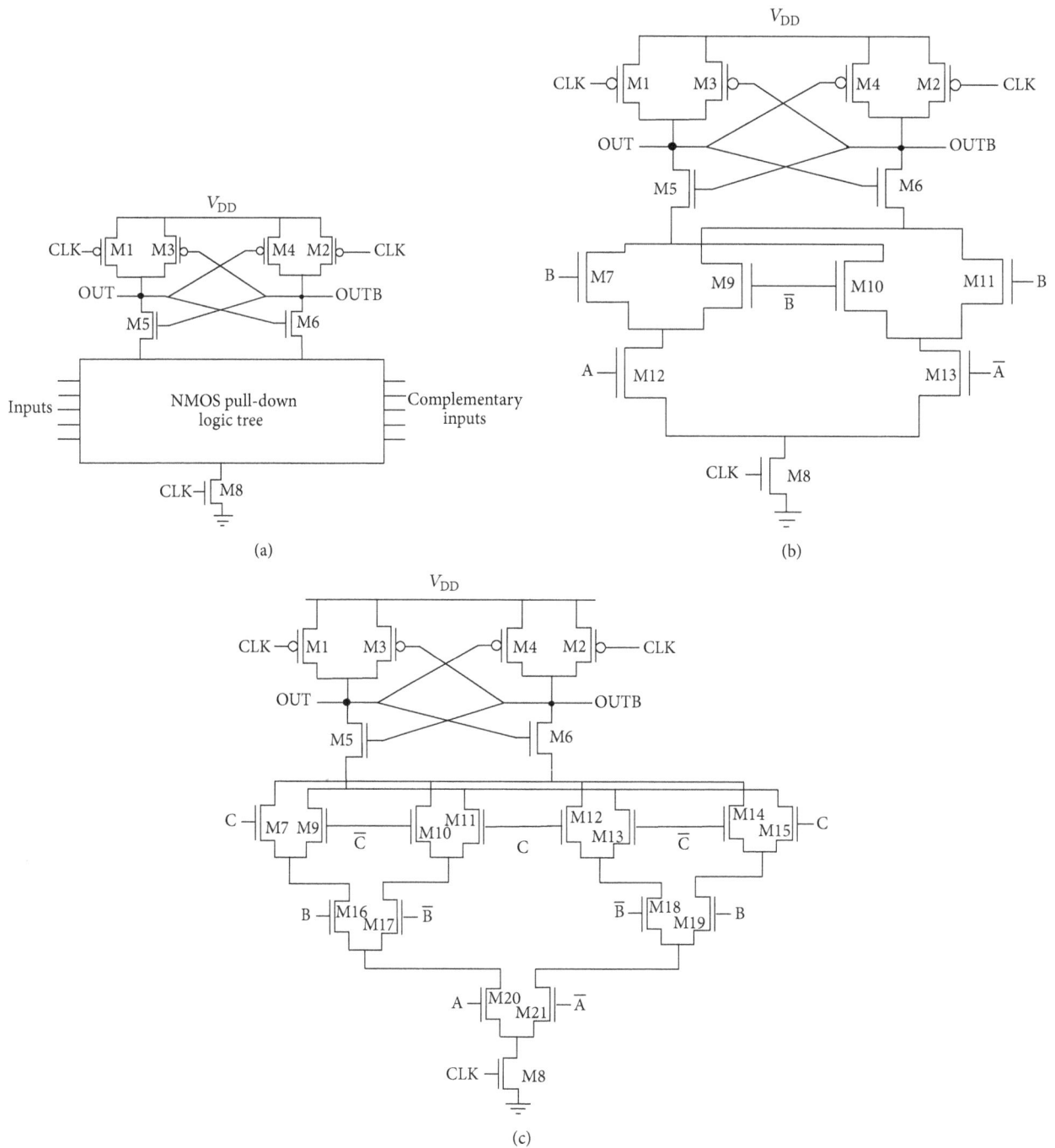

FIGURE 1: (a) Generic architecture of Dy-DCVSL circuit: (b) two- and (c) three-input Dy-DCVSL XOR-XNOR gate.

ratioed pseudo NMOS logic. It combines their advantages and provides a high speed, area efficient, and rail-to-rail logic design alternative. Both the static and the dynamic versions of the DCVSL style are available in open literature. This section briefly describes existing dynamic DCVSL styles (Dy-DCVSL).

2.1. Conventional Dy-DCVSL Architecture. The generic architecture of a Dy-DCVSL gate is shown in Figure 1(a). It consists of a pull-down network (PDN) that implements the

logic function (OUT) as well as its complement (OUTB) using NMOS transistors only. The operation of the circuit depends on clock signal (CLK). It works in precharge phase for low CLK signal and in evaluation phase otherwise. In the precharge phase, the transistors M1 and M2 are ON and transistors M3 and M4 remain OFF, so both output nodes are precharged to high (=V_{DD}) logic level. Any change in the inputs during this phase will not affect the output nodes' potential as a current path from the output nodes to the ground could not be established. A high CLK signal makes

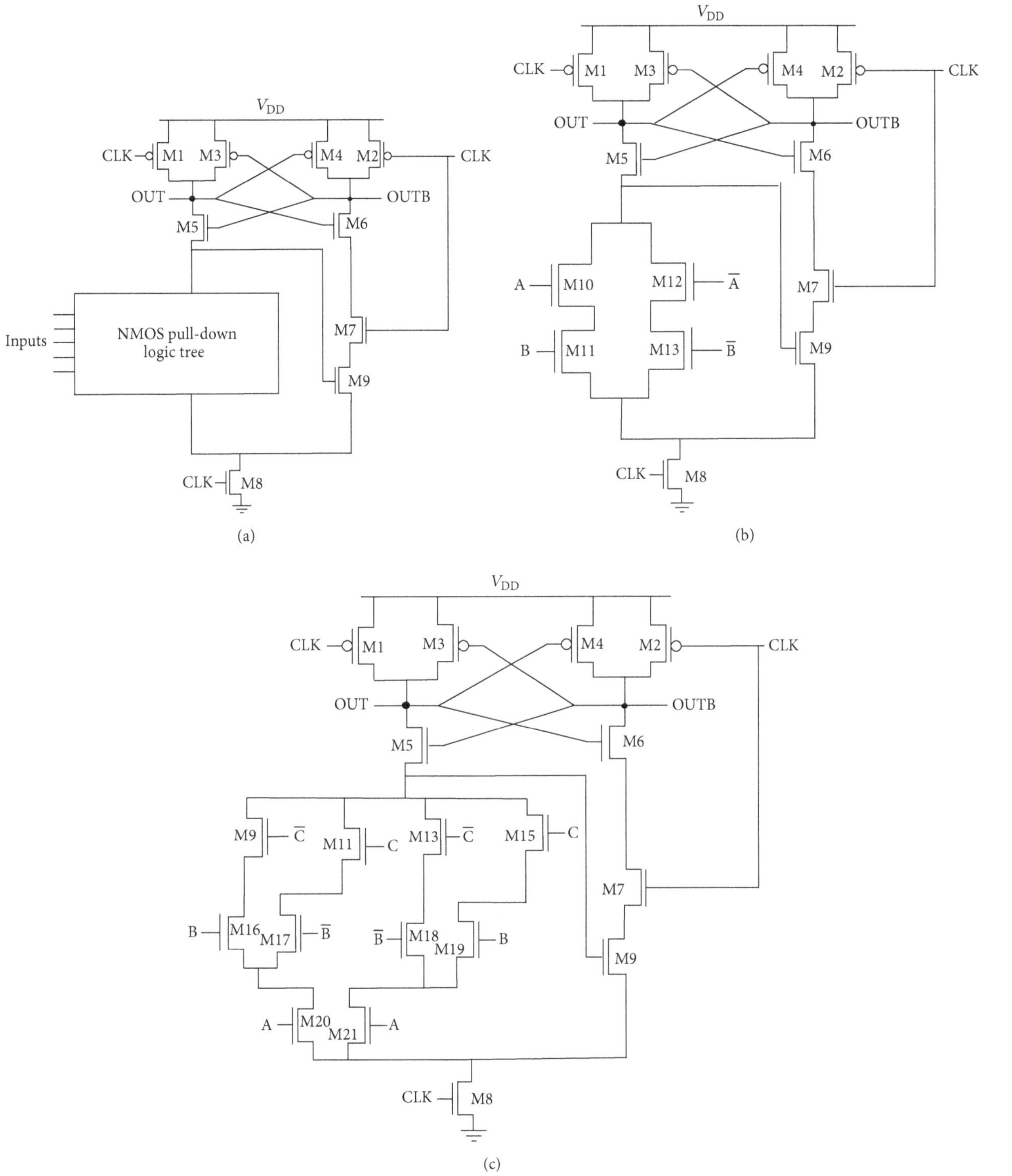

FIGURE 2: (a) Generic architecture of EDCVSL circuit: (b) two-input and (c) three-input EDCVSL XOR-XNOR gate.

transistors M1 and M2 OFF. One of the output nodes attains a low logic level depending on the value of inputs.

Cross-coupled transistors M3 and M4 help output to switch and transistors M5 and M6 accelerate this process [13]. The circuits of two- and three-input Dy-DCVSL XOR-XNOR gates are shown in Figures 1(b) and 1(c) as illustrations.

2.2. Conventional EDCVSL Architecture. The generic architecture of the EDCVSL circuits is drawn in Figure 2(a) and two/three-input EDCVSL XOR-XNOR gate is shown as an example in Figures 2(b) and 2(c). It is similar to Dy-DCVSL except for PDN where only one logic tree branch (OUT) is retained and the other is replaced by two stacked transistors

M7 and M9. The gate of transistor M7 is connected to CLK signal so it is activated in the evaluation phases only. The transistor M9 is connected to the output of the logic tree branch to achieve complementary operation. To elaborate the behavior, consider schematic of Figure 2(b). In the precharge phase, both outputs are precharged and the NMOS cross-coupled transistors M5 and M6 are ON but have no effect on output as M8 is OFF. In the evaluation phase, for low (high) values at both inputs, the transistors M12 and M13 (M10 and M11) are ON and therefore transistor M9 is OFF. So, OUTB remains high and M5 retains its ON state and finally OUT becomes low. For the case when one input is high and the other is low, no path exists between OUT and the ground so it remains high and OUTB goes low. The operation of three-input EDCVSL XOR-XNOR gate is similar to two-input EDCVSL XOR-XNOR gate and is omitted for the sake of brevity.

3. The Proposed Dynamic DCVSL Circuits

This section presents three new architectures to improve performance of existing Dy-DCVSL and EDCVSL. The first architecture aims at speed improvement, the second works on leakage reduction, and the third combines features of the first two architectures to see their combined effect on performance.

3.1. Proposed Architecture-1 (PA-1). Proposed architecture-1 is based on shifting the function realized by PDN logic tree to a separate block and using transmission gates (TG) logic for its implementation. The new Dy-DCVSL and EDCVSL circuits based on architecture-1 are named Dynamic TG based DCVSL (Dy-TG-DCVSL) and Dynamic TG based EDCVSL (TG-EDCVSL) circuits, respectively. A generic architecture of Dy-TG-DCVSL circuit along with two- and three-input XOR-XNOR realization is depicted in Figure 3. The PDN logic tree consists of two NMOS transistors which are controlled by the outputs of two separate blocks. The two blocks generate the complementary outputs such that either M9 or M10 is ON during evaluation.

The working of the proposed Dy-TG-DCVSL XOR2 gate can be explained for the two phases. The operation in precharge phase is the same as conventional Dy-DCVSL. Any changes in the inputs A and B may update the gate potential of M9 and M10 but will not affect the output, since M8 is OFF. Consequently, when CLK becomes high, the output gets evaluated according to the gate potential of M9 and M10. In comparison to the conventional Dy-DCVSL XOR-XNOR gate (Figure 1(b)), there is a speed advantage in terms of evaluation time due to the fact that the intermediate computation of the function is completed in the separate blocks just prior to the start of the evaluation phase. Also, a closer examination of the proposed architecture reveals a unique advantage of maintaining a constant evaluation time irrespective of the realized functionality. Similarly, placing logic functionality of EDCVSL in separate block logic leads to the proposed TG-EDCVSL architecture. Generic gate structure, two- and three-input XOR-XNOR gates are depicted in Figure 4.

3.2. Proposed Architecture-2 (PA-2). The differential nature of the DCVSL logic style has several advantages but in submicron regions it needs attention. For all input combinations, one of the two logic tree branches in the PDN will be conducting while the other would remain nonconducting. The nonconducting branch in submicron regions would have some amount of current due to OFF transistors in the path in both precharge and evaluation phases. This current can be classified as leakage current. To improve the performance in submicron region, these currents need to be minimized. Various leakage reduction techniques based on the use of sleep transistor [15] and high threshold voltage transistors [16] are available for static DCVSL circuits. These techniques require either routing of sleep signal [15] or a complex algorithm for selection of high threshold voltage transistors. A self-controlled technique named LECTOR [14] is presented for CMOS circuits, which reduces both types of currents and is adapted for dynamic DCVSL circuits, and the resulting topology is referred to as proposed architecture-2 (PA-2). LECTOR technique introduces two leakage control transistors (PMOS and NMOS) in between the PUN and the PDN of the logic gate with the gate terminal of each of the leakage control transistors (LCTs) controlled by the source of the other. This arrangement ensures that one of the LCTs is always in the "near-cut-off region" for any possible input combination. This results in an increase in the resistance of the path from the power supply to the ground, leading to a substantial drop in leakage currents through the path [14]. A further modification in achieving much more leakage control is to use high Vth LCTs. The architectures incorporating LECTOR technique in Dy-DCVSL and EDCVSL circuits are shown in Figure 5. The proposed architectures add four high Vth LCTs (LCT1–LCT4) in the basic architectures of the Dy-DCVSL (Figure 1(a)) and EDCVSL (Figure 2(a)) circuits and are called Dy-DCVSL-LCT and EDCVSL-LCT, respectively.

To understand the leakage control mechanism in Dy-DCVSL-LCT and EDCVSL-LCT circuits, the operating regions of LCTs during precharge and evaluation phases are examined. In the precharge phase, both the OUT and the OUTB are at V_{DD}. Under this condition, it can be observed that the transistor LCT2 is ON and LCT1 is in near-cut-off state. Thus, LCT1 offers more impedance along the path and reduces the leakage current. Similar behavior can be observed when a high voltage is obtained at the output in the evaluation phase.

3.3. Proposed Architecture-3 (PA-3). Proposed architecture-3 combines the features of proposed architecture-1 and proposed architecture-2. The resulting Dy-DCVSL and EDCVSL structures are called Dy-TG-DCVSL-LCT and TG-EDCVSL-LCT. The proposed architectures are shown in Figure 6.

4. Simulation Results

This section presents the simulation results for the new Dy-DCVSL circuits based on the proposed architectures. The simulations are performed using Symica tool and the PTM technology parameters for 90 nm, 65 nm, and 45 nm nodes. The frequencies of the inputs CLK, A, B, and C are 50 MHz,

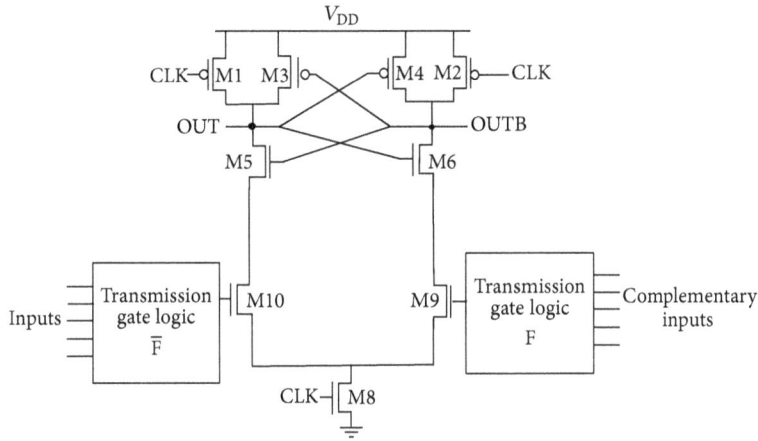

(a)

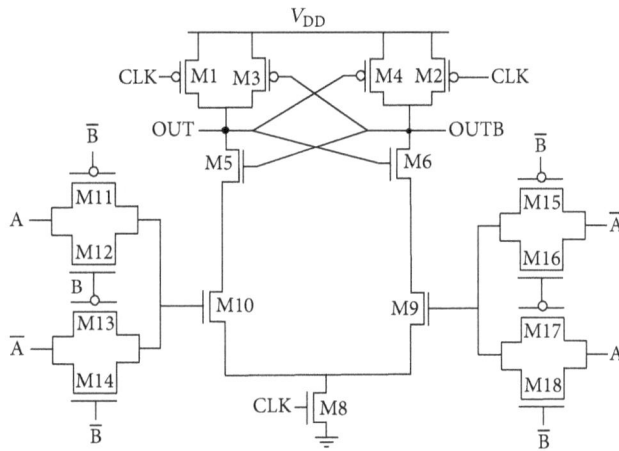

(b)

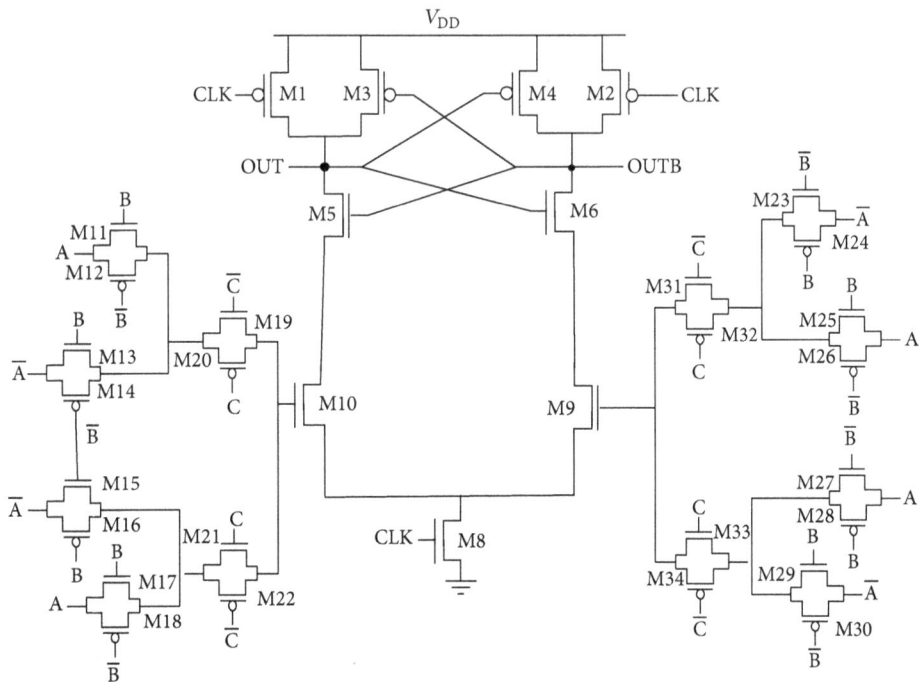

(c)

FIGURE 3: (a) Proposed architecture-1 for Dy-TG-DCVSL circuits. (b) Two-input and (c) three-input Dy-TG-DCVSL XOR-XNOR gate.

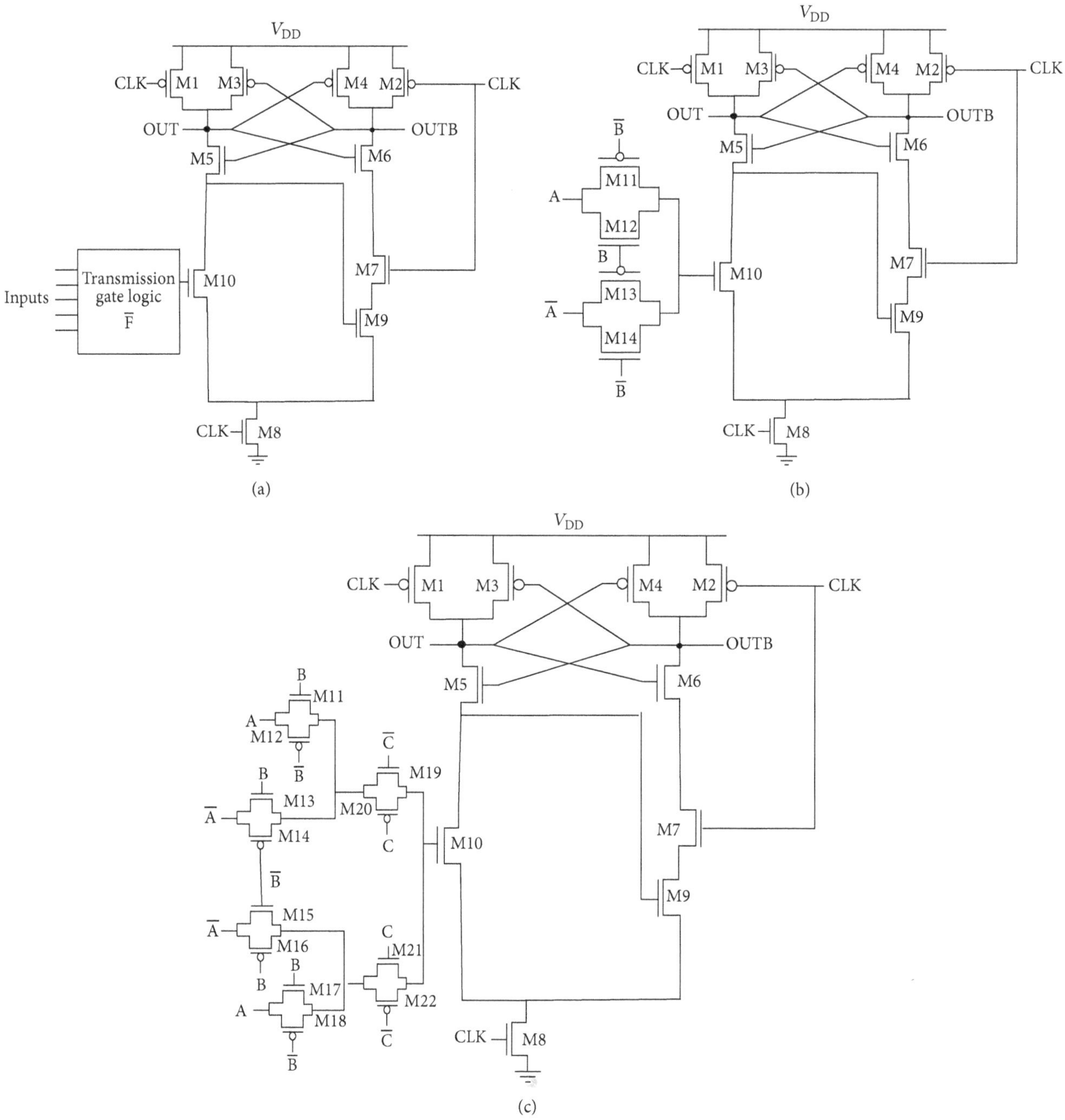

FIGURE 4: (a) Proposed architecture-1 for TG-EDCVSL circuits. (b) Two-input and (c) three-input TG-EDCVSL XOR-XNOR gates.

25 MHz, 12.5 MHz, and 6.25 MHz, respectively. FO4 of inverters is maintained as the load in all the gates. The results are categorized into three sections according to the proposed architectures. The leakage power is computed on the basis of leakage current and the power supply.

4.1. Simulation Results with PA-1. Dy-TG-DCVSL and TG-EDCVSL based two-input AND-NAND (AND-NAND2), three-input AND-NAND (AND-NAND3), two-input exclusive-OR (XOR-XNOR2), and three-input exclusive-OR

(XOR-XNOR3) circuits are simulated using 90 nm CMOS technology parameters. The simulation waveforms of the Dy-TG-DCVSL and TG-EDCVSL XOR-XNOR2 and XOR-XNOR3 gates are shown in Figure 7. For all the gates, it can be observed that, for low value of the CLK signal, both output nodes are precharged to V_{DD} (=1.8 V). The voltage changes in the input signals A and B during this phase do not affect the potential of the output nodes. In the evaluation phase, for the same value of the inputs A and B (Figure 7(a)), the output node OUT remains low. Similarly, when the inputs differ,

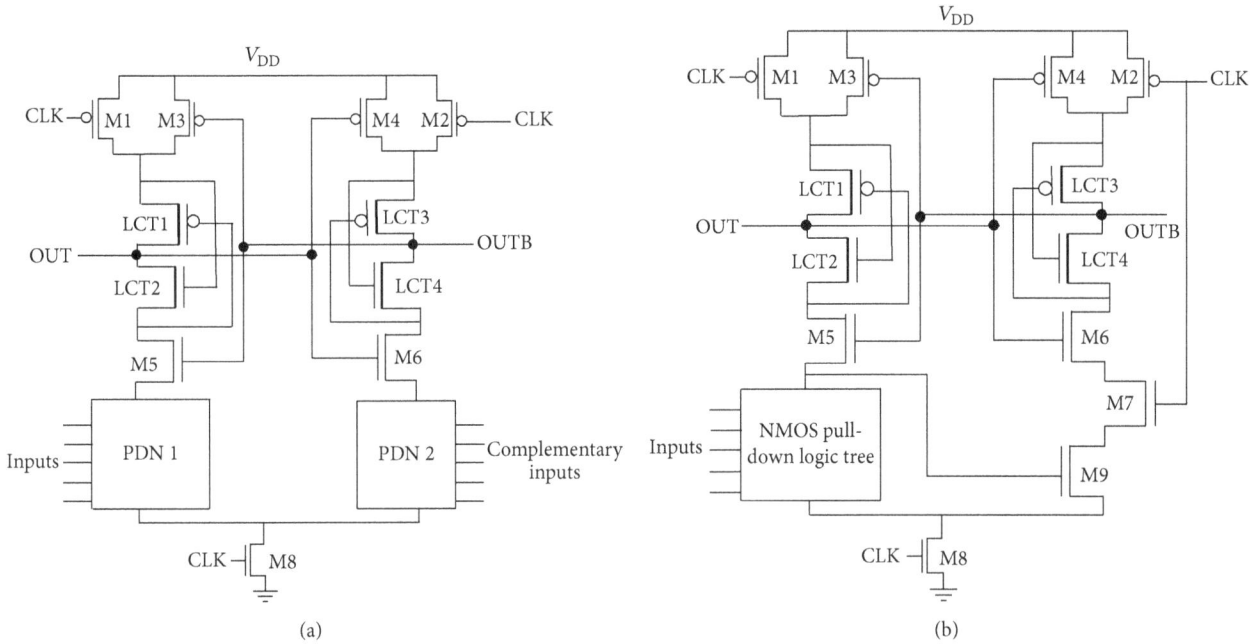

FIGURE 5: Proposed architecture-2 for (a) Dy-DCVSL-LCT circuits and (b) EDCVSL-LCT circuits.

a high voltage level is obtained at the output node OUT. Thus, the XOR functionality is achieved in the proposed Dy-TG-DCVSL and TG-EDCVSL XOR2 gates. The simulation waveform for Dy-TG-DCVSL and TG-EDCVSL XOR-XNOR3 gate is shown in Figure 7(b). Similar waveforms were achieved for the other gates and are omitted for the sake of brevity.

The gates are also designed in Dy-DCVSL and EDCVSL styles to analyze the speed advantage. The corresponding delay results are noted and enlisted in Table 1. The results clearly indicate the speed advantage of PA-1 based gates over the conventional counterparts. Also, the TG-EDCVSL gates are the fastest among all the logic styles. Lastly, the PA-1 based gates show almost equal delay values irrespective of the implemented functionality.

4.2. Simulation Results with PA-2. In this category, the leakage current reduction through incorporation of LECTOR technique in dynamic DCVSL circuits is demonstrated. An XOR-XNOR2 gate is chosen as the test bench due to its wide range of applications. The conventional Dy-DCVSL, EDCVSL, Dy-DCVSL-LCT, and EDCVSL-LCT XOR-XNOR2 gate circuits are simulated at various submicron technology nodes such as 90 nm, 65 nm, and 45 nm. Table 2 lists the leakage power for the conventional Dy-DCVSL, EDCVSL, Dy-DCVSL-LCT, and EDCVSL-LCT XOR-XNOR2 gate with $V_{DD} = 1.2$ V.

The following observations are made from Table 2:

(1) The percentage reduction ranges in the leakage power are 30.4%–56.6% for 90 nm, 32.2%–61% for 65 nm, and 33.8%–74.3% for 45 nm.

(2) Leakage power tends to follow an increasing trend with the scaling down of the technology.

(3) An increase in percentage reduction is seen as we dig down the lower technology nodes.

4.3. Simulation Results with PA-3. Dy-TG-DCVSL, TG-EDCVSL, Dy-TG-DCVSL-LCT, and TG-EDCVSL-LCT two-input XOR2 and three-input XOR3 gates are simulated at various technology nodes. Out of the two dynamic styles, the results pertaining to EDCVSL circuits are listed in Tables 3–5 for the leakage power and delay measurements. The delays reported in Table 5 are for 45 nm technology node. The findings can be summarized as follows:

(1) A percentage reduction range of 27%–64% for 90 nm, 30%–66% for 65 nm, and 38%–76% for 45 nm in leakage power is observed.

(2) The TG-EDCVSL-LCT XOR-XNOR2 gate shows less leakage power with respect to Dy-TG-DCVSL counterpart.

(3) Leakage power tends to follow an increasing trend with the scaling down of the technology.

(4) The percentage reduction in the leakage power increases with the lower technology nodes.

(5) The delay of the TG-EDCVSL-LCT XOR-XNOR2 gate is more than the Dy-TG-DCVSL due to the presence of the high resistance path for leakage current reduction, thus exhibiting a trade-off between the speed and leakage power reduction.

5. Conclusion

In this paper, three new architectures to enhance the performance of Dy-DCVSL and EDCVSL are proposed. The first

(a)

(b)

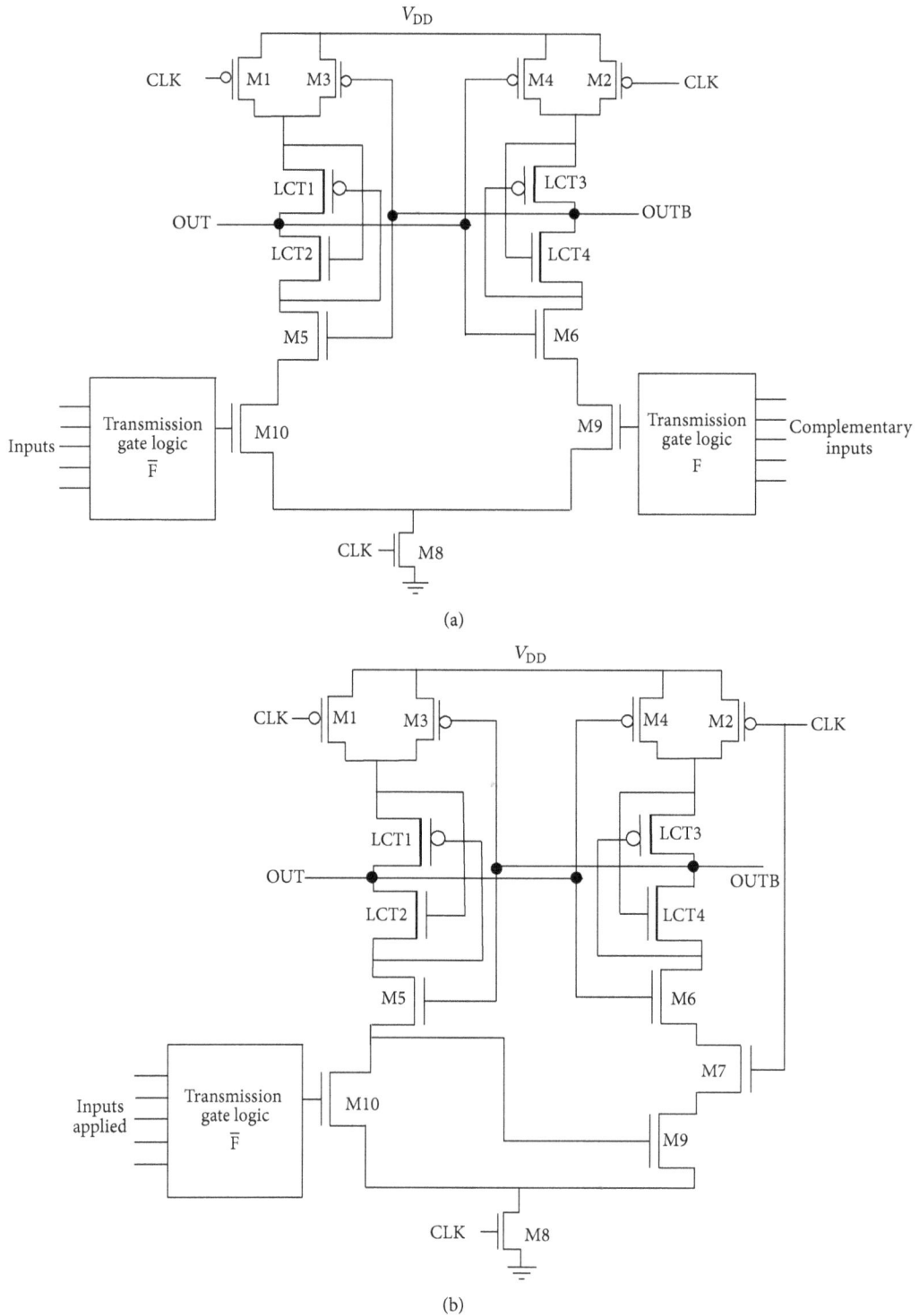

FIGURE 6: Proposed architecture-3 based (a) Dy-TG-DCVSL-LCT and (b) TG-EDCVSL-LCT circuits.

improves the speed by using transmission gates through logic tree depth reduction. The second architecture is derived to reduce leakage power at lower technology nodes. The incorporation of leakage control by incorporating LECTOR technique is proposed. The third architecture merges the two proposed architectural modifications to analyze their combined effect on the performance. Extensive simulations are done at various CNOS submicron technology nodes such as 90 nm, 65 nm, and 45 nm. It is observed that dynamic DCVSL gates based on the first architecture are 1.6 to 1.4 times faster than the conventional dynamic CVSL circuits. The gates based on the first architecture show almost equal delay values

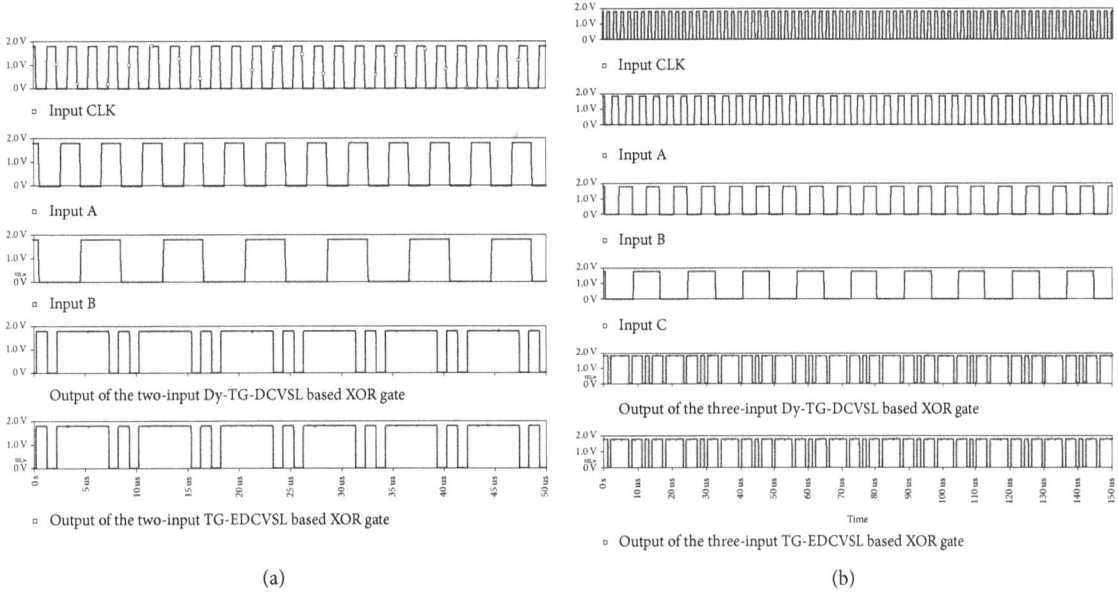

FIGURE 7: Simulation waveform of Dy-TG-DCVSL and TG-EDCVSL based (a) two-input XOR-XNOR gate and (b) three-input XOR-XNOR gate.

TABLE 1: Summary of delay results (ps).

Gate	Style			
	Dy-DCVSL	EDCVSL	Dy-TG-DCVSL	TG-EDCVSL
AND-NAND2	500	480	350	340
AND-NAND3	570	530	355	345
XOR-XNOR2	500	480	350	341
XOR-XNOR3	550	490	355	345

TABLE 2: Leakage power (nW) for the conventional Dy-DCVSL, EDCVSL-LCT and PA-2 Dy-DCVSL-LCT, EDCVSL-LCT based XOR-XNOR2 gate topologies.

Inputs		Architecture					
		Conventional		PA-2		Reduction with respect to conventional (%)	
A	B	Dy-DCVSL	EDCVSL	Dy-DCVSL-LCT	EDCVSL-LCT	Dy-DCVSL-LCT	EDCVSL-LCT
90 nm node							
0	0	9.27	8.9	4.02	4.06	56.6	54
0	1	0.23	0.21	0.16	0.14	30.4	33
1	0	0.23	0.21	0.16	0.14	30.4	33
1	1	9.27	8.9	4.02	4.06	56.6	54
65 nm node							
0	0	12.9	12.88	5.16	5	60	61
0	1	0.41	0.27	0.27	0.18	32.2	33.3
1	0	0.41	0.27	0.27	0.18	32.2	33.3
1	1	12.9	12.88	5.16	5	60	61
45 nm node							
0	0	22.48	22.48	5.79	5.79	74	74.3
0	1	0.41	0.28	0.28	0.18	33.8	35.7
1	0	0.41	0.28	0.28	0.18	33.8	35.7
1	1	22.48	22.48	5.79	5.79	74	74.3

TABLE 3: Leakage power (nW) for TG-EDCVSL and TG-EDCVSL-LCT based XOR-XNOR2 gate topologies.

Inputs		Architecture		
A	B	TG-EDCVSL	TG-EDCVSL-LCT	Reduction with respect to TG-EDCVSL (%)
		90 nm node		
0	0	8.8	4	55
0	1	0.18	0.13	27
1	0	0.18	0.13	27
1	1	8.8	4	55
		65 nm node		
0	0	12.7	5	61
0	1	0.2	0.14	30
1	0	0.2	0.14	30
1	1	12.7	5	61
		45 nm node		
0	0	22.24	5.7	74.3
0	1	0.21	0.13	38
1	0	0.21	0.13	38
1	1	22.24	5.7	74.3

TABLE 4: Leakage power (nW) in TG-EDCVSL and TG-EDCVSL-LCT three-input XOR-XNOR gates.

Inputs			Architecture		
A	B	C	TG-EDCVSL	TG-EDCVSL-LCT	Reduction with respect to TG-EDCVSL (%)
			90 nm node		
0	0	0	9.8	4	59
0	0	1	0.34	0.21	64
0	1	0	0.34	0.21	64
0	1	1	9.8	4	59
1	0	0	0.34	0.21	64
1	0	1	9.8	4	59
1	1	0	9.8	4	59
1	1	1	0.34	0.21	64
			65 nm node		
0	0	0	14	5.07	63
0	0	1	0.38	0.26	66
0	1	0	14	5.07	63
0	1	1	0.38	0.26	66
1	0	0	0.38	0.26	66
1	0	1	14	5.07	63
1	1	0	14	5.07	63
1	1	1	0.38	0.26	66
			45 nm node		
0	0	0	24.5	5.85	76
0	0	1	0.43	0.12	72
0	1	0	0.43	0.12	72
0	1	1	24.5	5.85	76
1	0	0	0.43	0.12	72
1	0	1	24.5	5.85	76
1	1	0	24.5	5.85	76
1	1	1	0.43	0.12	72

TABLE 5: Delay measurement for the PA-1 based and the PA-3 based ED-CVSL XOR/XNOR gate.

Mode of operation	A	B	Output	Delay
PA-1 based XOR gate				
Evaluation	1	1	1->0	300 ps
	0	0	1->0	299 ps
Precharge	1	1	0->1	184 ps
	0	0	0->1	187 ps
PA-3 based XOR gate				
Evaluation	1	1	1->0	355 ps
	0	0	1->0	354 ps
Precharge	1	1	0->1	598 ps
	0	0	0->1	598 ps

irrespective of the implemented functionality. A maximum leakage power reduction of 78.43% is achieved with the second architecture based DCVSL gates. An increasing trend in the leakage power with the scaling down of the technology is observed in the proposed circuits. Lastly, the third architecture achieves the same leakage power reduction values as the second one but is not able to exhibit the same speed advantage as achieved with the first architecture.

Conflicts of Interest

The authors declare that there are no conflicts of interest regarding the publication of this paper.

References

[1] J. M. Rabaey, A. P. Chandrakasan, and B. Nikolic, *Digital Integrated Circuits: A Design Perspective*, Pearson Education, 2003.

[2] N. Weste and D. Harris, *CMOS VLSI Design: A Circuits and Systems Perspective*, Addison-Wesley Publishing, 2010.

[3] C. Piguet, *Low-Power CMOS Circuits*, CRC Press, 2006.

[4] L. Bisdounis, D. Gouvetas, and O. Koufopavlou, "A comparative study of CMOS circuit design styles for low-power high-speed VLSI circuits," *International Journal of Electronics*, vol. 84, no. 6, pp. 599–613, 1998.

[5] K. M. Chu and D. L. Pulfrey, "A comparison of CMOS circuit techniques: differential cascode voltage switch logic versus conventional logic," *IEEE Journal of Solid-State Circuits*, vol. 22, no. 4, pp. 528–532, 1987.

[6] L. Heller, W. Griffin, J. Davis, and N. Thoma, "Cascode voltage switch logic: a differential CMOS logic family," in *Proceedings of the IEEE International Solid-State Circuits Conference*, pp. 16–17, San Francisco, Calif, USA, Feburary 1984.

[7] M. Stanisavljevic, A. Schmid, and Y. Leblebici, "Fault-tolerance of robust feed-forward architecture using single-ended and differential deep-submicron circuits under massive defect density," in *Proceedings of the International Joint Conference on Neural Networks (IJCNN '06)*, pp. 2771–2778, July 2006.

[8] R. Faghih Mirzaee, T. Nikoubin, K. Navi, and O. Hashemipour, "Differential Cascode Voltage Switch (DCVS) Strategies by CNTFET Technology for Standard Ternary Logic," *Microelectronics Journal*, vol. 44, no. 12, pp. 1238–1250, 2013.

[9] S. Mathew and R. Sridhar, "Data-driven micropipeline structure using DSDCVSL," in *Proceedings of the 21st IEEE Annual Custom Integrated Circuits Conference (CICC '99)*, pp. 295–298, May 1999.

[10] D. Z. Turker, S. P. Khatri, and E. Sánchez-Sinencio, "A DCVSL delay cell for fast low power frequency synthesis applications," *IEEE Transactions on Circuits and Systems. I. Regular Papers*, vol. 58, no. 6, pp. 1225–1238, 2011.

[11] Priyanka and A. K. Singh, "A low voltage high speed DCVSL based ring oscillator," in *Proceedings of the Annual IEEE India Conference (INDICON '15)*, pp. 1–5, New Delhi, India, December 2015.

[12] W. Y. Choi, "Miller effect suppression of tunnel field-effect transistors (TFETs) using capacitor neutralisation," *Electronics Letters*, vol. 52, no. 8, pp. 659–661, 2016.

[13] D. W. Kang and Y.-B. Kim, "Design of enhanced differential cascode voltage switch logic (EDCVSL) circuits for high fan-in gate," in *Proceedings of the 15th Annual IEEE International ASIC/SOC Conference (ASIC/SOC '02)*, pp. 309–313, September 2002.

[14] N. Hanchate and N. Ranganathan, "LECTOR: a technique for leakage reduction in CMOS circuits," *IEEE Transactions on Very Large Scale Integration (VLSI) Systems*, vol. 12, no. 2, pp. 196–205, 2004.

[15] P. Lakshmikanthan and A. Nuñez, "A novel methodology to reduce leakage power in differential cascode voltage switch logic circuits," in *Proceedings of the 3rd International Conference on Electrical and Electronics Engineering*, pp. 1–4, Veracruz, Mexico, September 2006.

[16] W. Chen, W. Hwang, P. Kudva, G. D. Gristede, S. Kosonocky, and R. V. Joshi, "Mixed multi-threshold differential cascode voltage switch (MT-DCVS) circuit styles and strategies for low power VLSI design," in *Proceedings of the International Symposium on Low Electronics and Design (ISLPED '01)*, pp. 263–266, August 2001.

8

Research on HILS Technology Applied on Aircraft Electric Braking System

Suying Zhou, Hui Lin, and Bingqiang Li

School of Automation, Northwest Polytechnical University, Xi'an, China

Correspondence should be addressed to Suying Zhou; nwpususu@nwpu.edu.cn

Academic Editor: Ephraim Suhir

On the basis of analyzing the real-time feature of hardware-in-the-loop simulation of aircraft braking system, a new simulation method based on MATLAB/RTW (Real-Time Workshop) and DSP is introduced. The purpose of this research is to develop a digital control unit with antilock brake system control algorithm for aircraft braking system using HILS. DSP is used as simulator. Using this method, a detailed mathematical modeling of system is proposed first. Studies on reducing sampling time with model simplification and modeling for applying to I/O interface of DSP and HILS are conducted. Compared with other methods, this method is low cost and convenient to implement. By using these methods, we can complete HIL simulation of aircraft braking under various experimental conditions, modify its control laws, and test its braking performance. The results have demonstrated that this platform has high reliability. The algorithm is verified by real-time closed loop test with HILS system and the results are presented.

1. Introduction

The Hardware In-the-Loop Simulation (HILS) is the technique used in the development and testing of complex and costly systems such as military tactical missiles, aircraft flight control system, satellite control systems, and automotive systems [1, 2]. It is a type of real-time simulation. HILS is an effective tool for design, performance evaluation, and test of vehicle subsystems such as antilock brake system (ABS), active suspension system, and steering system.

The HIL simulation uses model to simulate the parts which are not available or not easy to test, and the hardware of other parts can be connected to the system directly. HILS has the higher fidelity, so it is often used to verify the correctness and feasibility of the control system scheme. Here, the aim of HIL simulation is to verify the performance of electric braking system, so braking system is hardware, and the airframe of the aircraft is simulated by simulation software.

The HILS system consists of the hardware, software, and interface parts. The software part includes a virtual aircraft dynamic model and the hardware part includes the actual brake system including electromechanical actuator (EMA) and antibraking controller. Two parts are linked by the interface part in a real-time simulation environment.

By using HILS system, we can accomplish performance test and assessment of electric braking system. The control algorithm of braking system can be verified by real-time closed loop test. In the meantime, some certain faults appearing in the practical tests can also be simulated, so that the appropriate solutions and control algorithm can be given.

2. Composition of Aircraft Braking System

The aircraft braking system consists of aircraft frame, wheels, landing gears, runway, electric brake controller, motor drive controller, electromechanical actuator (EMA), sensor, and so on. In HILS system, the motor drive controller, electromechanical actuator, electric brake controller, and pressure sensor are the actual objects. The aircraft airframe, wheels, landing gears, runway, and speed sensors of aircraft are replaced by Simulink models. Its structure is shown in Figure 1.

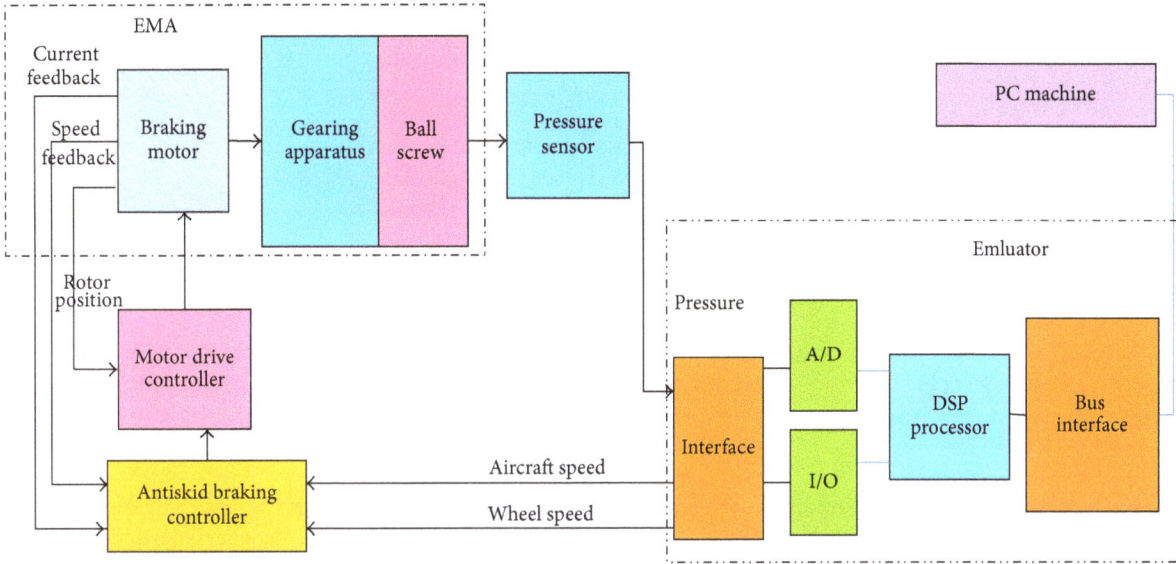

FIGURE 1: The structure of aircraft braking system.

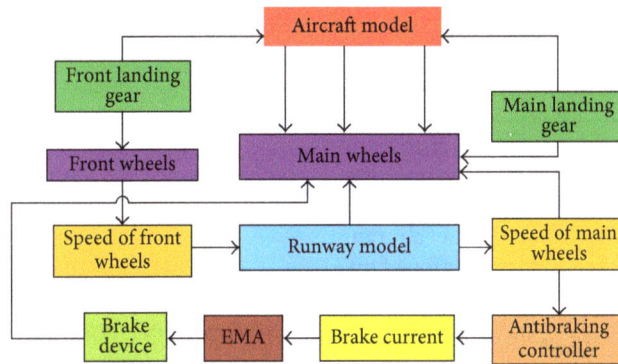

FIGURE 2: The functional relationship between the various parts.

In Figure 1, we are able to understand the working process of electric brake system clearly. We can obtain critical physical quantities, such as the braking distance, aircraft speed, wheel speed, and pitch angles of the aircraft according to the aircraft model. Based on the speed information obtained, braking control controller can calculate the corresponding braking current. Then, the motor generates the corresponding brake pressure, so as to realize the brake of the airplane.

In this paper, the developed all electric braking system model is uploaded and applied to brake experiment. Various brake control law is evaluated by controlling brake system according to different braking conditions.

3. Dynamic Simulink Model

According to the functional relationship between the various parts of the brake system, the system structure diagram is shown as in Figure 2.

The aircraft dynamics model is very important for evaluating the performance of antibraking controller, because the accuracy of dynamics model determines the performance of antibraking controller.

Here, we consider the landing gear and the wheels as the part of the airframe and give the main force analysis of the aircraft in the following.

3.1. Airframe Model. The main purpose here is to verify the brake control law by using HILS. The model is based on the following assumptions: first, the aircraft is considered as an ideal rigid body, without considering the elastic deformation. Second, only tackling the longitudinal, vertical, and pitching motion of the aircraft and the tire compression is not considered. The landing gear is regarded as a spring system. The wheel mechanism is consistent and synchronous control [3–6].

The force analysis of airframe is shown in Figure 3.

In Figure 3, V is the speed of the aircraft along the x-axis; the unit of V is m/s. m_A is the aircraft quality; the unit of m_A is kg. g is the acceleration of gravity; the unit of g is m/s^2. F_R is the engine thrust; the unit of F_R is N. Q is the pneumatic

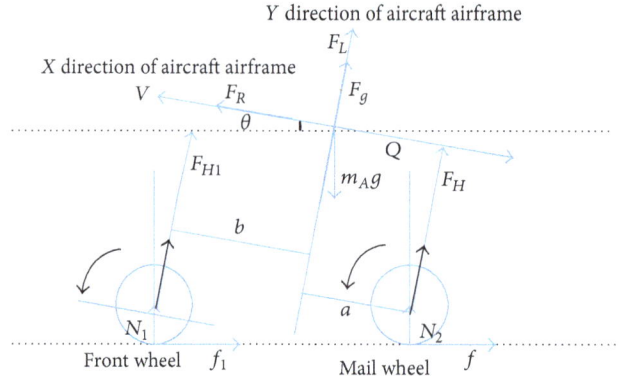

FIGURE 3: The force analysis of airframe.

resistance; the unit of Q is N. f_1 is the binding force between front wheel and runway; the unit of f_1 is N. f is the binding force between main wheel and runway; the unit of f is N. F_L is aircraft lift; the unit of F_L is N. F_g is the ground effect force of the airframe when aircraft touches the ground; the unit of F_g is N. F_{H1} is buffer support force of front landing gear; the unit of F_{H1} is N. F_H is buffer support force of main landing gear; the unit of F_H is N. N_1 is front wheel load; the unit of N_1 is N. N_2 is main wheel load; the unit of N_2 is N. b is the horizontal distance from the front wheel to the center of gravity; the unit of b is m. a is the horizontal distance from the main wheel to the aircraft center of gravity; the unit of a is m.

Considering the analysis of the force of the aircraft, according to Newton's second law, the balance equation of the aircraft can be obtained:

$$m_A \left(a_X + WV_Y\right) = F_R - f_1 - nf - m_A g \sin \theta - Q,$$

$$m_A \left(a_Y - WV_X\right) = F_L + F_{H1} + nF_H + F_g - m_A g \cos \theta,$$

$$Q = \frac{1}{2}C_d \rho_a S_A V^2, \tag{1}$$

$$F_L = \frac{1}{2}C_L \rho_a S_A V^2.$$

Here, n is the number of main wheels and C_d is the aerodynamic drag coefficient. ρ_a is the air density; the unit of ρ_a is kg sec^2/m^4. S_A is the wing area; the unit of S_A is m^2. C_L is the lift coefficient. V_X is the course speed of airframe; the unit of V_X is m/s. V_Y is the speed perpendicular to the course of the airframe; the unit of V_Y is m/s. a_X is the acceleration of aircraft along the X axis; the unit of a_X is m/s^2. a_Y is the acceleration of aircraft along the y-axis; the unit of a_Y is m/s^2. W is the pitching angular speed of aircraft; the unit of W is rad/s.

3.2. Wheels and Runway Models. According to the law of moment of inertia, the force equation of the wheel is listed as the following formula:

$$fR - T_b - B_w \omega = \dot{\omega} J_w. \tag{2}$$

In formula (2), R is the radius of main wheel; the unit of R is m. T_b is the brake torque; the unit of T_b is N·m. B_w is the

axle friction coefficient. ω is the main wheel angular velocity; the unit of ω is rad/s. J_w is the wheel inertia; the unit of J_w is kg sec^{2}m.

Here, the sliding of the aircraft wheel relative to the ground is defined as the relative slip rate; the formula is as follows:

$$\lambda = \frac{\left(V_{zx} - R\omega\right)}{V_{zx}}. \tag{3}$$

In formula (3), V_{zx} is wheel speed of aircraft.

Factor that affects the binding force f is called the friction coefficient. We can define it as the following formula:

$$\mu = \frac{f}{N_2}. \tag{4}$$

3.3. Design of Autobrake Control Law. Automatic brake system refers to the brake system that is slowed by electric brake to prevent the aircraft from flowing out of the runway. Automatic brake selector switch is used to select the brake stalls; then the system puts in braking performance according to the constant deceleration rate [7].

After the deceleration rate is selected, the reference speed of the aircraft can be obtained. The reference speed and the actual feedback of the aircraft can be used to adjust the brake current signal through control algorithm.

The control strategy adopted is PD + PBM antiskid control combined with a given rate of deceleration tracking control. The brake control part adopts the control scheme of given deceleration rate and PD + PBM control law is used in antiskid control. When the antiskid control is acting during braking, the output of the brake system maintains the output of previous moment. The output of control law is the difference between brake current and antiskid current. The realization of the algorithm is shown as in Figure 4.

In order to improve the robustness and adaptive ability of the automatic braking system, a quasi-sliding mode variable structure control based on fuzzy control is designed.

Here, we can define the error of speed as the following formula:

$$s = \int_0^t a_{\text{given}} - V_x. \tag{5}$$

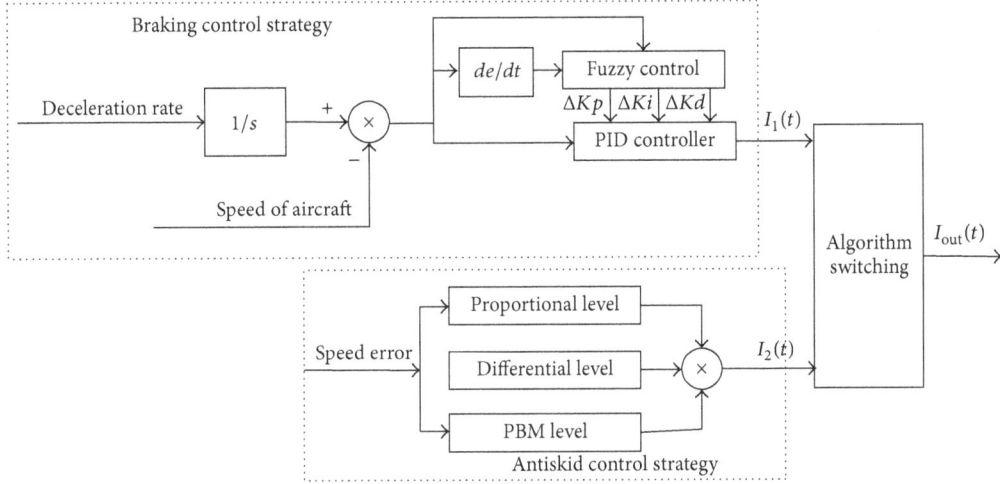

FIGURE 4: The control strategy of braking system.

In formula (5), a_{given} is deceleration rate.

$$I_1(t)$$

$$= \begin{cases} 0, & s > \varepsilon \text{ or } s < -\varepsilon, \\ k_p \cdot s + k_i \int s\, dt + k_d \cdot \dfrac{ds}{dt}, & -\varepsilon \le s \le \varepsilon, \end{cases}$$

$$I\text{brake}(t) = \begin{cases} \text{const}, & I_1(t) \ge \text{given value}, \\ I_1(t), & \text{others}, \\ 0, & I_1(t) < 0, \end{cases} \qquad (6)$$

$$I_{\text{out}} = \begin{cases} I\text{brake}, & \text{if } I_2 \le \zeta, \\ I\text{brake}_{k-1} - I_2, & \text{if } I_2 > \zeta. \end{cases}$$

Here, ζ and ε are very small boundary value. $I_1(t)$ is braking current. k_p, k_i, and k_d are proportion coefficient, differential coefficient, and integral coefficient, respectively.

And the proportion coefficient k_p, integral coefficient k_i, and differential coefficient k_d have important influence on the performance of the brake system. In order to improve the robustness and adaptive ability of the automatic braking system, fuzzy control is designed to adjust PID parameters online.

In the part of fuzzy parameter correction, the velocity error and the derivative of the velocity error are the input of the fuzzy controller, and the change of the three parameters of the PID is the output of the fuzzy controller [8].

According to the influence of control parameters on the output characteristics of the system, we can obtain adjustment rules of the parameter. When speed error is large, in order to improve the response speed of the system and prevent the differential overflow caused by the large transient, we should take a larger k_p and smaller k_d.

At the same time, in order to prevent integral saturation and avoid large overshoot, the integral separation PID regulation should be adopted. When the error is medium

TABLE 1: Fuzzy rules of Δk_p.

e	ec						
	NB	NM	NS	Z	PS	PM	PB
NB	PB	PB	PM	PM	PS	Z	Z
NM	PB	PB	PM	PS	PS	Z	NS
NS	PM	PM	PM	PS	Z	NS	NS
Z	PM	PM	PS	Z	NS	NM	NM
PS	PS	PS	Z	NS	NS	NM	NM
PM	PS	Z	NS	NM	NM	NM	NB
PB	Z	Z	NM	NM	NM	NB	NB

large, in order to reduce the overshoot of the system response and ensure response speed, the smaller k_p, smaller k_i, and moderate size k_d should be chosen. When speed error is the lowest, in order to make the system have steady performance and avoid the oscillation of the system in the vicinity of the reference speed, it is necessary to take a larger k_p, k_i, and proper k_d.

In addition, when $\dot{s}$ is smaller, the differential coefficient k_d should be larger. When $\dot{s}$ is larger, the differential coefficient k_d should be smaller. According to the regulation rules, the fuzzy rules can be designed and adjusted.

Fuzzy sets of fuzzy parameters are defined as

$$s = \{\text{NB NM NS Z PS PM PB}\},$$

$$\dot{s} = \{\text{NB NM NS Z PS PM PB}\},$$

$$\Delta k_p = \{\text{NB NM NS Z PS PM PB}\}, \qquad (7)$$

$$\Delta k_i = \{\text{NB NM NS Z PS PM PB}\},$$

$$\Delta k_d = \{\text{NB NM NS Z PS PM PB}\}.$$

For example, fuzzy rules of Δk_p can be obtained as in Table 1.

By using this method, the fuzzy rules of other parameters can also be obtained.

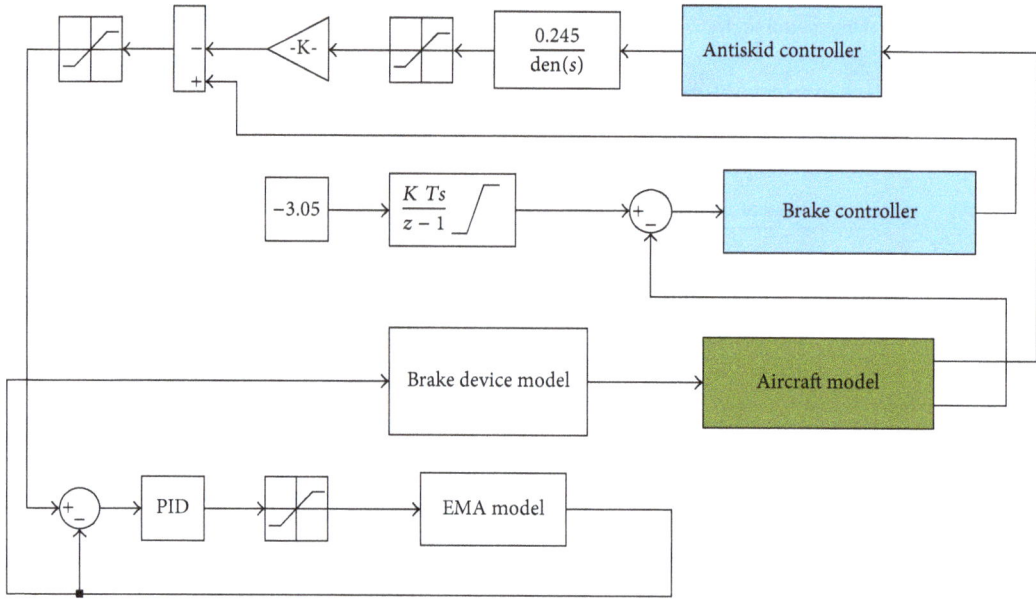

FIGURE 5: The integrated model of electric braking system.

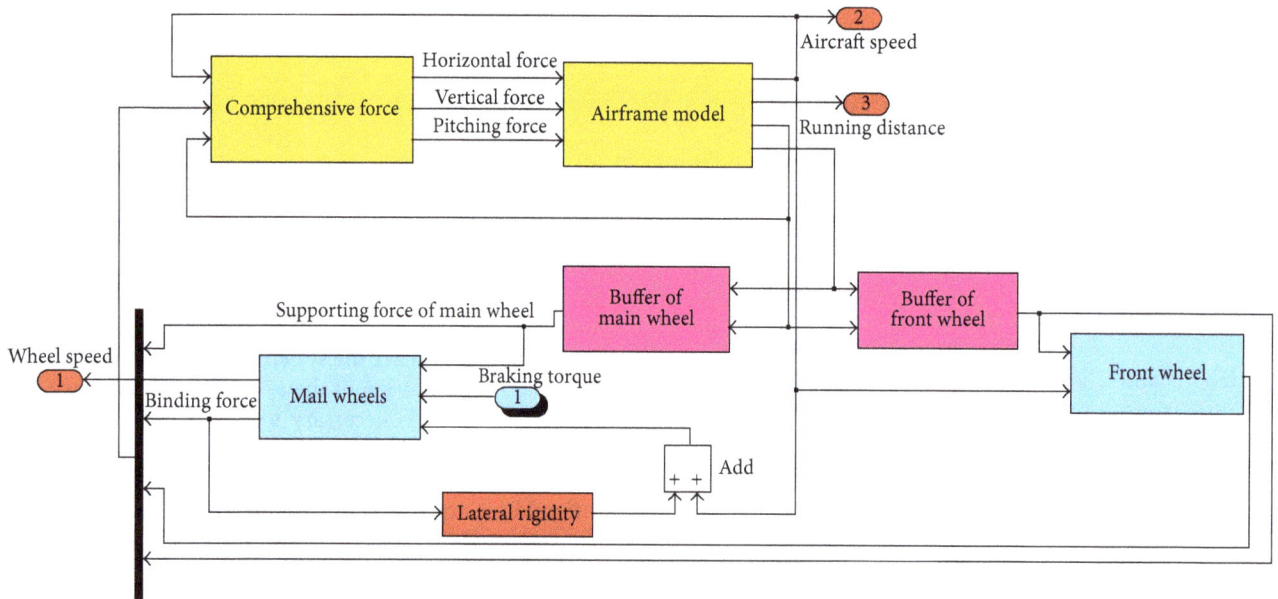

FIGURE 6: The simulation model of airframe.

3.4. Simulink Model of Aircraft Electric Braking System. The establishment of simulation model is one of the key technologies in HILS. The simulation model will be directly related to the authenticity of HILS. Therefore, in order to establish the simulation model of aircraft dynamics in MATLAB and verify the accuracy of the model, we must establish the whole mathematical model of aircraft. Then, landing gear, tires, and other factors must be considered.

According to the mathematical formula given in the previous section, the integrated model of aircraft electric braking system can be built as in Figure 5.

The brake system model mainly consists of airplane model, brake controller model, brake device model, and electromechanical actuator model.

According to the mathematical model of the aircraft, the simulation model of airframe can be built as in Figure 6.

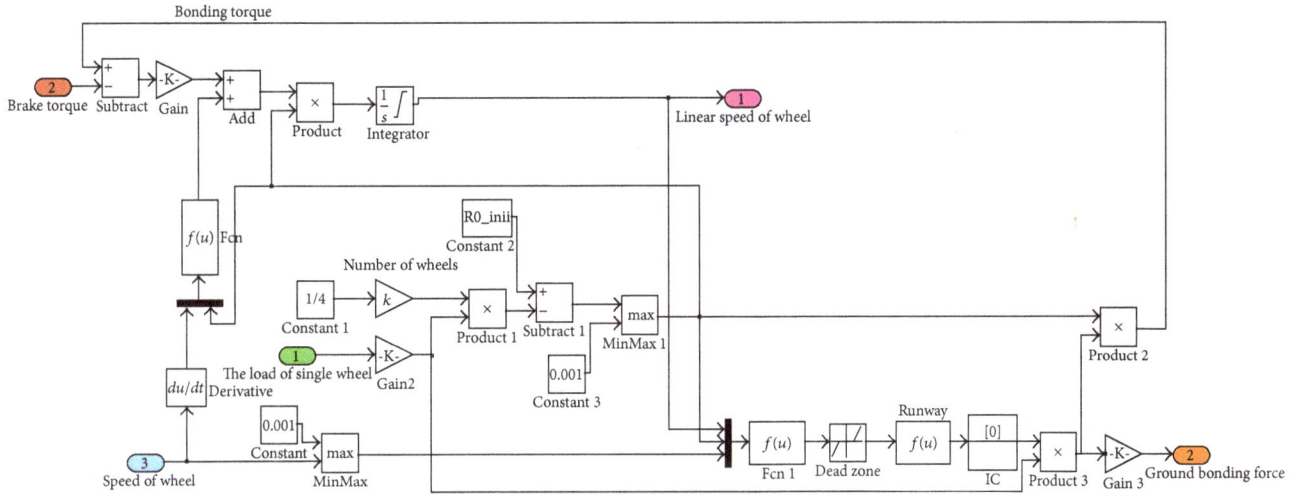

FIGURE 7: The model of main wheel and runway.

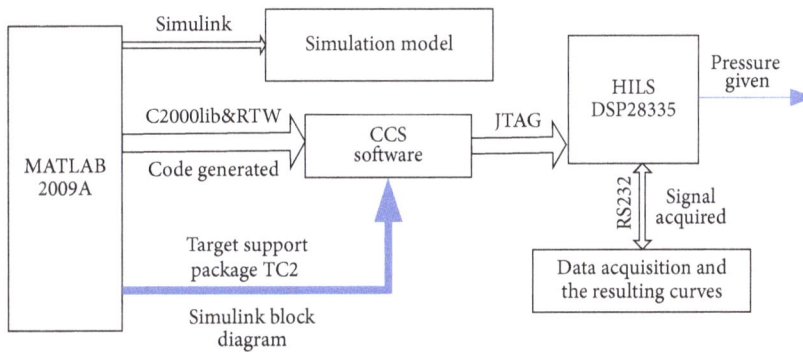

FIGURE 8: The structure of HILS system.

The model of main wheel and runway is shown as Figure 7.

In Figure 7, the brushless DC motor is chosen as the drive motor, and the corresponding mathematical model can be established.

According to the control law designed, the corresponding MATLAB model can be built.

In the same way, we can establish the mathematical model of the other control parts in electric braking system.

4. Design of All Aircraft Brake HILS System

The structure of HILS system is shown in Figure 8.

In HILS system, the simulation model should be discretized, and the module that needs to be replaced by the hardware should be determined. After we set the corresponding interface, then the Simulink simulation module can be converted into DSP code through the MATLAB toolbox and CCS software. Here, we call DSP code as digital model of HILS system.

In Figure 8, the simulation model is designed primarily first, just as we have done in the previous part. Then the DSP code can be generated automatically; the experimental results can be uploaded by using the interface part.

In HILS system, the electromechanical actuator and the brake controller are replaced by physical hardware. Other parts are replaced by simulation model.

(1) Simulation Model Discretization and Interface Setting. MATLAB/Simulink provides Target Support Package TC2 software, which supports Taxes Instrument (TI) C2000 series DSP and has various DSP function interface module. By using this tool, we can convert Simulink model of the aircraft braking system into executable code for DSP.

Simulation process based on MATLAB/RTW is listed as follows: an appropriate simulation model is established and discretized in the Simulink environment; then based on RTW simulation platform, real-time code for DSP can be generated and downloaded into the target machine (DSP). By using an external mode, we can adjust simulation parameters online

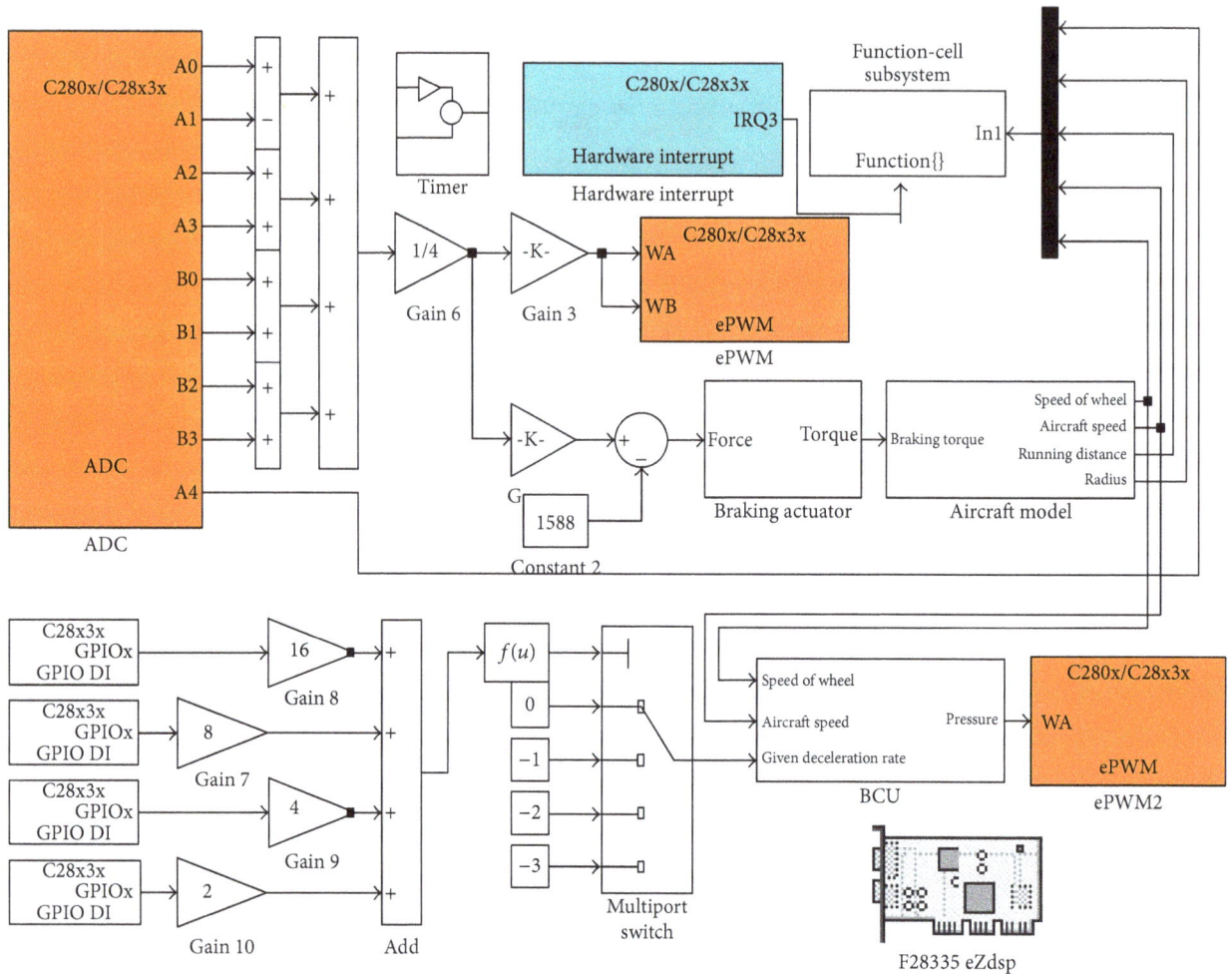

FIGURE 9: The discrete model of braking system.

and modify the simulation model until the results satisfy the design requirements.

In this paper, the discretized mathematical model of aircraft braking system is shown as in Figure 9.

On the basis of building the correct model of electric braking system, the corresponding interface of HILS is designed according to the function block of Target Support Package TC2. In Figure 9, the ADC module has 9 channels. Eight channels are used as pressure acquisition signal of pressure sensor. One channel is used as a trigger signal for PWM.

There are 4 GPIO modules; the position of automatic brake selector switch is determined by 4-bit binary code.

There is one hardware interrupt, SCI module, and timer. Timer is used to generate hardware interrupts. The SCI module uploads the brake signals to the host computer.

(2) The Design of the Hardware. The hardware of HILS system is composed of interface circuit, signal acquisition circuit, signal conditioning circuit, and braking motor. The diagram of the hardware is designed as in Figure 10. Digital signal processing is done by DSP and CPLD.

First, by use of powerful capabilities of DSP, functions such as given pressure regulation, SCI serial communication, brushless DC motor PWM control, the motor position, and speed of capture and signal feedback can be completed.

On the other hand, by using powerful logic computing power of CPLD, functions such as commutation control, enabling control, three phase full-bridge control signal conditioning, and other functions can be implemented.

5. Test Results

5.1. Simulation Results. In order to validate the validity of the simulation model and compare with HILS results, here, the performance of electric braking system is simulated under MATLAB/Simulink environment. The simulation studies are carried out when the reduction rate is high and medium, as well as emergency brake.

According to the actual working situation of the aircraft, when deceleration rate of automatic braking is high, the

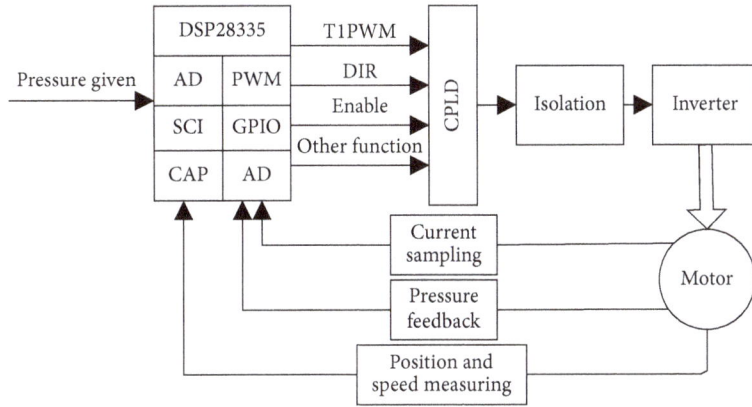

FIGURE 10: The hardware of HILS.

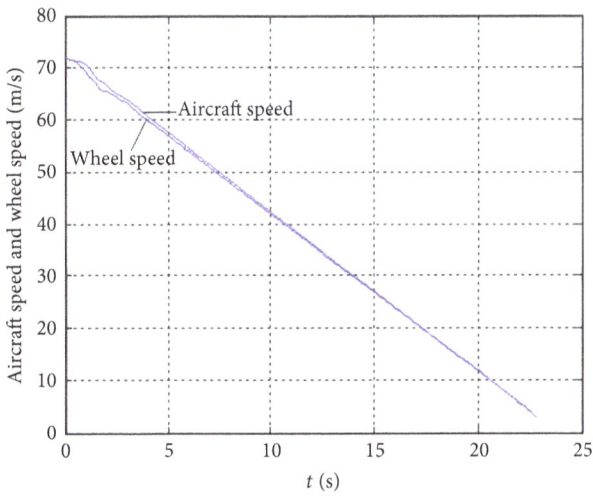

(a) Aircraft speed and wheel speed

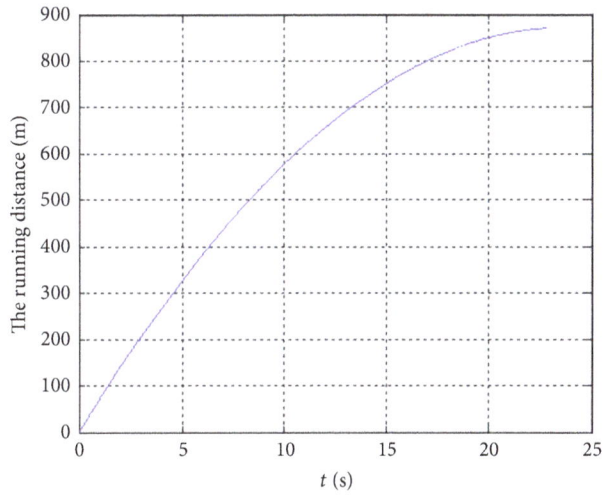

(b) Running distance

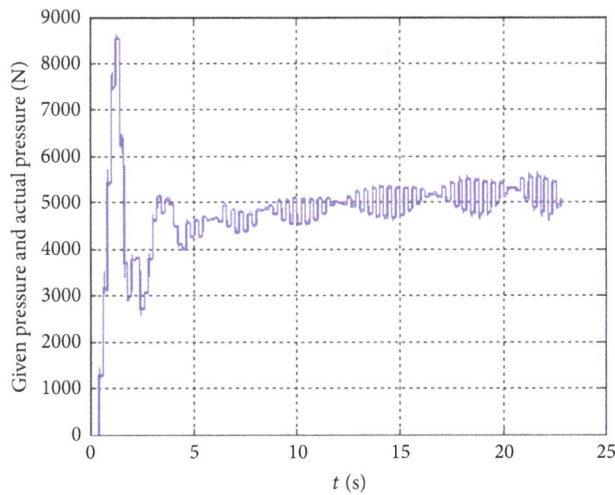

(c) Brake pressure

FIGURE 11: Simulation results when deceleration rate is 3.05.

(a) Aircraft speed and wheel speed

(b) Running distance

(c) Brake pressure

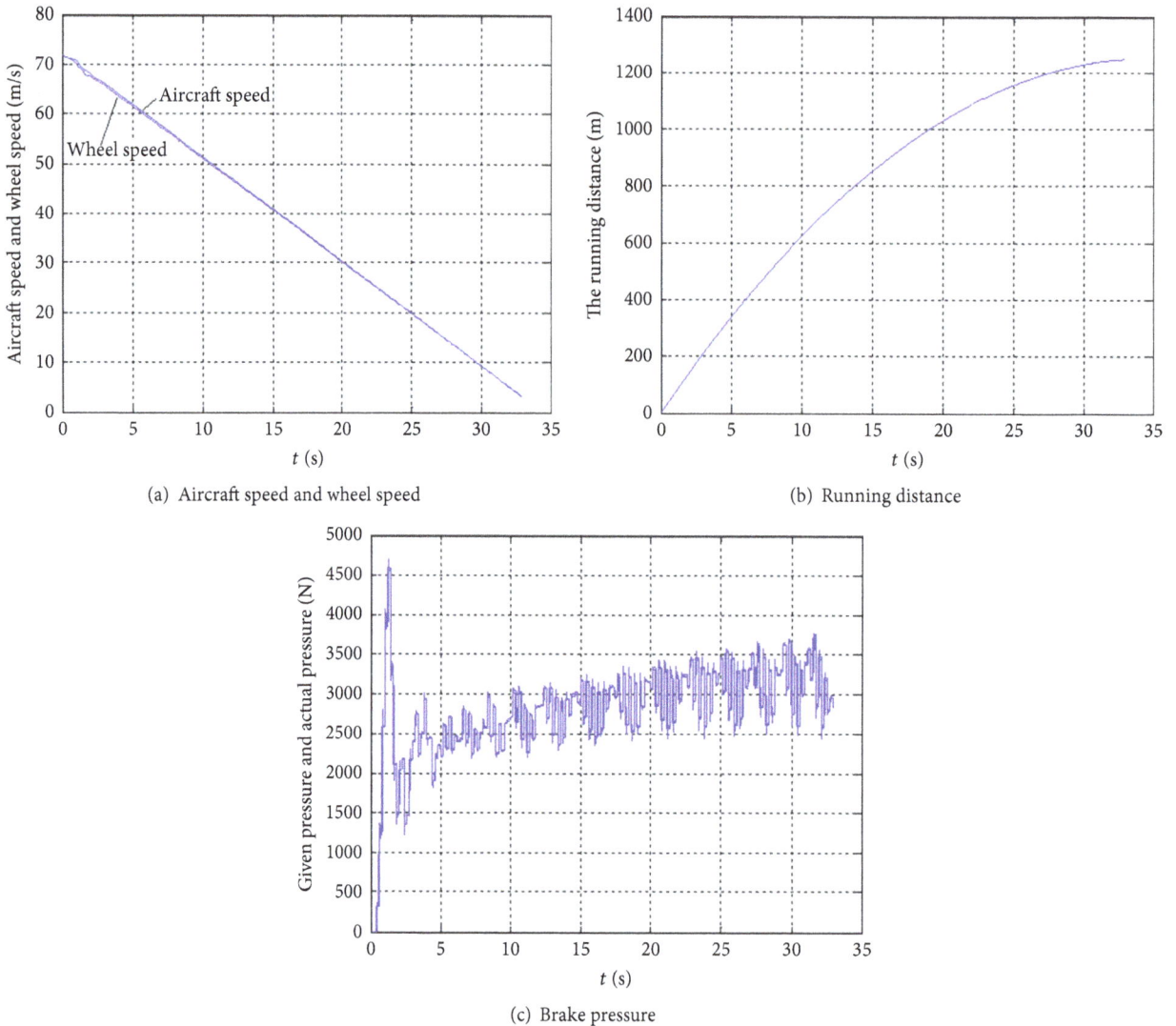

FIGURE 12: Simulation results when deceleration rate is 2.1.

actual deceleration rate is set to −3.05, and when the deceleration rate is medium, the actual deceleration rate is set to −2.1. In the case of emergency braking, the brake is controlled by the maximum braking force.

The simulation results of the system model are given below.

(1) Deceleration Rate Is 3.05. When the deceleration rate is 3.05, the digital simulation results are shown in Figures 11(a)–11(c).

(2) Deceleration Rate Is 2.1. When the deceleration rate is 2.1, the digital simulation results are shown in Figures 12(a)–12(c).

(3) Emergency Brake. In the case of emergency braking, manual braking mode is used. Maximum braking force is used to achieve emergency brake control.

Here, the digital simulation results are shown in Figures 13(a)–13(c).

5.2. HILS Results. In HILS system, the electromechanical actuator and the brake controller are physical hardware, and other parts are replaced by simulation model.

The experimental study is carried out by using HILS system. In order to compare the results, the simulations when the deceleration rate is 3.05 and 2.1 as well as in case of emergency brake are given.

The experiment results will be given in Figures 14, 15, and 16.

(1) Deceleration Rate Is 3.05. See Figure 14.

(2) Deceleration Rate Is 2.1. See Figure 15.

(3) Emergency Brake. See Figure 16.

(a) Aircraft speed and wheel speed

(b) Running distance

(c) Brake pressure

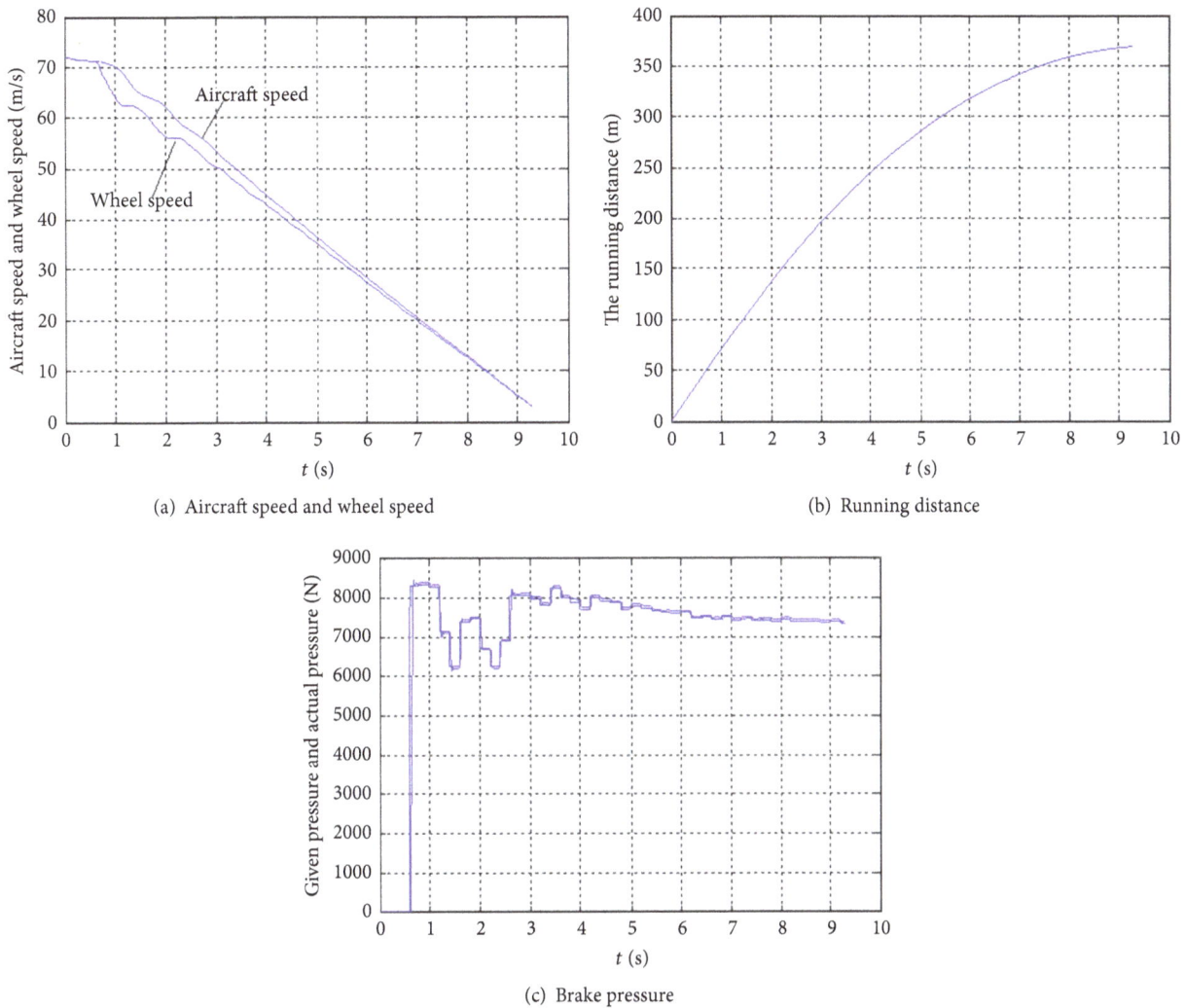

FIGURE 13: Simulation results in case of emergency brake.

5.3. Analysis of Results. According to the simulation results, the following conclusions can be obtained.

(1) With the increase of the deceleration rate, the brake pressure is bigger, the speed is faster, and the running distance is shorter. In case of the emergency brake, braking force is maximum, so the braking distance is the shortest.

(2) Taking the deceleration rate of 2.1 as an example, the ideal braking time is 72/2.1 (34.29) s, the actual simulation time is about 35 s, and the aircraft speed and wheel speed are in good agreement.

(3) When the brake pressure is simulated in MATLAB environment, the brake pressure is in good agreement with the actual brake pressure. When the deceleration rate is 2.1, brake pressure is about 3000 N.

According to the results of HILS, the following conclusions can be obtained.

(1) With the increase of the deceleration rate, the brake pressure is bigger, the speed is faster, and the running distance is shorter. In case of the emergency brake, braking force is maximum, so the braking distance is the shortest.

(2) Taking the deceleration rate of 2.1 as an example, the HILS time is about 34.95 s, and the aircraft speed and wheel speed are in good agreement. When the deceleration rate is 2.1, brake pressure is also about 3000 N in HILS.

(3) Since the start of the motor requires a certain reaction time, the actual brake pressure has a certain delay compared with the given pressure in HILS. But from the waveform and size of the HILS results, the actual brake pressure and the given pressure have good consistency.

By comparing the simulation results with the results of HILS, we can know that the simulation results are in good agreement with the HILS results. The results of HILS can reflect the actual working state of the brake system.

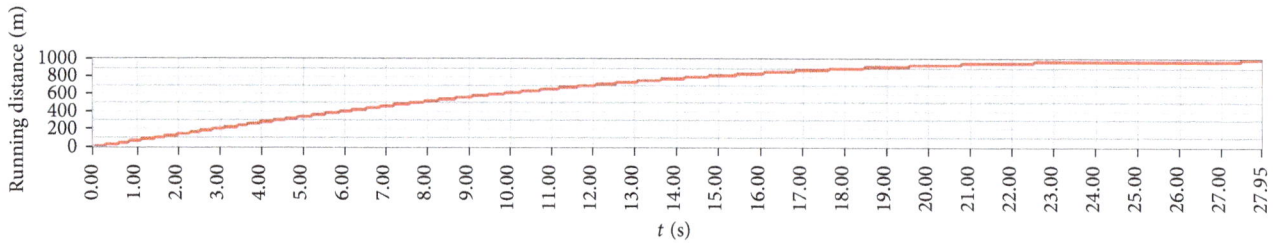

(a) Running distance

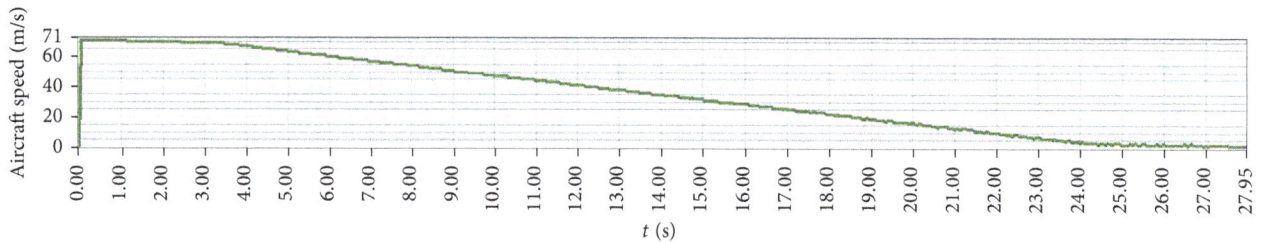

(b) Aircraft speed

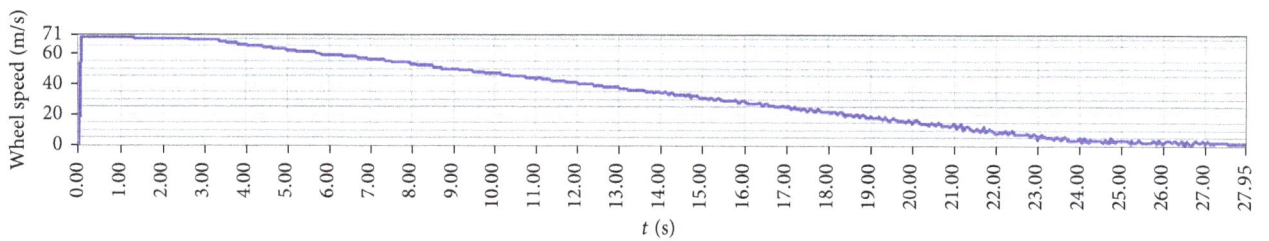

(c) Wheel speed

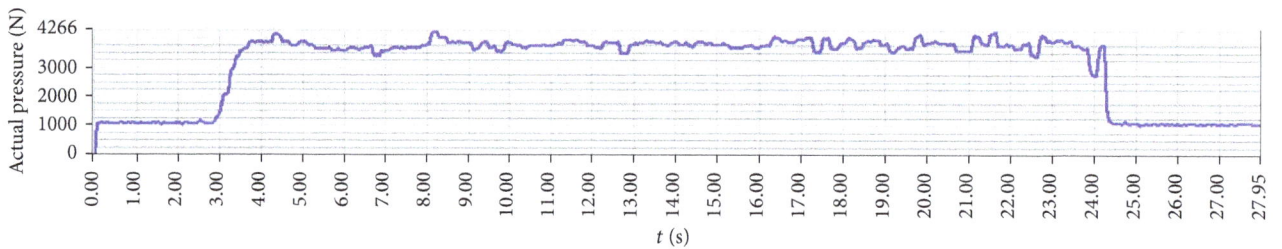

(d) Actual brake pressure

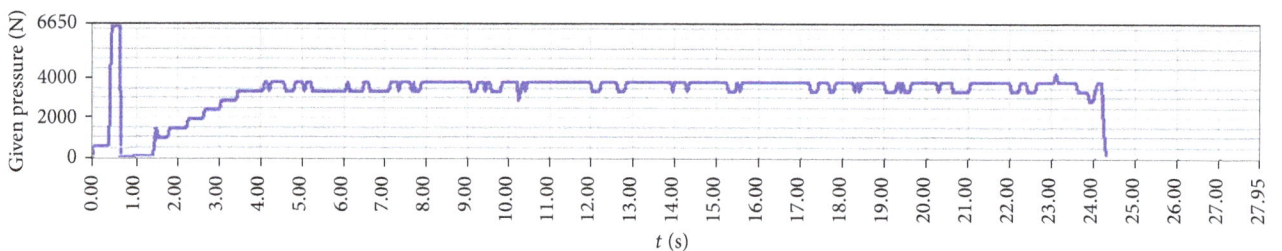

(e) Given brake pressure

FIGURE 14: HILS results when deceleration rate is 3.05.

6. Conclusion

In this paper, the HIL simulation technology is introduced into the aircraft electric braking system. By making full use of

the RTW toolbox of MATLAB and DSP, the HIL simulation platform of the aircraft brake system is completed.

The automatic generation of the simulation model code, the online adjustment of the control law parameters in

(a) Running distance

(b) Aircraft speed

(c) Wheel speed

(d) Actual brake pressure

(e) Given brake pressure

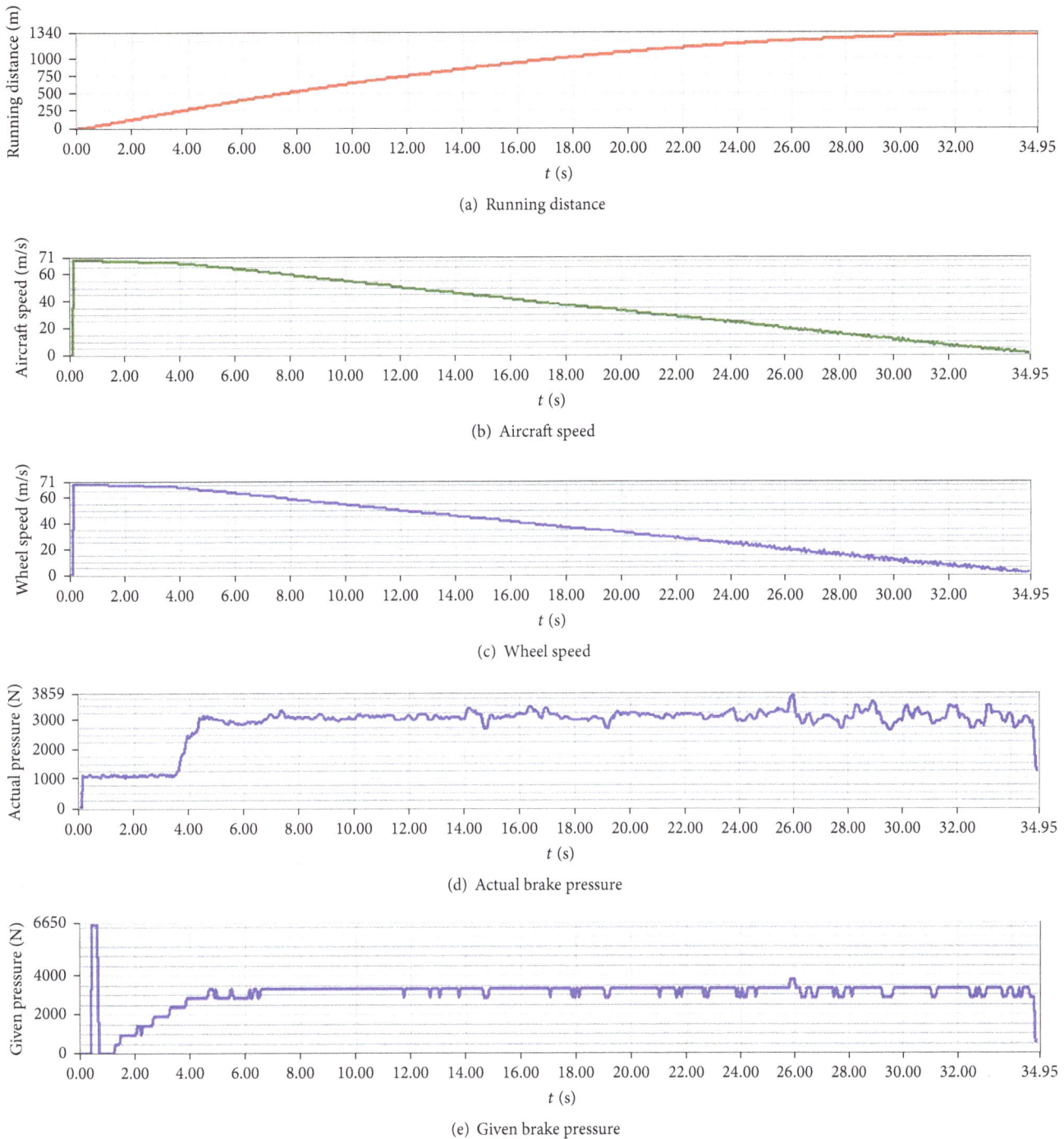

FIGURE 15: HILS results when deceleration rate is 2.1.

the simulation process, and the online monitoring of the simulation data in the simulation process are realized.

From the results of simulation and HILS, the simulation curves are very close to the HILS results, which shows that the results in HILS environment are realistic and reliable.

Based on the HILS system, the brake control law can be easily modified or designed. It provides a platform for the design of brake system and the design of control law.

Conflicts of Interest

The authors declare that there are no conflicts of interest regarding the publication of this paper.

Acknowledgments

The work is supported by Foundation for Shaanxi Key Laboratory of Small and Special Electrical Machines and

(a) Running distance

(b) Aircraft speed

(c) Wheel speed

(d) Actual brake pressure

(e) Given brake pressure

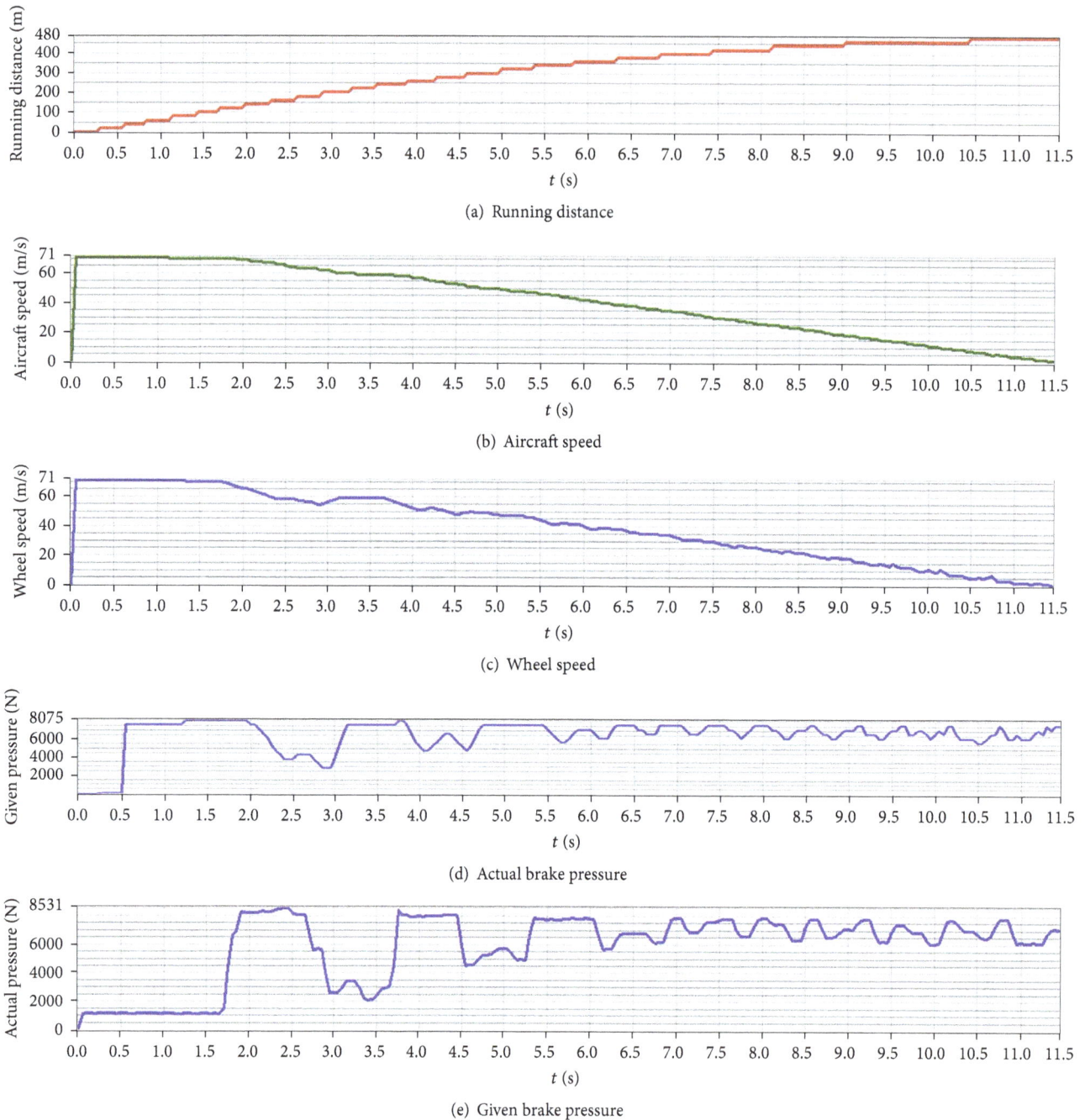

FIGURE 16: HILS results in case of emergency brake.

Drive Technology (2014SSJ1004). The work is also supported by National Natural Science Foundation of China (51407143).

References

[1] J. Wang and L. Xie, "Real-time software design of hardware in-the-loop simulation of aircraft braking system," *Computer Measurement & Control*, pp. 773–775, 2009.

[2] M. Yoon, J. Lee, and J. Hong, "The simulink model of motor system for HEV using HILS (hardware-in-the-loop)," in *Proceedings of the 2012 Sixth International Conference on Electromagnetic Field Problems and Applications (ICEF)*, pp. 1–4, Dalian, China, June 2012.

[3] J. W. Jeon, G. A. Woo, K. C. Lee, D. H. Hwang, and Y. J. Kim, "Real-time test of aircraft brake-by-wire system with HILS & dynamometer system," in *Proceedings of the IEEE International Conference on Mechatronics 2004, (ICM '04)*, pp. 322–327, June 2004.

[4] N. E. Daidzic and J. Shrestha, "Airplane landing performance on contaminated runways in adverse conditions," *Journal of Aircraft*, vol. 45, no. 6, pp. 2131–2144, 2008.

[5] H. B. Pacejka and E. Bakker, "The magic formula tyre model," *Vehicle System Dynamics*, vol. 21, supplement 001, pp. 1–18, 1993.

[6] D. Xu, Y. Li, and L. Xie, "Research on Modeling and Simulation of Aircraft Anti-Skid Braking System," *Measurement & Control Technology*, vol. 23, pp. 66–68, 2004.

[7] J. I. Miller and D. Cebon, "A high performance pneumatic braking system for heavy vehicles," *Vehicle System Dynamics*, vol. 48, no. 1, pp. 373–392, 2010.

[8] H. Zhang, G. Xu, W. Li, and M. Zhou, "Fuzzy logic control in regenerative braking system for electric vehicle," in *Proceedings of the 2012 IEEE International Conference on Information and Automation, ICIA 2012*, pp. 588–591, June 2012.

9

DPFFs: C²MOS Direct Path Flip-Flops for Process-Resilient Ultradynamic Voltage Scaling

Myeong-Eun Hwang[1] and Sungoh Kwon[2]

[1]*Memory Division, Samsung Electronics Inc., Hwaseong, Gyeonggi 18448, Republic of Korea*
[2]*School of Electrical Engineering, University of Ulsan, Nam-gu, Ulsan 44610, Republic of Korea*

Correspondence should be addressed to Sungoh Kwon; nj0324@gmail.com

Academic Editor: Ahmed M. Soliman

We propose two master-slave flip-flops (FFs) that utilize the clocked CMOS (C²MOS) technique with an internal direct connection along the main signal propagation path between the master and slave latches and adopt an adaptive body bias technique to improve circuit robustness. C²MOS structure improves the setup margin and robustness while providing full compatibility with the standard cell characterization flow. Further, the direct path shortens the logic depth and thus speeds up signal propagation, which can be optimized for less power and smaller area. Measurements from test circuits fabricated in 130 nm technology show that the proposed FF operates down to 60 mV, consuming 24.7 pW while improving the propagation delay, dynamic power, and leakage by 22%, 9%, and 13%, respectively, compared with conventional FFs at the iso-output-load condition. The proposed FFs are integrated into an 8×8 FIR filter which successfully operates all the way down to 85 mV.

1. Introduction

The rapidly growing volume of data processing and transferring in contemporary clocked electronics has constantly drawn considerable attention to the design of high speed low-power yet robust sequential timing elements. With a clock network, these sequential components account for 30% to 60% of total power consumption in VLSI systems [1]. Moreover, the propagation delay and latency of the timing units are responsible for a large portion of the cycle time as the operating frequency increases. As a result, FF selection and design have a great effect on both reducing power dissipation and providing more slack time for easier time-budgeting in high-performance applications.

FFs and latches are the major building blocks of digital circuits, and their primary function is to store binary data. Many commercial digital applications selectively use master-slave and pulse-triggered FFs. Examples of the master-slave flip-flop (MSFF) include the transmission gate-based FF [2], push-pull D-type FF [3], and true single phase clocked FF (TSPC) [4]. Despite popularity of high-performance design, the TSPC can malfunction when the slope of the clock is insufficiently steep. Slow clocks can cause both the clocked pull-up and pull-down networks to be on simultaneously, resulting in undefined value of the state and race condition. Further, transistor sizing is critical to achieve correct functionality in the TSPC. With improper sizing, glitches may also occur at the output due to a race condition when the clock transits [5]. Ishikawa et al. proposed a MSFF in which a hysteresis characteristic of the input circuit of the slave latch prevents the output which was once inverted from being inverted again in the metastable state [6]. Another edge-triggered FF is the sense amplifier based FF [7]. All these hard-edged FFs are characterized by positive setup time, causing a large cycle time. Alternatively, the pulse-based FFs (PBFFs) [8–10] have been widely used to decrease the data-to-output (D-to-Q) delay. In scan logic of the PBFF, however, the scan control becomes too complex and incompatible with the conventional MSFF for more enhancement of the D-to-Q speed of the FF. The PBFF requires the addition of pulse generators internally or externally, which can cause an increase in area and power and routing congestion, even though external pulse generation scheme would provide several advantages such as shareability among the neighboring

FFs and availability of the dual-edge triggering. Other various types of FFs and their analysis at normal or high voltages are found in [11–20]. Considering circuit robustness against race condition and signal integrity in the presence of clock skew and power consumption and performance in aiming at commercial products, we will focus on the fully static MSFFs [21] for ultralow as well as normal voltage applications in this paper.

Based on the conventional MSFFs which are widely used especially in commercial products, we propose two C^2MOS asymmetric direct path master-slave FFs (DPFFs) which use the C^2MOS technique at the primary input stage and have direct signal path between master and slave latches. The proposed DPFFs adopt a low-power technique, called ABRM (adaptive β-ratio modulation), to dynamically adjust the skewness in PMOS and NMOS transistors in different regions of operation. Highlights include that the proposed FFs provide (1) high-performance in speed, which can be traded off to reduce power and area; (2) full compatibility with the most widely used commercial tools for characterization while inheriting main advantages of the conventional FFs, such as high signal integrity and noise immunity against clock skew; and (3) compensation over the device mismatches and skewed P/N-ratios as a supply voltage changes from the normal voltage all the way down to the deep subthreshold region (say, 60 mV).

The remaining of this paper is organized as follows. Section 2 explores the FF design metrics of interest. Conventional MSFFs used as references in the paper are presented in Section 3 and DPFFs are proposed in Section 4. Section 5 briefly explains a body bias technique applied to improve circuit robustness especially at ultralow voltages. Section 6 discusses simulation and measurement results, and the conclusions are drawn in Section 7.

2. Flip-Flop Design Metrics

In general, gate-level static timing analysis requires that the sequential elements are characterized for three important design metrics: setup and hold times and propagation delay. These design metrics affect the system-level features such as performance in speed, signal integrity, and noise immunity under noisy and race conditions. The setup time t_{setup} is defined as the minimum amount of time for which the input data D should remain steady before the active clock edge so that the data is reliably sampled by the clock. Any violation may cause incorrect data to be captured, which is known as a setup violation. The hold time t_{hold} is the minimum amount of time for which the input D should remain stable after the active clock edge so that D is correctly sampled. Otherwise, violation may cause incorrect data to be latched, which is known as a hold violation. Finally, the clock-to-Q delay is the propagation delay that a FF takes to compute its correct output Q after the active clock edge. Considering the polarity of propagated data, the clock-to-Q delay is defined by $t_{clk2q} = \max(t_{clk2qh}, t_{clk2ql})$, where t_{clk2qh} and t_{clk2ql} are the clock-to-Q low-to-high and high-to-low delays, respectively.

These design metrics are determined as time difference between data and clock signals and generally

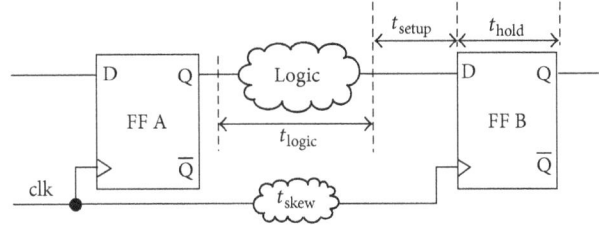

FIGURE 1: Flip-flop environment in a digital system.

precharacterized and stored in a table indexed by the input slope, clock slope, and output load. Typically, each sequential gate is characterized using commercial library delivery methodologies. The characterization procedure is repeated for a rising or falling edge with various combinations of input slopes and output loads. Figure 1 illustrates a condition which FF environment in a digital circuit has to satisfy for correct operation. The clock period T must be greater than or equal to the sum of the clock-to-Q delay t_{clk2q}, setup time t_{setup}, maximum combinational logic delay t_{logic}, and relative clock skew t_{skew}. Then, the FF delay has to meet the maximum delay limit given by

$$\text{Delay} = 1.05 \times t_{clk2q} + t_{setup} \leq T - t_{logic} - t_{skew}, \quad (1)$$

where 5% margin of the clock-to-Q delay is considered to avoid the metastable state where setup and hold violations occur and thus the output is unpredictable. The worst race condition happens when there is no logic between the two FFs in Figure 1. The internal race immunity of a FF is given by

$$R = t_{clk2q} - t_{hold} \geq t_{setup}. \quad (2)$$

3. Conventional Flip-Flops

The master-slave flip-flop (MSFF) typically consists of two gated latches connected in series and with an inverted enable input to one of them. Clocking causes the FF to either change or retain its output based upon the value of the input signals at transition. It is known that the transmission gate-based FF (TGFF) presents the best power-performance trade-off with the total delay ($t_{clk2q} + t_{setup}$) among the fully static FFs.

Figure 2(a) shows the TGFF which was originally designed for the IBM PowerPC microprocessor [2, 11]. The main advantages of the TGFF include a short signal path and a low-power feedback. The butterfly-structured low-power feedback in the C^2MOS cross-coupled inverters is usually insensitive to overlap of the clocks. The TGFF covers a relatively wide range of total energy-delay space [11] and presents the least amount of total leakage as an average across all states compared with all other FFs even with limited performance and positive setup time [21]. On the other hand, the use of the transmission gates not only degrades signal integrity at the presence of output noise but also increases the sensitivity to race condition when two phases overlap. The transmission gate T1 at the primary input stage is normally vulnerable to the output noise due to its generic characteristics of bidirectional signal transferring capability which can cause output

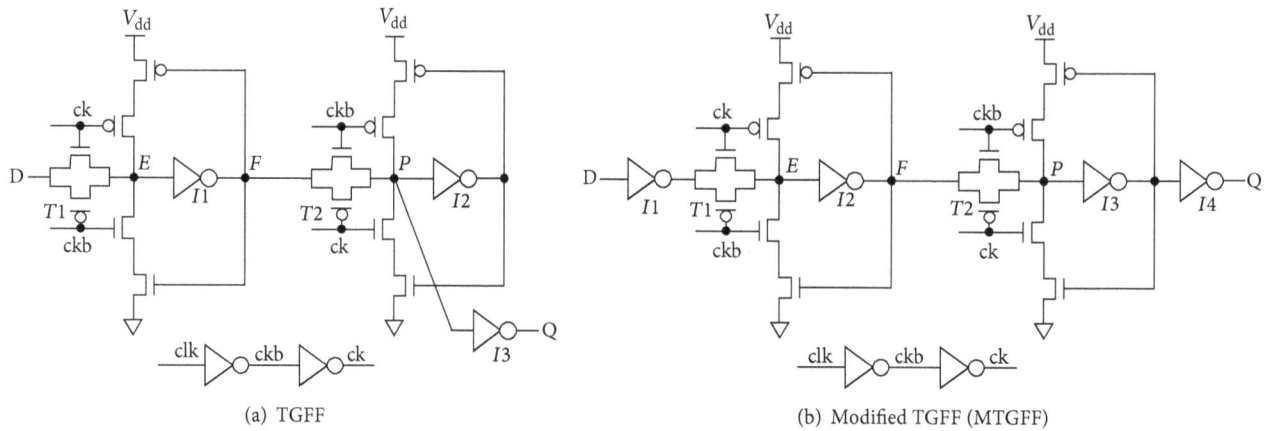

FIGURE 2: Transmission gate flip-flops (TGFFs) [2, 11].

noise to flow back to and disturb the input stage. Moreover, nonbuffered (or bare) input directly applied to the transmission gate can be limited by the standard cell library characterization flow since power consumed by a FF should partially be delivered through the input data D terminal. The authors believe that design constraints of the characterization flow even with leading commercialized EDA (Electronic Design Automation) tools [22–25] come with a restricted capability in characterizing library cells which, besides the power source V_{dd}, require power delivery from the input D as well. From the performance characterization perspective, the current drive of the previous stage can also cause inaccuracy in FF delay measurements.

The modified TGFF (MTGFF) shown in Figure 2(b) addresses the noise immunity issue of the TGFF by adding an inverter buffer at the primary input stage which keeps the output noise from propagating back to and interfering with the input [11]. In addition, the added inverter ensures compatibility with the characterization constraints by delivering power from a single voltage source V_{dd} to the entire cell including the transmission gate. The MTGFF inherits the main advantages of the TGFF such as a low-power feedback to store the cell value. The C^2MOS technique along with the transmission gate separates the hold mode from the transparent mode. In general, large transmission gates are used to speed up signal propagation in the transparent mode, resulting in increased area overhead. Unlike the TGFF, the addition of the inverter $I1$ at the primary input stage enables the MTGFF to achieve high noise immunity against the output noise and provides full compatibility with the primitive cell characterization methodology. The added inverter, however, requires earlier data arrival, which increases the setup time by the inverter delay $t_{p,inv}$. Further, inverter insertion now needs another inverter $I4$ at the output stage in order to keep the same polarity with the input, which may increase the propagation delay by the inverter delay $t_{p,I4}$.

Despite these unfavorable aspects, high robustness and low-power features and full compatibility with the underlying characterization process allow the MTGFF to be successfully embedded in numerous commercial applications such as Intel's mainstream microprocessors and Samsung's SSD

(Solid State Drive) controllers. We will use the MTGFF as a reference to evaluate the proposed FFs.

Figure 3 shows the C^2MOS FF (C2FF) as another approach to resolve the issues associated with the TGFF. Unlike the MTGFF, the C2FF utilizes the C^2MOS inverter that combines the inverter and transmission gate in the MTGFF at the input stage for both master and slave latches. The use of the C^2MOS inverter as an input buffer at the input stage shortens the logic depth along the main signal propagation path, reducing the setup time roughly by the inverter delay $t_{p,I1}$. The reduced setup time can increase system performance and robustness by relaxing the timing constraint of the maximum FF delay in (1) and improving the internal race immunity in (2). Note that, for a speedup in charging or discharging the output of the FF, the input data D is applied to the outer PMOS and NMOS transistors, whereas the clock signals of ck and ckb are applied to the inner transistors, where we assume that input data arrives and becomes stable earlier than the clock signals.

4. Proposed Direct Path Flip-Flops

As basic and common building blocks of digital systems, the FFs are required to have high-performance and low-power consumption while providing high robustness under data and clock skews and compatibility with a characterization flow of primitive cells. In this paper, we propose two C^2MOS direct path master-slave FFs with an internal direct connection between the asymmetric master and slave latches.

Figure 4 shows the first proposed FF, called C^2MOS DPFF (C2DPFF), which utilizes the C^2MOS scheme at the primary input stage and takes over the main advantages of the C2FF, addressing the output noise and noncompatibility issues of the TGFF while reducing the setup time. Unlike the C2FF, direct interconnection at node E between the master and slave latches enables prompt signal propagation along the main signal path. Performance improvement in time can be traded off with power so that the C2DPFF can achieve further power saving with area reduction. Due to the transistor stacking effect, however, the size of the C^2MOS inverter at the input stage may need to be enlarged to offer the current

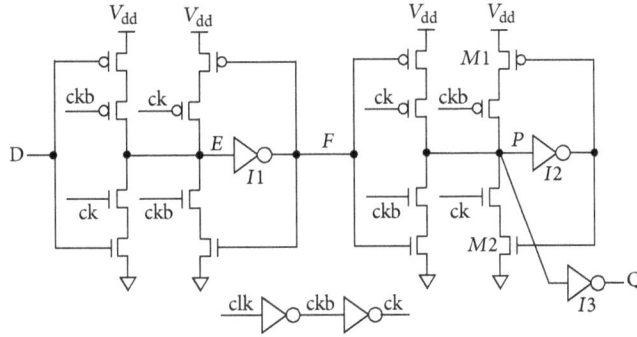

FIGURE 3: C^{2}MOS master-slave flip-flop (C2FF).

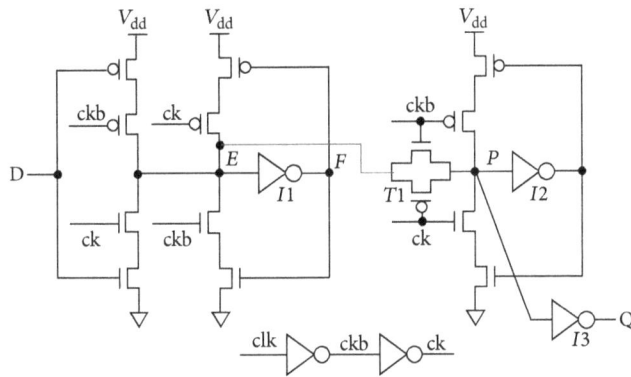

FIGURE 4: Proposed C^{2}MOS direct path flip-flop (C2DPFF).

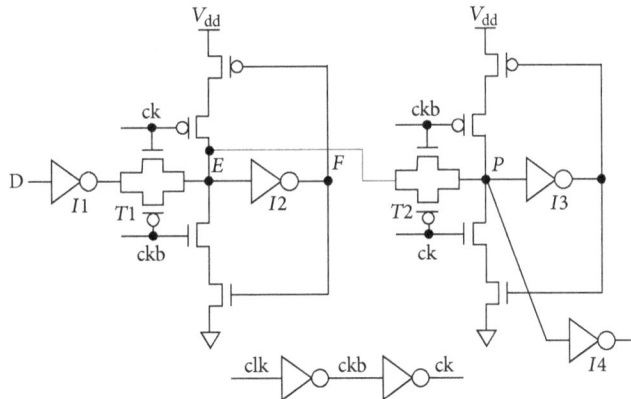

FIGURE 5: Proposed transmission gate direct path FF (TGDPFF).

drive or strength comparable with that of the nonstacked counterpart. Larger input capacitance of a gate requires larger current drive of the previous driver, resulting in more power dissipation.

The second proposed FF, called transmission gate DPFF (TGDPFF), shown in Figure 5, uses the transmission gate with reduced gate input capacitance at the input stage while leveraging the key advantages of the MTGFF. With the employment of butterfly-structured C^{2}MOS cross-coupled inverters to store the cell value, the TGDPFF also presents good low-power properties, assuring fully static operation.

The direct path structure may, however, have write-back glitches between storage nodes E and P in both Figures 4 and 5 due to charge sharing through or bidirectional signal transferring capability of the transmission gate [26]. The write-back issue is a kind of contention which can happen when the clock transitions high the value stored in the slave node P which writes back into the nonprotected master node E, resulting in incorrect bit flip because of reduced noise margins especially at lower voltages. At lower supply voltages, the issue is getting more serious since degradation in the transistor ON/OFF current ratio, random and systematic

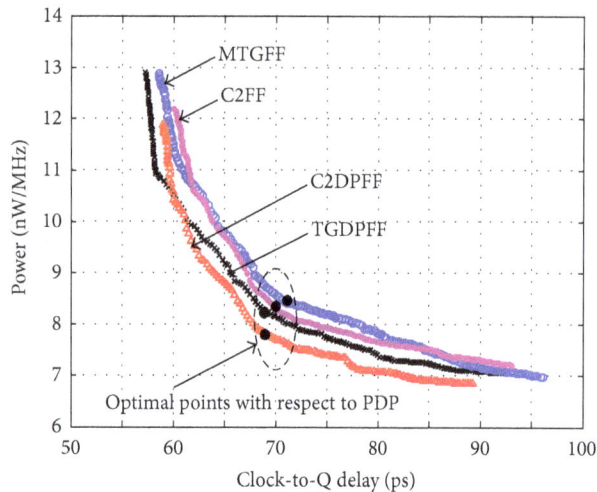

FIGURE 6: Power and delay with different sizes (simulated).

process variations, affects stability of the storage nodes. In order to address the issue, the keepers need to be upsized to improve the stage of state retention and made interruptible to avoid write contention. During retention phase, the on-current of the keepers can hence fully contend with the off-current of the transmission gates and thus avoid incorrect bit flipping. A clocked CMOS style flip-flop implementation of the proposed DPFFs replaces master and slave transmission gates in the conventional circuit topologies with pass-gate free clocked inverter, thereby eliminating the risk of data write-back through the transmission gate.

On the other hand, the proposed direct path scheme may cause an increase in load capacitance at node E due to the directly connected transmission gate of $T1$ or $T2$ in Figure 4 or Figure 5, respectively. Hence, the size of the C²MOS inverter needs to be enlarged to secure enough current drive of the inverter, which may increase area and power consumption accordingly. This side effect can be compensated by the shortened signal path and thus increased performance which in turn allows the use of smaller sized DPFFs during synthesis while meeting a given performance constraint.

Asymmetric structure may cause an unbalanced timing specification for positive or negative edge-triggered FFs, which desires careful sizing and optimization for target edge-triggered systems.

Figure 6 shows the power and delay profile of the conventional and proposed FFs with different sizes. Size optimization is made with an in-house tool which varies individual transistor size in the FFs. Boundaries of size variation and initial sizes of the transistors are set by using the theory of Logical Effort [27] and prelayout simulations without layout-extracted parasitics are performed with various transistor sizes. The optimal points with respect to both power consumption and delay are marked with black dots which have the minimum power and delay product (PDP) for the FFs. On the other hand, 3000 Monte-Carlo simulations at isoarea conditions show that the proposed DPFFs have

similar variations in key design metrics compared with their corresponding conventional counterparts (not shown).

Both proposed DPFFs can be extended to include scan logic. For example, Figure 7 shows one possible implementation of the scanned C2DPFF which, same as the scanned TGDPFF, includes total of 36 transistors and functions as a scanned asynchronous reset D-type FF.

5. Adaptive Body Bias Technique for Ultralow Voltage Operation

Due to the impact of process variation and skewness between the PMOS and NMOS transistors, circuit robustness can severely degrade especially for subthreshold operation. This limits supply voltage scaling while providing proper logic functionality with limited voltage headroom under process variation. Hence, it is of primary essence to keep an equal device strength ratio between the transistors in the FFs as well as logic cells to minimize the impact of process variation [28].

We proposed a circuit technique, ABRM (adaptive β-ratio modulation) [29, 30], which dynamically adjusts the P/N-ratio (or β-ratio) in the current drive between the PMOS and NMOS transistors and thus maximizes noise margin and circuit robustness for ultradynamic voltage operation. For reader's convenience, we restate a brief explanation of ABRM as follows.

The body bias technique is used to equalize the strength of pull-up and pull-down networks when switching back and forth between different regions of operation (Figure 8(a)). Forward body bias (FBB) lowers V_T, whereas reverse body bias (RBB) increases V_T. Body biases are implemented with additional body-biasing rails for PMOS and NMOS transistors (namely, V_{pbody} and V_{nbody}) and a body bias generating circuit.

Figure 8(b) shows the body bias circuit for ABRM. The proposed adaptive body bias generator (BBG) consists of two comparators, switch logic, body bias voltage sources, two reference voltage sources, and an inverter to monitor the logical threshold voltage V_M. The monitored V_M of the inverter is compared against the reference voltages of V_{ref1} and V_{ref2}. If V_M is below a predetermined reference potential (V_{ref1}), indicating that the NMOS transistor is stronger than the PMOS transistor, we apply a FBB to the pull-up network (PUN) and/or a RBB to the pull-down network (PDN) to make them equally strong. Conversely, if the monitored V_M is higher than V_{ref2}, the β-ratio is too large compared to the optimal value due to strong PMOS. We apply a FBB to the PDN (and/or a RBB to the PUN). If V_M is monitored to be between the two reference levels, zero body bias (ZBB) is applied to the target system. The generated BB voltages are fed to the inverter, and then the updated V_M is again compared against the reference voltages. With more voltage references (i.e., fine-grained levels), this loop repeats until the best BB voltages are found.

6. Results and Discussions

To compare FF features among the conventional FFs and proposed DPFFs in terms of design metrics, we implement

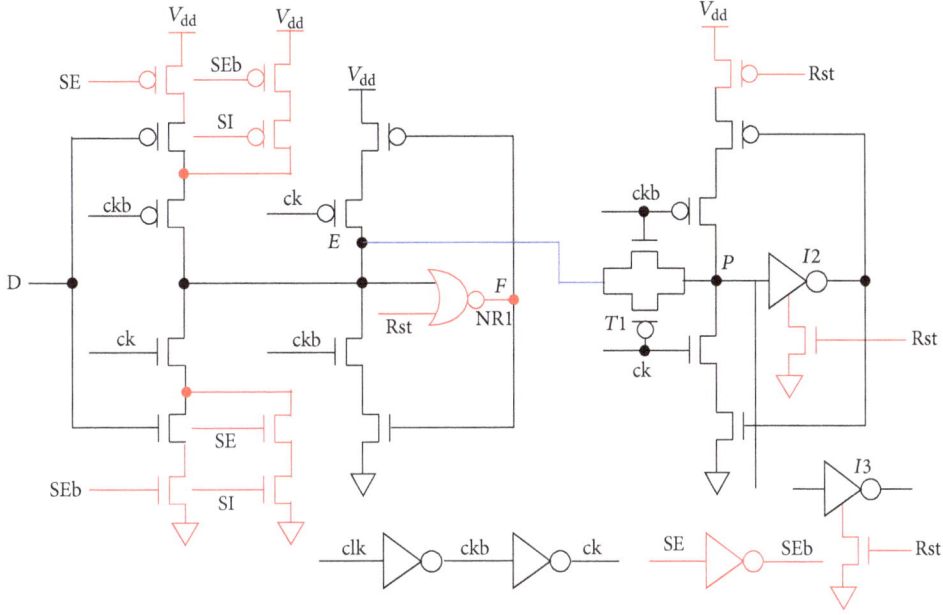

FIGURE 7: Proposed C2DPFF with scan logic.

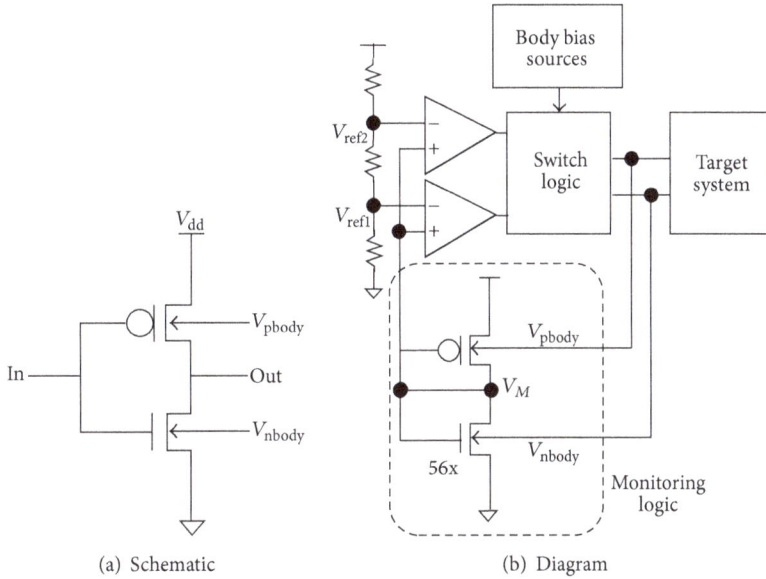

(a) Schematic

(b) Diagram

FIGURE 8: Adaptive β-ratio modulation and variation tolerant body bias circuit.

isolated FF test circuits with a direct probing capability. Figure 9(a) shows the photograph and GDS views of a test chip and isolated FF test circuits fabricated in 130 nm process. The DPFFs are integrated into an 8-tap, 8-bit FIR filter as an example to demonstrate highly robust low-power operation at the circuit level. Figure 9(b) shows the diagram of the FF test circuits in the test chip and FF design metrics as defined in Section 2. The FF cells in the test circuits are drawn, respectively, according to the transistor sizes at the PDP optimal points marked in Figure 6. Note that the FF test circuits include two inverters connected in series which are used to shape the waveforms of both input and clock

signals, whereas four inverters connected in parallel are used for an output load. The same structure of the test circuits is used for FF simulations where the sizes of the wave-shaping and output-load inverters are modulated to change the slew rate of input or clock waveforms and the value of an output capacitive load, respectively.

Conventionally, the setup and hold times are independently characterized as a skew so that an increase in the clock-to-Q delay remains within a certain amount of percentage (say, 10%). The basic concept behind setup and hold time characterization is to sample and propagate the data in the stable region of operation. Otherwise, if the data-to-clock

(a) Photograph and GDS layout view

Wave shaping   Output load

(b) FF test circuits

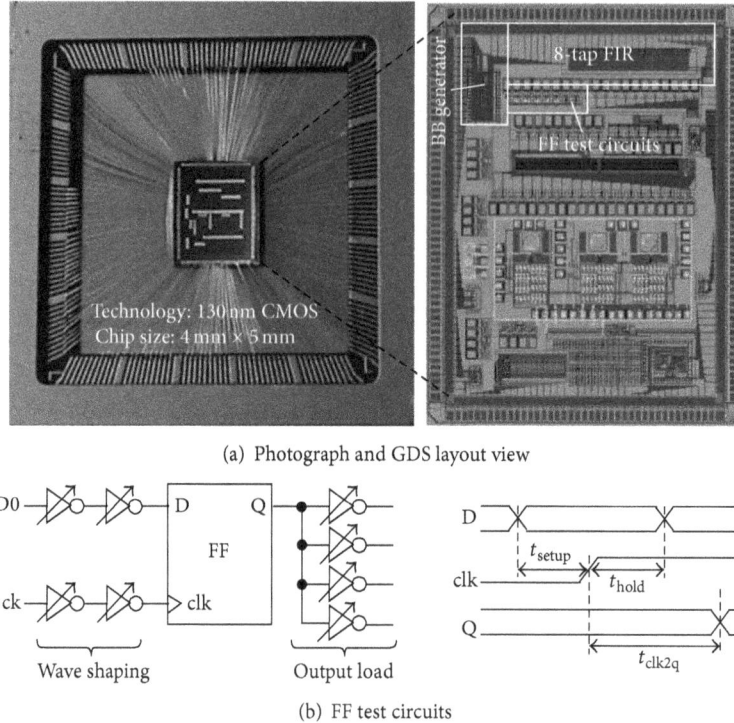

FIGURE 9: Test chip with isolated FF test circuits in 130 nm technology.

TABLE 1: Conventional FFs versus proposed DPFFs (measured).

Param.	MTGFF	C2FF	C2DPFF	TGDPFF
t_{clk2qh} [ps]	185	172	136	142
t_{clk2ql} [ps]	174	169	130	139
t_{setup} [ps]	57	43	46	54
t_{hold} [ps]	49	44	45	48
P_{tot} [μW]	2.290	2.041	1.846	1.815
P_{leakh} [nW]	3.060	1.885	1.766	1.757
P_{leakl} [nW]	2.868	2.252	1.804	1.714
Area [μm^2]	122.21	77.91	77.65	74.21

Test condition: SS/1.15 V/25°C for delay and power.
Test condition: FF/1.25 V/125°C for leakage.

TABLE 2: Delay sensitivity over V_{dd} variation (measured).

| Voltage [V] | Clock-to-Q rise/fall delay [ns] | | | |
	MTGFF	C2FF	C2DPFF	TGDPFF
1.15	0.185/0.174	0.172/0.1692	0.136/0.130	0.142/0.139
1.10	0.208/0.197	0.198/0.1911	0.155/0.149	0.165/0.162
1.05	0.237/0.227	0.229/0.2199	0.179/0.174	0.195/0.191
1.00	0.279/0.265	0.262/0.2580	0.210/0.208	0.236/0.229
0.95	0.331/0.317	0.318/0.3018	0.250/0.240	0.288/0.281
0.90	0.411/0.393	0.392/0.3825	0.316/0.322	0.365/0.358
0.85	0.525/0.514	0.532/0.5094	0.417/0.431	0.482/0.479
0.80	0.726/0.716	0.746/0.7058	0.658/0.674	0.671/0.681
0.75	1.091/1.086	1.126/1.0961	0.962/1.011	0.978/1.025
0.70	1.916/1.864	1.975/1.8429	1.674/1.704	1.649/1.711
0.65	3.543/3.448	3.657/3.4009	2.954/2.807	2.905/2.853
Ratio	19.1/19.8	21.2/20.1	21.7/21.5	20.4/20.5

Test condition: SS/25°C.
Ratio (%) = delay at 0.65 V/delay at 1.15 V.

skew (or time difference) is too small then a FF fails to capture the data or fails to correctly transfer the data. The window of data-to-clock skew is termed as the failure region. During timing analysis the constraints ensure that the FF does not fall into the failure region. In the stable region, the nominal clock-to-Q delay is named t^0_{clk2q}.

Table 1 summarizes the average values of 97 sets of the referential and proposed FFs, measured from the isolated test circuits shown in Figure 9 at the slow-slow (SS) corner with a supply voltage of 1.15 V and 25°C for delay and power consumption and at the fast-fast (FF) corner with a supply voltage of 1.25 V and 125°C for leakage characterization. The area is calculated without scan logic. It can be seen that, compared with the MTGFF, both proposed DPFFs achieve a considerable improvement in the clock-to-Q delay by more

than 20% due to the reduced logic depth, which in turn may allow the direct path applied FFs to achieve further reduction in area and power consumption. Note that the use of C²MOS inverter at the primary input stage of the C2FF and C2DPFF lowers the setup time by 25% and 19%, respectively, over the MTGFF. Improvement in the delay and setup time apparently relaxes the timing constraints and improves performance of target systems.

Table 2 shows delay variation over supply voltage scaling of the conventional and proposed FFs, measured at the SS

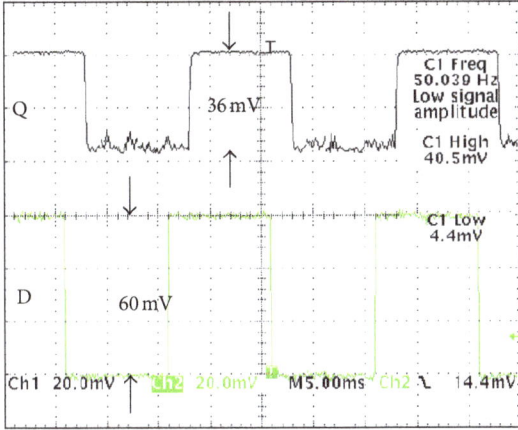

FIGURE 10: Waveforms of the C2DPFF at 60 mV (measured).

corner with a supply voltage range of 1.15 V to 0.65 V and $-25°C$, where the delay ratio or sensitivity is calculated as a ratio of the delay at 0.65 V over the delay at 1.15 V. It can be observed that the proposed DPFFs have comparable delay sensitivity over supply voltage scaling with the conventional FFs especially at low voltages.

Figure 10 shows the input D and output Q waveforms measured from the C2DPFF in the deep subthreshold region. The measured minimum supply voltage of $V_{dd,min} = 60$ mV results in the dynamic switching power of 24.7 pW, five orders of magnitude smaller than normal voltage operation, with the minimum-sized design at an operating frequency of 50 Hz during ten complete binary cycles (i.e., low-to-high and high-to-low transitions). On the other hand, $V_{dd,min}$ of 80 mV and 85 mV is measured for the conventional C2FF and MTGFF, respectively. It is worthwhile to mention that the output of a 36 mV swing is observed at a supply voltage of $V_{dd} = 60$ mV. This voltage diminution is due to the fact that, unlike in the normal V_{dd} region, the OFF leakage current is not negligible anymore compared to the (operating) subthreshold current in the ultralow V_{dd} region. That is, $I_{sub}/I_{off} \sim 10^2$ for $V_{dd} < V_T$, whereas $I_{on}/I_{off} \sim 10^5$ for $V_{dd} > V_T$, where I_{on} is the (normal) ON current, I_{sub} is the subthreshold current, I_{off} is the leakage current, and V_T is the threshold voltage of the device. The main conduction current at high or normal V_{dd}'s can be explained by the *drift* mechanism while the subthreshold current at ultralow V_{dd}'s is mainly governed by the *diffusion* mechanism. For example, assume that the input D is set to "1" and a "0" value is driven to the internal node P in Figure 4. Then, the PMOS transistor of the output driver $I3$ must pull up the output to V_{dd} by overriding the (idle) OFF current of the NMOS transistor of $I3$. If the PMOS operating current is not strong enough to overcome the NMOS OFF current (unlike at high V_{dd}'s), a supply voltage is divided resistively across the transistors and, as a result, the output will not rise all the way to V_{dd} in deep subthreshold operations.

Many studies have been reported regarding the theoretical and practical limit of CMOS logic operation [26, 31, 32]. Presentation of astonishing circuit operation at aggressively

scaled supply voltages (i.e., 36 mV swing at a supply voltage of 60 mV) does not necessarily mean that it is recommended to operate the system at the voltage level but to provide measurement results as an evidence of increased circuit robustness with the proposed technique and as a possible advantage of the body bias technique to lower $V_{dd,min}$ and salvage a silicon chip; otherwise, a circuit would fail to operate due to the presence of various process variations.

Comparison of the FF types in the clock-to-Q delay requires consideration of negative as well as positive influences. As discussed in Section 4, the reduced logic depth along the main signal propagation path usually decreases the propagation delay, whereas the use of the C^2MOS inverter generally comes with the stacking effect which results in higher threshold voltage and less current drive and thus increases the propagation delay. In comparison of C2FF with MTGFF, delay decrease tendency thanks to the reduced logic depth compensates delay increase due to the stacking effect and hence the delay of the C2FF is comparable to or slightly better than that of the MTGFF.

In the proposed DPFFs, however, much reduced logic depth further lowers the clock-to-Q delay by roughly one and two inverter delays over C2FF and MTGFF, respectively, surpassing a feasible delay increase caused by the stacking effect. This is the best advantage of the DPFFs where the direct path connection considerably improves the clock-to-Q delay by 17% over the conventional FFs. From the power and area perspective, this gate delay improvement can provide an opportunity to save more power at the isoperformance condition even with less area. Moreover, direct path connection allows the use of minimum-sized transistors in the data-storing units (e.g., $I1$ and $I2$ in Figure 4) that are now on the noncritical path, providing a further reduction in area, dynamic, and leakage powers by more than 36% (0.3%), 19% (9%), and 37% (6%) over the MTGFF (C2FF), respectively, with the same output load. Note that, in the C2FF, the output driver $I2$ drives the feedback transistors, $M1$ and $M2$, as well as the underlying output load, which decreases the output slope especially in the presence of large fanout. The C2DPFF and TGDPFF address this fanout issue by using the additional driver $I3$ and $I4$, respectively, dedicated to drive output as in MTGFF, which improves the output slope and current drive of the cell, covering a wide range of fanout.

Figure 11 now provides the measurement results with various signal polarities and driving strengths. The normalized average values of the MTGFF, C2FF, TGDPFF, and C2DPFF are plotted in red, black, magenta, and blue, respectively. As the setup skew becomes smaller, the contamination delay, the amount of time needed for a change in a logic input to cause an initial change at an output, dramatically increases. Consequently, there is a radical push out in the clock-to-Q delay as shown in the figure. Note that, for a certain clock-to-Q delay, the hold time increases with a decrease in the setup time to keep the internal race immunity in (2). One may argue that the DPFFs cause an increase in the setup and hold times as side effects. This is due to the fact that the decreased nominal clock-to-Q delay lowers the 10% constraint as well, which increases the setup and hold times by their definition; even the absolute values almost remain

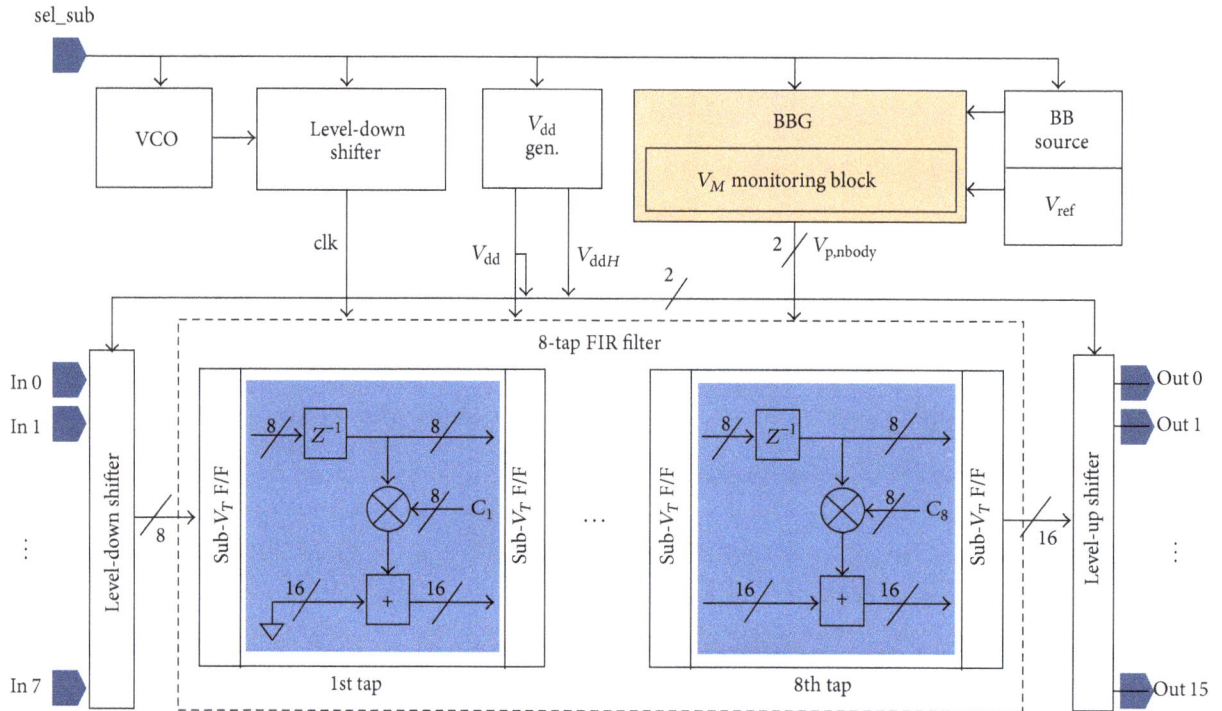

FIGURE 11: The FIR filter.

the same. The authors believe that the FFs should be designed to operate with margin in the nominal delay region (i.e., the flat region in Figure 11) for stable and prompt operation. The increased setup and hold times are acceptable and fully redeemed by considerable advantages in propagation delay, power consumption, and area.

The proposed DPFFs are fully integrated into an 8×8 FIR filter fabricated in 130 nm technology. Figure 11 shows the architecture of the 8-tap, 8-bit FIR filter. Detailed discussions of the filter are found in [29, 30]. With the application of ABRM by which the optimal β-ratio value of the filter is automatically driven by the BBG (body bias generator) as shown in Figure 9, the filter successfully operates all the way down to 85 mV, consuming 40 nW of power at an operating frequency of 240 Hz. This ultralow voltage operation proves high circuit robustness of the DPFFs since relative variations are significantly higher with voltage scaling and the circuit becomes much more vulnerable to noise disturbance with limited voltage headroom.

7. Conclusions

Design metrics are of primary importance for the FFs to be used as primitive library cells. We proposed two direct path master-slave FFs which adopt a C^2MOS style input buffer to improve performance while providing full compatibility with widely used EDA characterization tools. Internal direct path between the master and slave latches reduces the logic depth along the main signal path, achieving a further speedup in the propagation delay. Measurements from the ABRM-applied test circuits fabricated in 130 nm demonstrated potential advantages of the proposed FFs in design metrics for ultralow as well as normal voltage applications.

Competing Interests

The authors declare that they have no competing interests.

Acknowledgments

This work has been supported by the 2014 Research Fund of University of Ulsan.

References

[1] H. Kawaguchi and T. Sakurai, "A reduced clock-swing flip-flop (RCSFF) for 63% power reduction," *IEEE Journal of Solid-State Circuits*, vol. 33, no. 5, pp. 807–811, 1998.

[2] G. Gerosa, S. Gary, C. Dietz et al., "A 2.2W 80 MHz superscalar RISC microprocessor,," *IEEE Journal of Solid-State Circuits*, vol. 29, no. 12, pp. 1440–1454, 1994.

[3] U. Ko and P. T. Balsara, "High-performance energy-efficient D-flip-flop circuits," *IEEE Transactions on Very Large Scale Integration (VLSI) Systems*, vol. 8, no. 1, pp. 94–98, 2000.

[4] J. Yuan and C. Svensson, "High-performance energy-efficient D-flip-flop circuits," *IEEE Journal of Solid-State Circuits*, vol. 24, no. 1, pp. 62–70, 1989.

[5] J. M. Rabaey, A. P. Chandrakasan, and B. Nikolic, *Digital Integrated Circuits: A Design Perspective*, chapter 7, Prentice Hall, Upper Saddle River, NJ, USA, 2nd edition, 2003.

[6] T. Ishikawa, M. Kinugasa, and M. Kida, "Master-slave clocked CMOS flip-flop with hysteresis," US Patent, No. US5107137A, April 1992.

[7] B. Nikolić, V. G. Oklobdžija, V. Stojanović, W. Jia, J. K.-S. Chiu, and M. M.-T. Leung, "Improved sense-amplifier-based flip-flop: design and measurements," *IEEE Journal of Solid-State Circuits*, vol. 35, no. 6, pp. 876–884, 2000.

[8] V. Zyuban, "Optimization of scannable latches for low energy," *IEEE Transactions on Very Large Scale Integration (VLSI) Systems*, vol. 11, no. 5, pp. 778–788, 2003.

[9] F. Ricci, L. T. Clark, T. Beatty et al., "A 1.5 GHz 90 nm embedded microprocessor core," in *Proceednigs of the VLSI Circuits, Digest of Technical Papers*, vol. 15, pp. 12–15, Kyoto, Japan, June 2005.

[10] K. Absel, L. Manuel, and R. K. Kavitha, "Low-power dual dynamic node pulsed hybrid flip-flop featuring efficient embedded logic," *IEEE Transactions on Very Large Scale Integration (VLSI) Systems*, vol. 21, no. 9, pp. 1693–1704, 2013.

[11] V. Stojanovic and V. G. Oklobdzija, "Comparative analysis of master-slave latches and flip-flops for high-performance and low-power systems," *IEEE Journal of Solid-State Circuits*, vol. 34, no. 4, pp. 536–548, 1999.

[12] M. Alioto, E. Consoli, and G. Palumbo, "Analysis and comparison in the energy-delay-area domain of nanometer CMOS Flip-Flops: part I—methodology and design strategies," *IEEE Transactions on Very Large Scale Integration (VLSI) Systems*, vol. 19, no. 5, pp. 725–736, 2011.

[13] H. Partovi, R. Burd, U. Salim, F. Weber, L. DiGregorio, and D. Draper, "Flow-through latch and edge-triggered flip-flop hybrid elements," in *Proceedings of the IEEE International Solid-State Circuits Conference*, Digest of Technical Papers, pp. 138–139, San Francisco, Calif, USA, February 1996.

[14] F. Klass, "Semi-dynamic and dynamic flip-flops with embedded logic," in *Proceedings of the Symposium on VLSI Circuits. Digest of Technical Papers*, pp. 108–109, Honolulu, Hawaii, USA, June 1998.

[15] S. Hesley, V. Andrade, B. Burd et al., "A 7th-generation x86 microprocessor," in *Proceedings of the IEEE International Solid-State Circuits Conference*, Digest of Technical Papers, pp. 92–93, San Francisco, Calif, USA, February 1999.

[16] C. F. Webb, C. J. Anderson, L. Sigal et al., "A 400-MHz S/390 microprocessor," *IEEE Journal of Solid-State Circuits*, vol. 32, no. 11, pp. 1665–1675, 1997.

[17] Y. Zhang, H. Yang, and H. Wang, "Low clock-swing conditional-precharge flip-flop for more than 30% power reduction," *Electronics Letters*, vol. 36, no. 9, pp. 785–786, 2000.

[18] B.-S. Kong, S.-S. Kim, and Y.-H. Jun, "Conditional-capture flip-flop for statistical power reduction," *IEEE Journal of Solid-State Circuits*, vol. 36, no. 8, pp. 1263–1271, 2001.

[19] N. Nedovic, M. Aleksic, and V. Oklobdzija, "Conditional techniques for small power consumption flip-flops," in *Proceedings of the IEEE International Conference on Electronics, Circuits & Systems (CAS '01)*, pp. 803–806, Malta, Spain, September 2001.

[20] M. Tokumasu, H. Fujii, M. Ohta, T. Fuse, and A. Kameyama, "A new reduced clock-swing flip-flop: NAND-type Keeper flip-flop (NDKFF)," in *Proceedings of the IEEE Custom Integrated Circuits Conference*, pp. 129–132, Orlando, Fla, USA, May 2002.

[21] A. Ma and K. Asanovic, "A double-pulsed set-conditional-reset flip-flop," MIT LCS Technical Report MIT-LCS-TR-844, 2002.

[22] Cadence, *Encounter Library Characterizer*, Cadence, 2013.

[23] Synopsys, "SiliconSmart," July 2013, http://www.synopsys.com.

[24] Synopsys, "Liberty NCX," 2013, http://www.synopsys.com.

[25] Altos Design Automation, *Liberate User Guide*, Ver.2.4, 2009.

[26] A. Bryant, J. Brown, P. Cottrell, M. Ketchen, J. Ellis-Monaghan, and E. J. Nowak, "Low-power CMOS at Vdd = 4kT/q," in *Proceedings of the Device Research Conference*, pp. 22–23, Notre Dame, Ind, USA, June 2001.

[27] M. E. Hwang, S. O. Jung, and K. Roy, "Slope interconnect effort: gate-interconnect interdependent delay modeling for early CMOS circuit simulation," *IEEE Transactions on Circuits and Systems I: Regular Papers*, vol. 56, no. 7, pp. 1428–1441, 2008.

[28] A. Bryant, J. Brown, P. Cottrell, M. Ketchen, J. Ellis-Monaghan, and E. J. Nowak, "Low-power CMOS at Vdd = 4kT/q," in *Device Research Conference*, pp. 22–23, Notre Dame, Ind, USA, June 2001.

[29] M.-E. Hwang, "ABRM: adaptive β-ratio modulation for process-tolerant ultradynamic voltage scaling," *IEEE Transactions on Very Large Scale Integration (VLSI) Systems*, vol. 18, no. 2, pp. 281–290, 2010.

[30] M.-E. Hwang, A. Raychowdhury, K. Kim, and K. Roy, "A 85 mV 40 nW process-tolerant subthreshold 8x8 FIR filter in 130 nm technology," in *Proceedings of the IEEE Symposium on VLSI Circuits (VLSI '07)*, pp. 154–155, Kyoto, Japan, June 2007.

[31] S. Vangal and S. Jain, "Claremont: a solar-powered near-threshold voltage IA-32 processor," in *Design Technologies for Green and Sustainable Computing Systems*, P. P. Pande, A. Ganguly, and K. Chakrabarty, Eds., chapter 9, pp. 229–239, Springer, New York, NY, USA, 2013.

[32] E. J. Nowak, "Maintaining the benefits of CMOS scaling when scaling bogs down," *IBM Journal of Research and Development*, vol. 46, no. 2-3, pp. 169–180, 2002.

10

Data Mining for Material Feeding Optimization of Printed Circuit Board Template Production

Shengping Lv ⓘ,[1,2] **Binbin Zheng,**[1] **Hoyeol Kim,**[2] **and Qiangsheng Yue**[1]

[1]College of Engineering, South China Agricultural University, Guangzhou 510642, China
[2]Department of Industrial, Manufacturing and Systems Engineering, Texas Tech University, Lubbock, TX 79409, USA

Correspondence should be addressed to Shengping Lv; lvshengping@scau.edu.cn

Academic Editor: Ephraim Suhir

Improving the accuracy of material feeding for printed circuit board (PCB) template orders can reduce the overall cost for factories. In this paper, a data mining approach based on multivariate boxplot, multiple structural change model (MSCM), neighborhood component feature selection (NCFS), and artificial neural networks (ANN) was developed for the prediction of scrap rate and material feeding optimization. Scrap rate related variables were specified and 30,117 samples of the orders were exported from a PCB template production company. Multivariate boxplot was developed for outlier detection. MSCM was employed to explore the structural change of the samples that were finally partitioned into six groups. NCFS and ANN were utilized to select scrap rate related features and construct prediction models for each group of the samples, respectively. Performances of the proposed model were compared to manual feeding, ANN, and the results indicate that the approach exhibits obvious superiority to the other two methods by reducing surplus rate and supplemental feeding rate simultaneously and thereby reduces the comprehensive cost of raw material, production, logistics, inventory, disposal, and delivery tardiness compensation.

1. Introduction

Printed circuit board (PCB) is found in practically all electrical and electronic equipment (EEE), being the base of the electronics industry [1]. Due to increased competition and market volatility, demand for highly individualized products promotes a rapid growth of PCB orders designed with many specialized features but short delivery time [2, 3]. Customer-oriented small batch production is always employed by a factory with lots of PCB template orders, which is different from mass production, and therefore causes companies to face serious challenges. Optimization of material feeding is one of the critical problems.

The scrap rate and material feeding area of each PCB template order are difficult to be accurately determined in advance of the production. Many factories undergo the violent fluctuation in both surplus rate and supplemental feeding rate due to empirical manual feeding in practice by heavily depending on their experience and knowledge. Individualized surplus template products can be placed in inventory or directly destroyed while frequent material feeding brings supplemental production cost and delivery tardiness compensation. This motivates us to explore the pattern of historical orders through data mining (DM) approach to facilitate more reasonable material feeding for the orders automatically and therefore reduce the comprehensive cost caused by excessive or underestimated material feeding before production.

The general process of DM also known as knowledge discovery in databases (KDD) includes problem clarification, data preparation, preprocessing, DM in the narrow sense, and interpretation and evaluation of results [4]. DM in the narrow sense as a step in the KDD process consists of applying data analysis and particular discovery algorithm within an acceptable computational efficiency limit [4]. DM tasks can be classified into descriptive and predictive two groups [4, 5]. Descriptive function of DM mainly aims to explore the potential or recessive rules, characteristics, and relationships (dependency, similarity, etc.) that exist in the data, such as generalization, association, sequence pattern mining, and clustering [4–8]. As to the predictive functions of DM, they are usually selected to analyze the relevant trends of the data or the relevant laws to predict the future state. It includes classification, prediction, time series, and analysis [4–8]. To achieve these goals, DM solutions employ a wide variety of enabling techniques and specific mining techniques to both predict and describe interpretable and valuable information [4, 5, 8–10]. The enabling techniques mainly refer to the methods for data cleaning, data integration, data transformation, and data reduction that can support the implementation of DM in the narrow sense, while specific mining techniques, like regression, support vector machine, and artificial neural network (ANN), are the approaches used to explore useful knowledge from massive data [4, 9]. The scrap rate prediction and material feeding optimization of PCB template production can be taken as an application of predictive function of DM; and the specification of scrap rate related features, identifying features which affect scrap rate significantly and related mining techniques, should be carefully studied. Moreover, many features (e.g., required panel) have structural change influence for the scrap rate according to empirical knowledge, and therefore the enabling and mining techniques, interpretation, and evaluation steps in DM should also be adjusted accordingly.

The details of enabling techniques, DM applications for different manufacturing task and different manufacturing industry, patterns in the use of specific mining techniques, application performance, and software used in these applications have been widely studied, and one can refer to [4–10] for comprehensive review. Electronic product manufacturing industries also exploited several DM methods with the purpose of summarization, clustering, association, classification, prediction, and so on [5, 8]. And many are closely related to PCB manufacturing industry. Tseng et al. employed Kohonen neural networks, decision tree (DT), and multiple regression to improve accuracy of work hours estimation based on the PCB design data, and the performance clearly exceeds the conventional method of regression equations [11]. Tsai et al. developed three hybrid approaches including ANN-genetic algorithm (GA), fuzzy logic-Taguchi, and regression analysis-response surface methodology (RSM) to predict the volume and centroid offset two responses and optimize parameters for the micro ball grid array (BGA) packages during the stencil printing process (SPP) for components assembly on PCB, and the confirmation experiments show that the proposed fuzzy logic-based Taguchi method outperforms the other two methods in terms of the signal-to-noise ratios and process capability index [12, 13]. Some other approaches, like support vector regression (SVR) and mixed-integer linear programming, have also been developed for the parameter optimization of SPP [14]. Haneda et al. [15] employed variable cluster analysis and K-means approach to help engineers determine appropriate drilling condition and parameter for PCB manufacturing. DM-based defect (faults) diagnosis or quality control during manufacturing has also been widely studied [5, 7], and many algorithms like adaptive genetic algorithm-ANN [16] and DT [17] have been developed for the defect (faults) diagnosis of PCB manufacturing.

The marketing and sales is another widely investigated direction of DM application in PCB industry. Success in forecasting and analyzing sales for given goods or services can

mean the difference between profit and loss for an accounting period [18]. Many DM-based methods like K-mean cluster and fuzzy neural network, fuzzy case-based reasoning, and weighted evolving fuzzy neural network have been developed by Chang et al. [19–21] to select a combination of key factors which have the greatest influence on PCB marketing and then forecasts the future PCB sales. Tavakkoli et al. [22] combined SVR, Bat metaheuristic, and Taguchi method to predict the future PCB sales, and performance comparison indicates that the accuracy of the proposed hybrid model is better than the GA-SVR, particle swarm optimization-SVR, and classical SVR. Hadavandi et al. hybridize fuzzy logic with GA and K-means to extract useful information patterns from sales data, and results show that the proposed approach outperforms the other previous approaches [18].

However, the quality related research for PCB manufacturing mainly focuses on one operation of the manufacturing process for the purpose of yield improvement [12–15], and there are few studies on material feeding optimization especially for PCB production using DM mechanism to the best of our knowledge. Meanwhile, the change structure of the studied problem and corresponding change of relevant features have seldom been considered during the mining procedure. Meanwhile, ANN-based approach, as a most frequently used DM method that will also be employed in the study, tries to exploit nonlinear patterns in different problems demonstrating reasonable results; however, problem divided into different subproblems according to structural change always requires different ANN architecture and different learned link weights based on different input features, while this is difficult to learn by the ANN without reasonable preprocessing.

In this paper, a data mining approach (MSCM-ANN) is presented to establish the scrap rate prediction model and optimize material feeding of PCB orders considering the structural change influence based on the use of multivariate boxplot, multiple structural change model (MSCM), neighborhood component feature selection (NCFS), and ANN. The comparison of MSCM-ANN to ANN and manual feeding will be conducted to verify the performance of the proposed approaches. The rest of the paper is organized as follows. In Section 2, variables specification and sample data are described. Methodology, including multivariate boxplot, MSCM, NCFS, ANN, and performance indicators, is presented in Section 3, followed by experimental results and discussion in Section 4. Lastly, conclusions are drawn in Section 5.

2. Variables and Sample

The data used in this study were collected from Guangzhou FastPrint Technology Co., Ltd. A total of 56 variables inherited from enterprise resource planning system combined with the derived variables were selected and specified in Table 1, in which variables 40 to 56 are the statistic results of manual feeding adopted by FastPrint. The unit in a panel, required quantity/panel/area, and delivery unit area can not only be taken as statistic items, but also feature candidates for MSCM-ANN and ANN model establishment.

Set and unit are two types of delivery unit, whereas panel is production unit that will be partitioned into either set

TABLE 1: Variables specification.

Number	Variable name	Symbol	Description	Value range
Overall characteristics				
1	PCB thickness	Pt	Thickness of the ordered PCB	0.3–8
2	Layer number	Ln	Number of copper layers	4–20
3	Rogers material	Ro	Whether substrate material is Rogers	0/1
4	Plating frequency	$Plfr$	Number of plating operations	0–4
5	Number of operations	Noo	Number of operations to produce the order	16–71
6	Number of Prepregs	NPP	Number of Prepregs for lamination	1–50
7	Scrap units in a set	Sus	Allowed maximum scrap units in a set	0–8
8	Photoelectric board	$Photb$	Whether the order is the specified board	
9	High frequency board	$Highfb$		
10	Test board	$Semictb$		
11	Negative film plating	$Nflp$	Whether the order takes negative film plating	0/1
12	Tinning copper	$Tinc$	Whether the order has tinning copper	
13	IPCIII standard	$IPCIII$	Whether the order takes IPCIII or Huawei standard	
14	Huawei standard	Huawei		
Feature of internal/outer layer line				
15	Minimum line width in internal layer (mil)	$Mwil$	Minimum line width or space in core boards	0.1–100
16	Minimum line space in internal layer (mil)	$Mlsil$		0.1–137.66
17	Minimum line width in outer layer (mil)	$Mwol$	Minimum line width or space in outer layer	1–157.5
18	Minimum line space in outer layer (mil)	$Mlsol$		1.2–290
19	Average residual rate	$Arcr$	Average residual rate of copper layer	0.15%–94.75%
Feature and operation information of hole				
20	Solder resist plug hole	$Srph$	Whether the order has the specified hole related operation	0/1
21	Plug hole with resin	$Phwr$		
22	Second drilling	$Secd$		
23	Back drilling	$Bcdr$		
Operation information of character/solder mask				
24	Character print	$Chaprt$	Whether the order has the specified character/solder mask related operation	0/1
25	White oil solder mask	$White$		
26	Blue oil solder mask	$Blue$		
27	Black oil solder mask	$Black$		
Surface finishing operation options				
28	Hot air solder leveling	$Hasl$		
29	Lead-free hot air solder leveling	$Lfhasl$		
30	Entek	Osp		
31	Cu/Ni/Au pattern plating	$Cnapp$	Whether the order takes the specified surface finishing operation	0/1
32	Gold finger plating	$Gfig$		
33	Gold plating	$Godp$		
34	Soft Ni/Au plating	$Snap$		
35	Immersion Ag/Sn/Au	$Iasa$		
Statistic items				
36	Delivery unit in a panel	$Duap$	Number of delivery units in a panel	1–262

TABLE 1: Continued.

Number	Variable name	Symbol	Description	Value range
37	Supplemental feeding frequency	$Supff$	Material feeding frequency minus 1	0–14
38	Required quantity	$Reqq$	Demand quantity of delivery unit minus delivery unit in inventory for the same order number	1–3,000
39	Required panel	$Reqp$	$Reqq/Duap$ rounded up to the nearest integer	1–225
40	Feeding quantity	$Fedq$	Feeding number of delivery units	2–6296
41	Least feeding panel	Lfp	$Reqq/(1 - scrap\ rate)$ rounded up to the nearest integer	1–245
42	Feeding panel	$Fedp$	Number of feeding panels	1–308
43	Scrap quantity	$Scraq$	Scrap number of delivery units	0–712
44	Qualified quantity	$Qualq$	Qualified number of delivery units	1–6,226
45	Surplus quantity	$Surpq$	$Qualq$-$Fedq$	0–3,226
46	Delivery unit area (m^2)	$Dunita$	Area of a delivery unit	0.001–0.393
47	Required area (m^2)	$Reqa$	$Reqq \times Dunita$	0.001–25.74
48	Feeding area (m^2)	$Feda$	$Fedq \times Dunita$	0.011–42.63
49	Scrap area (m^2)	$Scraa$	$Scraq \times Dunita$	0–15.39
50	Qualified area (m^2)	$Quala$	$Qualq \times Dunita$	0.009–37.49
51	Surplus area (m^2)	$Surpa$	$Surpq \times Dunita$	0–25.45
52	Supplemental feeding rate	$Supfr$	$Supff$ in a certain period/number of orders $\times 100\%$	19.84%
53	Scrap rate	$Scrar$	$Scraa/Feda \times 100\%$	0%–72.58%
54	Qualified rate	$Qualr$	$Quala/Feda \times 100\%$	27.42%–100%
55	Surplus rate	$Surpr$	$Surpa/Reqa \times 100\%$	0–827.18%
56	Historical qualified rate	$Hquar$	The $Qualr$ for the same order *number* in the past 2 years	8.824%–100%

Note. New orders having no $Hquar$ are replaced by the $Qualr$ for orders having the same layer number and surface finishing operation during the past 2 years.

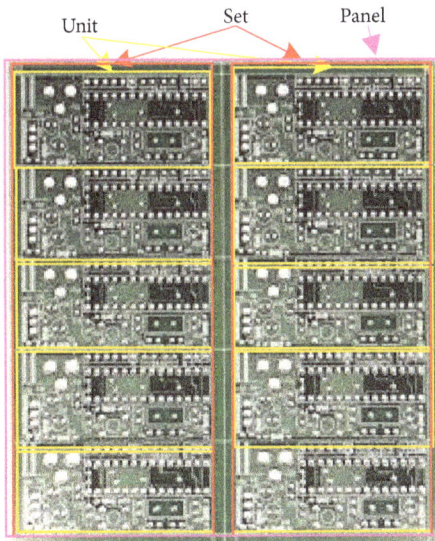

FIGURE 1: Structure of a PCB panel.

or unit before delivery depending on the requirement of customers. The relation between set, unit, and panel specified in Table 1 is illustrated in Figure 1, in which each panel consists of 10 units in the PCB order. Suppose the customer's required quantity and required panel of the PCB order given in Figure 1 are 90 units and 9 panels, respectively. If the initial feeding is 100 units (10 panels) but finally ended up with 95 qualified units due to scrap rate (i.e., $(100 - 95) \times Dunita/100 \times Dunita \times 100\% = 5\%$ in this example) after production, then the surplus quantity is 5 units (= 95 – 90) and therefore feeding 10 panels is more reasonable to reduce the redundancy of the customized orders. Conversely, it will result in supplemental feeding if we feed only 9 panels initially due to the scrap rate.

On this basis, 30,117 samples of the orders placed between October 31, 2015, and October 31, 2016, were exported with careful audit for erroneous and missing values. The number of the orders for each required panel is illustrated in Figure 2. It can be seen that the required panel is less than 30 in most of the case, which represents a typical customer-oriented small batch template production in PCB industry.

3. Methodology

The main flow of the proposed approach (MSCM-ANN) is presented in Figure 3, and various aspects of MSCM-ANN are discussed in detail in the following subsections. The multivariate boxplot, MSCM-based partition, and neighborhood

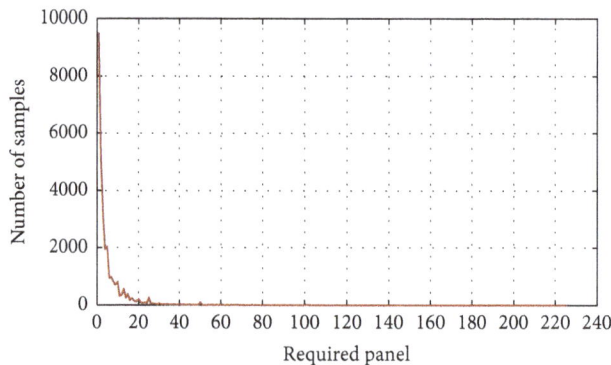

FIGURE 2: Number of orders for each required panel.

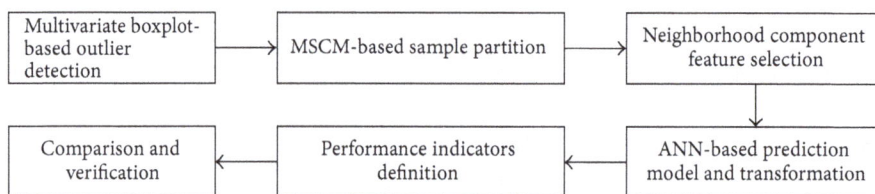

FIGURE 3: Main flow of the proposed approach.

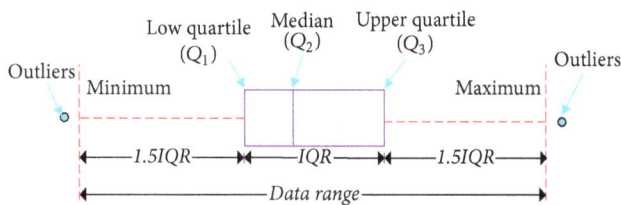

FIGURE 4: Description of boxplot.

component feature selection are the enabling technique of DM considering structural change influence; and the ANN-based prediction model is the mining technique to predict scrap rate; meanwhile, the transformation of scrap rate to surplus rate and supplemental feeding rate is conducted. Performances of the proposed MSCM-ANN will be compared to the ANN and manual feeding based on the same 29,157 samples after outlier detection of original 30,117 samples. Some statistic results of manual feeding are given in Table 1 by the variables 40 to 56. MSCM-ANN and ANN were implemented in Matlab® version 2017a.

3.1. Multivariate Boxplot-Based Outlier Detection. Identification of outliers and the consequent removal are a part of the data screening process which should be done routinely before analyses [23, 24]. There are various methods of outlier detection. Some are graphical such as normal probability plot, and others are model-based approaches which assume that the data are from a normal distribution [24]. Boxplot is a hybrid of the above two mechanisms for exploring both symmetric and skewed quantitative data, and it can also identify infrequent values from categorical data. Figure 4 shows a description of boxplot, and one could define an

outlier as any observation outside the range $[Q_1 - 1.5IQR, Q_1 + 1.5IQR]$, where $IQR = Q_2 - Q_1$ and is the interquartile range which contains 50% of the data. The value is a lower outlier, if $x < Q_1 - 1.5IQR$, and an upper outlier, if $x > Q_1 - 1.5IQR$.

Detection of outlier sample according to scrap rate, as the target variable of the prediction here, can reduce the impact of accidents that may be caused by machine break, wrong operation, and so on. However, scrap rate related outlier detection influenced by multivariable with structural change does not guarantee they are subject to normal distribution. The modification of the boxplot, called multivariate boxplot here, is developed to identify and discard outliers. The main procedure can be described in Algorithm 1.

3.2. Multiple Structural Change Model-Based Sample Partition. The required panel has significant influence on the scrap rate according to expert experience and initial analysis. The average scrap rate of the orders with different required panel is illustrated in Figure 5. The curve shows declining tendency when the required panel is less than 9 but presents great fluctuations when the required panel is larger than 30. The multiple structural change of average scrap rate versus the required panel may require separate features and prediction models to improve the prediction accuracy.

MSCM was employed to explore multiple structural changes of samples and partition samples. MSCM was initially developed by Bai and Perron [25, 26] to address problem of online (time serial related) multiple linear regression (MLR) with multiple structural change along with time. MSCM takes least squares method to detect the number of break points and estimate the change position. Here, the required panel in ascending sort order is considered as time serial related date, and the scrap rate is taken as the online

---

Input: N samples; categorical variable set $fset = \{f_1, f_2, \ldots, f_u\}$
Output: Outlier frequency for each sample $fq = \{fq_1, \ldots, fq_N\}$
Initialize: $fq = \{fq_1, \ldots, fq_N\} = \{0, 0, \ldots, 0\}$
    for $i = 1 : u$ **do**
        set $Va_i = \{v_{i1}, v_{i2}, \ldots, v_{is}\}$ as the values of feature f_i
        Plot a boxplot of scrap rate for each values in Va_i, and detect samples with scrap rate out of
$[Q_1 - 1.5\text{IQR}, Q_1 + 1.5\text{IQR}]$ for each boxplot; and outlier frequency of these samples are added by 1.
    End for
    Remove the samples corresponding to fq with $fq_k \geq 2$, $1 \leq k \leq N$. Return $fq = \{fq_1, \ldots, fq_N\}$

ALGORITHM 1: Multivariate boxplot-based outlier detection of scrap rate.

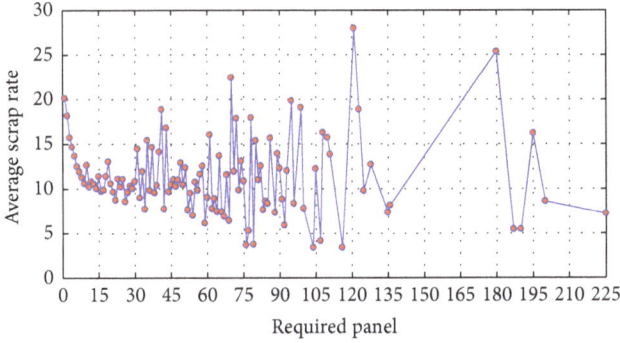

FIGURE 5: Average scrap rates with different required panel.

regression objective. Then the online MLR of the scrap rate with m breaks ($m + 1$ regimes) can be expressed as $scrapr_t = z_t'\delta_j + u_t$, $t = N_{j-1} + 1, \ldots, N_j$, for $j = 1, \ldots, m + 1$ with the convention $N_0 = 0$ and $N_{m+1} = N$. In this model, $scrapr_t$ is the observed scrap rate, and z_t is vectors of independent variable, but only the layer number, required panel, and number of operations are considered here. δ_j ($j = 1, \ldots, m + 1$) is the corresponding vectors of coefficients, and u_t is the disturbance. The break points, $N_1, \ldots, N_m$, are explicitly treated as unknowns. The purpose is to estimate the unknown coefficients together with the break points when N observations (samples) on $scrapr_t$, z_t are available.

For each m-partition $(N_1, \ldots, N_m)$, $(\hat{\delta}_1, \ldots, \hat{\delta}_j)$ as the estimated results of δ_j are obtained by minimizing sum of squared residuals $S = \sum_{i=1}^{m+1} \sum_{t=N_{i-1}+1}^{N_i} [scrapr_t - z_t'\delta_i]^2$. Substituting them in the objective function and denoting the resulting as $S_N(N_1, \ldots, N_m)$, the estimated partition to determine the break points and related change position are such that $(\hat{N}_1, \ldots, \hat{N}_m) = \arg\min_{N_1, \ldots, N_m} S_N(N_1, \ldots, N_m)$ [26].

Significance test for the structural changes based on newly introduced statistic items $\sup F(l)$, UD_{max}, WD_{max}, and $\sup F_T(l + 1/l)$ is conducted. The $\sup F(l)$ is the F test with original hypothesis $m = 0$ and alternative hypothesis $m = l$. Then $UD_{max} = \max_{1 \leq l \leq L} \sup F_T(l)$ and $WD_{max} = \max_{1 \leq l \leq L} w_l \sup F_T(l)$ are introduced to check whether there is the structural change in the model, in which w_l is set as the weight of p value based on $\sup F_T(l)$ hypothesis test, and L is the maximum number of l. The number of the break points

for the model is determined according to $\sup F_T(l + 1/l)$ ($l \geq 1$), and $F_T(l + 1/l)$ is the F test with original hypothesis having l break points and alternative hypothesis having $l + 1$ break points. Therefore the samples can be partitioned into subgroups according to the break points and corresponding change position [25, 26]. All the estimation and hypothesis tests are conducted based on Matlab code provided by Qu [27].

3.3. Neighborhood Component Feature Selection. It is necessary to employ some feature selection methods to remove irrelevant and redundant features to reduce the complexity of analysis and the generated models and also improve the efficiency of the whole modeling processes [28–30]. Wrappers, embedded, and filter are three types of the approaches developed for feature selection [31]. In this study, neighborhood component feature selection (NCFS) was applied for each group of the samples. NCFS is an embedded method for feature selection with regularization to learn feature weights for minimization of an objective function that measures the average leave-one-out regression loss over the training data [32].

Given n observations, $S = \{(x_i, y_i), i = 1, 2, \ldots, n\}$, where $x_i \in R^p$ are the feature vectors and $y_i \in R$ are response (scrap rate). In this study, the aim is to predict the response given the training set S. Consider a randomized regression model that randomly picks a point ($\text{Ref}(x)$) from S as the "reference point" for x and sets the response value at x equal to the response value of the reference point $\text{Ref}(x)$. Now consider the leave-one-out application of this randomized regression model, that is, predicting the response for x_i using the data in $S^{-i} = S/(x_i, y_i)$. The probability that point x_j is picked as the reference point for x_i is

$$p_{ij} = P\left(\text{Ref}(x) = x_j \mid S^{-i}\right) = \frac{k\left(d_w\left(x_i, x_j\right)\right)}{\sum_{j=1, j \neq i}^{N} k\left(d_w\left(x_i, x_j\right)\right)}, \quad (1)$$

where $d_w(x_i, x_j) = \sum_{r=1}^{m} w_r^2 |x_{ir} - x_{jr}|$, w_r, $r = 1, 2, \ldots, m$, are the feature weights, and k is the kernel function. Suppose $k(z) = \exp(-z/\sigma)$ as suggested in [32], where σ is set to 1 after standardizing the dependent value to have zero mean and unit standard deviation.

Let $\hat{y}_i$ be the response value of the randomized regression model and let y_i be the actual response for x_i. And let l be a

---

Procedure $NCFS(N, \alpha, \lambda, \eta)$, N: samples size; α: initial step length; λ: regularization parameter; η small positive constant

    Initialization: Standardize features to have zero mean and unit standard deviation; $w^{(0)} = (1, \ldots, 1)$, $\varepsilon^{(0)} = -\infty$, $t = 0$

        for $i = 1, \ldots, N$ **do**

          Compute p_{ij} and l_i according to (1) and (2)

          for $r = 1, \ldots, m$ **do**

$$\frac{\partial f(w)}{\partial(w_r)} = \frac{1}{N}\sum_{i=1}^{N} l_i \left[2w_r p_{ij} \left(\sum_{k \neq i} p_{ik} |x_{ir} - x_{kr}| - |x_{ir} - x_{jr}| \right) \right] + 2\lambda w_r,$$

          $t = t + 1,$
          $w^{(t)} = w^{(t-1)} + \alpha\Delta,$
          $\varepsilon^{(t)} = f(w^{(t-1)})$

          If $\varepsilon^{(t)} > \varepsilon^{(t-1)}$ **then**

              $\alpha = 1.01\alpha$

          else

              $\alpha = 0.4\alpha$

          Until $|\varepsilon^{(t)} - \varepsilon^{(t-1)}| < \eta$

             $w = w^{(t)}$

    return w

ALGORITHM 2: Neighborhood component feature selection.

loss function that measures the disagreement between $\hat{y}_i$ and y_i. Then, the average value of $l(y_i, \hat{y}_i)$ is

$$l_i = E\left(l\left(y_i, \hat{y}_i \right) \mid S^{-i} \right) = \sum_{j=1, j \neq i}^{N} p_{ij} \times l\left(y_i, y_j \right). \qquad (2)$$

After adding the regularization term $\lambda \sum_{r=1}^{m} w_r^2$, the objective function for minimization is

$$
\begin{aligned}
f(w) &= \frac{1}{N}\sum_{i=1}^{N} l_i + \lambda \sum_{r=1}^{m} w_r^2 \\
&= \frac{1}{N}\sum_{i=1}^{N} \sum_{j=1, j \neq i}^{N} p_{ij} \times l\left(y_i, y_j \right) + \lambda \sum_{r=1}^{m} w_r^2.
\end{aligned}
\qquad (3)
$$

The loss function for $l(y_i, y_j)$ here is the mean absolute deviation defined as $\sum_{i=1}^{N} |y_i - y_j|/N$. The main procedure of NCFS for regression feature selection can be summarized in Algorithm 2.

3.4. Neural Network-Based Prediction Model and Transformation. The most frequently used DM method for prediction is ANN. An ANN is network neurons that consist of propagation function and activation function, which receives the input, changes their internal state (activation) according to the input, and produces the output depending on the input and activation [33]. Despite the black box mechanism of ANN, it has been widely used in prediction problems demonstrating reasonable results as scrutinized in the literature [34]. ANN with their successful experience in forecasting diverse problems are among the most accurate and trustworthy used models. Their ability to learn from incomplete datasets in

order to predict the unseen section of data besides their capability of modeling the problem with the least available data and estimating almost all continuous functions has made them attractive enough to be used in prediction problems [34]. Köksal et al. [5] reviewed the reported performance of the DM methods and also pointed out that the ANN performance is mostly compared to the performance of the classical statistical modeling method such as multiple linear regressions (MLR), and better performance of ANN can naturally be observed in multidimensional data since these are powerful tools in modeling nonlinear relationships.

In this study, three-layer back propagation ANN was taken to predict the scrap rate; then it was transformed to determine the predicted surplus rate and supplemental feeding rate, two most concerned performances for material feeding optimization. The architecture of ANN is set by trial and error, and the number of nodes in hidden layer is set to $\max(3, (n_i + 1)/20)$, in which n_i is the number of input features. The ANN-based architecture for the scrap rate prediction and material feeding optimization is illustrated in Figure 6, in which a neuron (in the hidden layer or the output layer) j receives the outputs $a_1, a_2, \ldots, a_{n^{[l]}}$ of other neurons $1, \ldots, n^{[l]}$ which are connected to j with bias θ_j, and the propagation function of neuron j is defined as $z_j^{[l]} = \sum_{i=1}^{n^{[l-1]}} w_{ij} a_i + \theta_j$, in which the superscript $[l]$ denotes the lth layer and $n^{[l-1]}$ is the number of units of the $(l-1)$th layer. The results of propagation function are further processed by sigmoid activation function; that is, $a_j = f(z_j) = 1/(1 + e^{-z_j})$.

The transformation can be conducted according to (4)-(5) following the hypothesis that each feeding panel of an order has the same scrap probability, and scrap rate for an order will not change along with the predicted feeding area. Therefore,

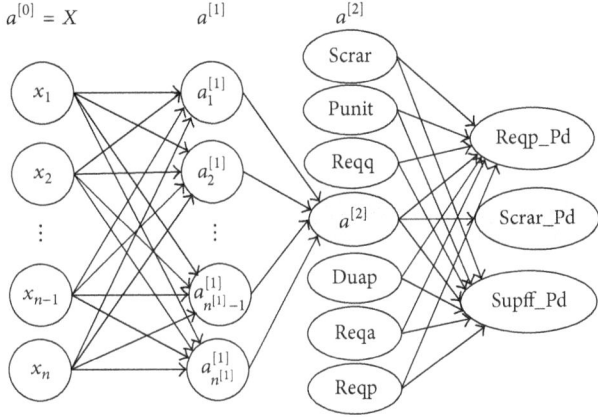

FIGURE 6: ANN-based architecture for the prediction of scrap rate and material feeding optimization (one can refer to Tables 1 and 2 for the notations).

TABLE 2: Prediction results related variables.

Variable	Symbol
Predicted scrap rate	$Scrar\_Pd$
Predicted feeding quantity	$Fedq\_Pd$
Predicted feeding panel	$Fedp\_Pd$
Predicted feeding area	$Feda\_Pd$
Predicted scrap area	$Scraa\_Pd$
Predicted surplus rate	$Surpr\_Pd$
Predicted supplemental feeding frequency for an order	$Supff\_Pd$
Predicted supplemental feeding rate	$Supfr\_Pd$

the predicted surplus rate and supplemental feeding rate can be calculated according to (6)–(8), and related variables are presented in Table 2.

The predicted feeding quantity and panel for each order are described as

$$Fedq\_Pd = \frac{Reqq/(100 - Scrar\_Pd)}{100},$$
$$Fedp\_Pd = \left\lceil \frac{Fedq\_Pd}{Duap} \right\rceil, \tag{4}$$

in which $Duap$ is the delivery unit in a panel given in Table 1. Then the predicted feeding quantity, area, and scrap area should be revised accordingly:

$$Fedq\_Pd = Fedp\_Pd \times Duap,$$
$$Feda\_Pd = Fedq\_Pd \times Punita, \tag{5}$$
$$Scraa\_Pd = Feda\_Pd \times Scrar,$$

Thus the predicted surplus rate of each order can be defined as

$$Surp\_Pd = \frac{(Feda\_Pd - Reqa - Scraa\_Pd)}{Reqa} \times 100\%. \tag{6}$$

The predicted supplemental feeding frequency for each order is specified as

$$Supff\_Pd = \begin{cases} 1, & Feda\_Pd - Reqa - Scraa\_Pd < 0 \\ 0, & \text{otherwise.} \end{cases} \tag{7}$$

$Reqa$ is the required area defined in Table 1. Therefore, the predicted surplus rate for these orders with $Supff\_Pd_i = 0$ can be calculated by

$$Surpr\_Pd = \frac{\sum_{i=1}^{N'} \left(Feda\_Pd_i - Reqa_i - Scraa\_Pd_i \right)}{\sum_{i=1}^{N'} Reqa_i}. \tag{8}$$

N' is the number of samples with $Supff\_Pd_i = 0$, $1 \le i \le N$. The $Surpr\_Pd$ for these orders with $Supff\_Pd_i = 1$ is not considered here because the surplus area cannot be determined before the supplemental feeding is finished. But their surplus rate is always lower than the $Surpr\_Pd$ defined in (10) in practice.

The supplemental feeding rate for all the samples can be defined as

$$Supfr\_Pd = \frac{\sum_{i=1}^{N} Supff\_Pd_i}{N} \times 100\%. \tag{9}$$

The surplus rate and supplemental feeding rate of the manual feeding can be computed by

$$Surpr = \frac{\sum_{i=1}^{N} \left(Feda_i - Reqa_i - Scraa_i \right)}{\sum_{i=1}^{N} Reqa_i}, \tag{10}$$

$$Supfr = \frac{\sum_{i=1}^{N} Supff_i}{N} \times 100\%. \tag{11}$$

3.5. Performance Indicators. In order to evaluate the effectiveness of the model, the following evaluation indicators are used [35]. The mean squared error (MSE) is the average of square sums between predicted data $\hat{y}_i$ and original data y_i, which can be described as

$$\text{MSE} = \frac{\sum_{i=1}^{N} \left(\hat{y}_i - y_i \right)^2}{N}. \tag{12}$$

The mean absolute error (MAE) is the average of the sum of the absolute difference between observed values and estimated values. It can be expressed as

$$\text{MAE} = \frac{\sum_{i=1}^{N} \left| \hat{y}_i - y_i \right|}{N}. \tag{13}$$

The mean absolute percentage error (MAPE) is the average of the sum of the normalized absolute difference between observed values and estimated values. The formula is written as

$$\text{MAPE} = \frac{1}{N} \sum_{i=1}^{N} \left| \frac{\hat{y}_i - y_i}{y_i} \right| \times 100, \tag{14}$$

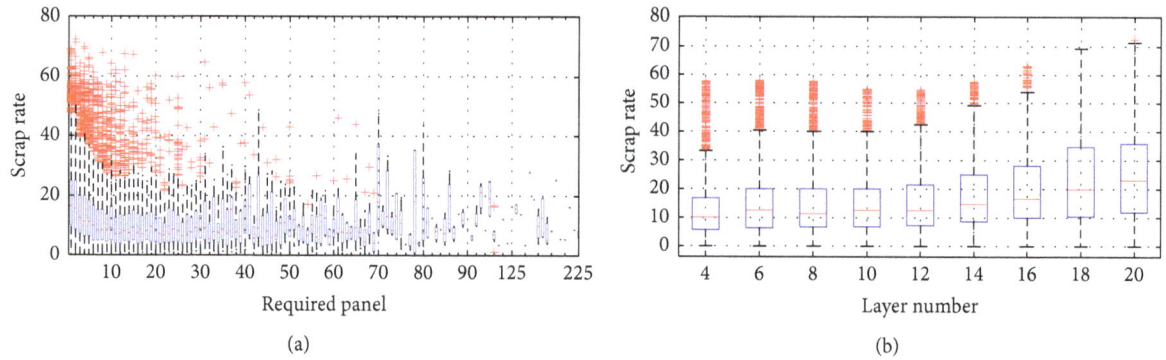

FIGURE 7: (a) Boxplot for different value of required panel. (b) Boxplot for different value of layer number.

TABLE 3: Significance test of break points.

Test variables	Test value	Critical value [18]
UD_{max}	8,543.4119*	10.17
WD_{max}	8,543.4119*	10.91
sup $F_T(2 \mid 1)$	4,030.0921*	11.14
sup $F_T(3 \mid 2)$	3,297.9988*	12.16
sup $F_T(4 \mid 3)$	1,437.4137*	12.83
sup $F_T(5 \mid 4)$	793.6005	13.45

5 break points corresponding to the required panel intervals 1, 2, 3, 4–6, 7–19, and greater than 19

*A statistic significance at the 5% level.

where N is the number of samples. The final purpose is to determine the feeding panel for each order on the basis of the predicted scrap rate, and y_i and $\hat{y}_i$ are replaced by the least feeding panel and predicted feeding panel, respectively. Then the deviation of the predicted feeding and the manual feeding can be computed according to $\hat{y}_i - y_i$ and $Fedp_i - y_i$ for sample i, $1 \leq i \leq N$, respectively. The error diagram can be drawn as a distribution of the deviation for all samples. Combined with the aforementioned predicted surplus rate and supplemental feeding rate for material feeding optimization, the final performance will be evaluated by the five indicators.

4. Results and Discussion

According to the multivariate boxplot approach described in Section 3.1, 960 outliers were trimmed and 29,157 samples left. Figure 7 shows the boxplot of the scrap rate for different value of the required panel and the layer number. Figure 7 illustrates that the outliers are shifted by the values of the required panel and the layer number, and therefore outlier detection considering different feature values is more reasonable.

Significance test for break points of the samples according to UD_{max}, WD_{max}, and sup $F_T(l + 1/l)$ was conducted based on default parameters given in [27], and the results are given in Table 3. The values of UD_{max} and WD_{max} indicated that the samples have significant structural change at 5% level, and the values of sup $F_T(l + 1/l)$ showed that 5 break

points are significant. The final break position of ascending sorted samples according to the required panel was 8,935, 13,995, 17,003, 21,791, and 27,491. Therefore the samples were partitioned into 6 groups with indexes 1–6 which corresponds to the samples with the required panels 1, 2, 3, 4–6, 7–19, and greater than 19, respectively. The samples in group 6 can still be segmented for sup $F_T(5 \mid 4)$ greater than the critical value at 5% level [25]. However, the sample size in group 6 was small (1666 samples) and the average of scrap rate greatly fluctuated which can be seen from Figures 1 and 6(a). Thus further partition for the samples in group 6 was not conducted.

Then NCFS was conducted for each group of the samples in which the initial step length was set to 0.9 and small positive constant η was set to 10^{-4}. 5-fold cross-validation instead of a single test was conducted to optimize regularization parameter λ initialized with 20 randomly selected values between 0–1.2×10^{-3} according to [32], and the λ value that minimizes the mean loss across the cross-validation was selected to fit NCFS. Figure 8(a) shows that the loss performance for 20 different λ values for the group of the samples with the required panels between 7 and 19 and the fourth λ corresponding to the lowest mean loss was selected as the regularization parameter for NCFS. Figure 8(b) illustrates the indexes of the selected features based on the selected λ. Final selected features for different group of the samples and all samples as a whole are given in Table 4. Difference of the selected features indicates that the samples with the different features may distribute in different regimes. However, layer number, Rogers material, number of operations, Huawei standard, plug hole with resin, second drilling, back drilling, Cu/Ni/Au pattern plating, gold finger plating, gold plating, delivery unit area, and historical qualified rate are critical features for most of the samples, which means that different values of these features will influence the scrap rate greatly in general, and these selected features also match well with the experience of experts from the factory.

The 70%, 15%, and 15% of the mutually exclusive samples were randomly selected as training, validation, and test data for each partitioned group of the samples, and the sample sizes for each group are given in Table 5. Prediction models of MSCM-ANN were trained, validated, and tested for each group of the samples with 5 runs based on the corresponding selected features while the ANN was trained, validated, and

TABLE 4: Selected features for different group of the samples.

Index	Feature symbol	Sample group						
		1	2	3	4	5	6	All
1	Pt						◊	◊
2	Ln	◊	◊	◊	◊	◊	◊	◊
3	Ro		◊	◊		◊	◊	◊
4	Plfr						◊	◊
5	Noo				◊	◊	◊	◊
6	NPP							◊
7	Sus							
8	Photb	◊	◊					
9	Highfb	◊				◊		
10	Semictb	◊	◊	◊				
11	Nflp							
12	Tinc					◊	◊	◊
13	IPCIII							
14	Huawei		◊	◊	◊	◊	◊	◊
15	Mwil							
16	Mlsil							
17	Mwol							
18	Mlsol							
19	Srph					◊		
20	Phwr		◊		◊	◊		
21	Secd	◊	◊	◊	◊	◊	◊	◊
22	Bcdr	◊			◊	◊	◊	◊
23	Chaprt	◊			◊	◊	◊	◊
24	White						◊	◊
25	Blue							
26	Black							
27	Chaprt			◊				
28	Hasl					◊	◊	◊
29	Lfhasl	◊						
30	Osp	◊				◊		
31	Cnapp	◊		◊	◊		◊	◊
32	Gfig			◊	◊	◊	◊	
33	Godp	◊	◊	◊	◊	◊		
34	Snap							
35	Iasa					◊	◊	◊
36	Reqq	◊						
37	Reqpl							◊
38	Dunita	◊	◊	◊	◊	◊	◊	◊
39	Reqa							
40	Hquar	◊	◊	◊	◊	◊	◊	◊

Note. Selected features are marked with ◊. The description of features (variables) has been specified in Table 1.

tested from all samples based on the selected features as listed in the last column of Table 4. Comparison of average MSE, MAE, and MAPE of 5 runs for MSCM-ANN, ANN, and the manual feeding is given in Table 6. The results indicate that both MSCM-ANN and ANN have obvious superiority in reducing the three indicators. However, MSCM-ANN can achieve smaller MSE, MAE, and MAPE compared to ANN, which means that the established models considering structural change can further improve the precision.

The average results of each group of the samples achieved by MSCM-ANN, ANN, and manual feeding are given in Table 7. The surplus rate and supplemental feeding rate obtained by the three approaches are given in Table 8. The following results can be drawn accordingly:

(1) Both MSCM-ANN and ANN can reduce the surplus rate and supplemental feeding rate performances simultaneously compared to the manual feeding as shown in Table 8. MSCM-ANN obtained lower values with 11.96%

(a)

(b)

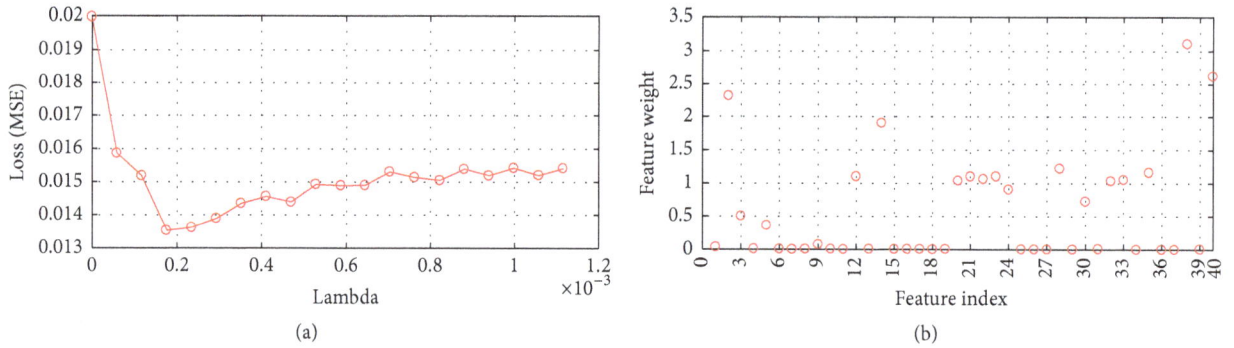

FIGURE 8: (a) Mean loss performance for 20 different lambda (λ) values for samples with required panel between 7 and 19. (b) Feature selection based on NCFS with the lowest loss lambda for samples with required panel between 7 and 19.

TABLE 5: Samples sizes of training, validation, and test data.

Sample group	Panel intervals	Training	Validation	Test	Total number
1	1	6,255	1,340	1,340	8,935
2	2	3,542	759	759	5,060
3	3	2,105	451	451	3,007
4	4–6	3,353	718	718	4,789
5	7–19	3,990	855	855	5,700
6	≥20	1,166	250	250	1,666
All	All	20,409	4,374	4,374	29,157

TABLE 6: Performance indicators achieved by different approaches.

Performance indicators	Samples	MSCM-ANN	ANN	Manual feeding
MSE	Training	0.719	1.803	24.272
	Validation	1.186	1.639	20.830
	Test	1.272	1.847	14.849
	All	0.872	1.786	22.342
MAE	Training	0.330	0.502	1.417
	Validation	0.391	0.509	1.458
	Test	0.396	0.515	1.409
	All	0.349	0.505	28.591
MAPE	Training	5.402	5.784	29.088
	Validation	5.817	6.089	28.582
	Test	5.687	5.872	28.664
	All	5.522	5.783	28.591

predicted surplus rate and 11.91% predicted supplemental feeding rate while ANN achieved 15.16% and 12.69% for the two performance indicators, respectively. Better performance of MSCM-ANN may be influenced by more precisely selected features, more reasonable ANN architecture, and well-trained models for each partitioned sample group based on MSCM considering the structural change influence compared to ANN which could not explore the pattern in each partitioned group.

(2) The results in Table 8 indicate that MSCM-ANN and ANN achieved the lower surplus rate but relatively higher supplemental feeding for the samples when the sample group corresponding to the intervals of required panel increases. The main reason is that the required panel that was rounded up to the nearest integer based on the required quantity

resulted in high redundancy when the number of the required panels was small, which therefore caused lower supplemental feeding rate. Taking the PCB order in Figure 1, for example, if the required quantity is only 4 units, then it will cause 100% ((10-2-4)/4 × 100%) surplus rate for feeding one panel with 20% scrap rate; the supplemental feeding should not be conducted until the scrap rate is greater than 60%. In contrast, lower surplus rate but relatively higher supplemental feeding rate was obtained when the required panel increased with lower surplus, but great fluctuation of the scrap rate may cause insufficient feeding panel and therefore bring about high supplemental feeding frequency.

The predicted scrap rate and predicted supplemental feeding rate on average obtained by MSCM-ANN for the training, validation, and test sample in each group are

TABLE 7: Predicted and real results of each group of samples.

Group	Supplemental feeding frequency			Required area			Surplus area		
	MSCM-ANN	ANN	Manual feeding	MSCM-ANN	ANN	Manual feeding	MSCM-ANN	ANN	Manual feeding
1	345	494	347	721.78	702.69	760.36	567.88	528.97	983.96
2	678	801	407	1,039.63	1,005.84	1,200.13	325.61	302.35	726.15
3	476	509	301	981.31	965.91	1,157.35	198.09	201.12	515.11
4	551	790	776	2,895.08	2,725.32	3,259.56	348.94	427.56	974.55
5	953	849	1810	7,417.77	7,600.13	8,956.11	585.65	908.73	1702.68
6	470	256	1581	6,184.03	7,310.21	8,754.72	301.72	710.30	1708.27
All	3,473	3,699	5222	19,471.17	20,310.11	24,088.22	2,327.89	3,079.03	6610.72

Note. The required area and surplus area of MSCM-ANN and ANN are the sum of required area and predicted surplus area of these orders with $Supff\_Pd_i = 0, 1 \leq i \leq N$, respectively.

TABLE 8: Comparison of surplus rate and supplemental feeding rate.

Group	MSCM-ANN		ANN		Manual feeding	
	Surpr_Pd (%)	Supfr_Pd (%)	Surpr_Pd (%)	Supfr_Pd (%)	Surpr (%)	Supfr (%)
1	78.68	3.86	75.28	5.53	129.41	3.88
2	31.32	13.40	30.59	15.83	60.51	8.04
3	20.19	15.83	20.82	16.93	44.51	10.01
4	12.05	11.51	15.69	16.49	29.90	16.20
5	7.90	16.72	11.96	14.89	14.15	31.75
6	4.88	28.21	9.72	15.37	19.51	94.90
ALL	11.96	11.91	15.16	12.69	27.44	17.91

Note. $Surpr\_Pd$ and $Supfr\_Pd$ are the predicted surplus rate and supplemental feeding rate, respectively, and they can be obtained according to the definition specified in Section 3.5 and the data provided in Table 7.

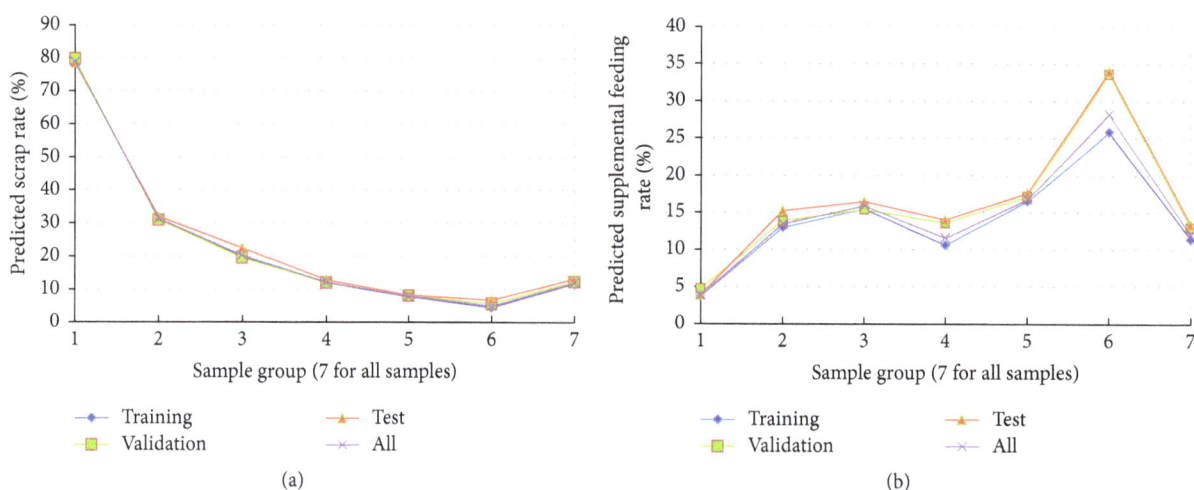

(a)

(b)

FIGURE 9: (a) Predicted scrap rate of MSCM-ANN for different samples. (b) Predicted supplemental feeding rate for different samples.

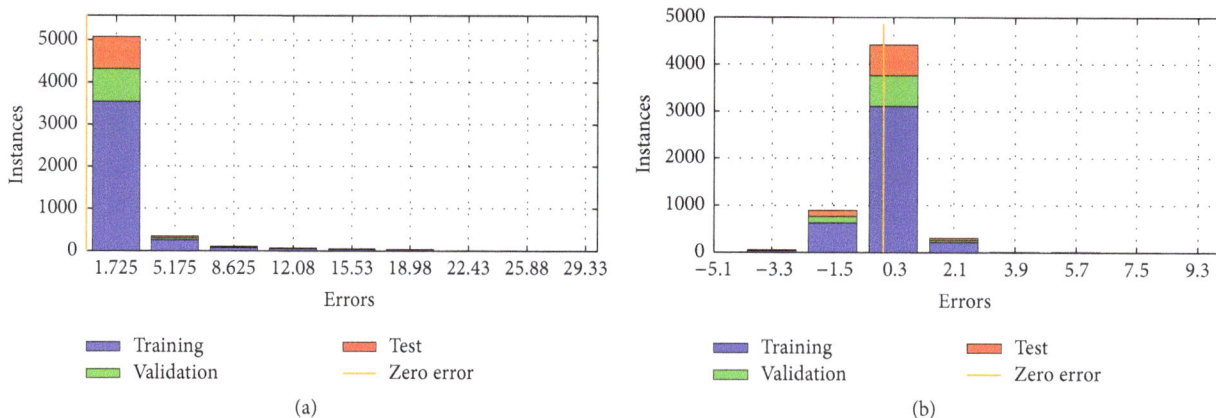

(a)

(b)

FIGURE 10: (a) Error diagram of the results from manual feeding for the samples in group 5. (b) Error diagram of predicted result obtained by MSCM-ANN for the samples in group 5.

illustrated by Figure 9, which indicates that the MSCM-ANN is stable to determine the surplus rate and supplemental feeding rate for each group of samples in most of the case. The relatively large deviation of the predicted supplemental feeding rate between training and test for the samples in group 6 may be caused by the large fluctuation of the scrap

rate for different orders. Meanwhile, relatively small sample size is harmful to maintain the stability of the model.

Figures 10(a) and 10(b) present the error diagrams of the results obtained from the manual feeding and run predicted results of MSCM-ANN, respectively, for the samples in group 5. Figure 10(b) illustrates that the errors obtained by

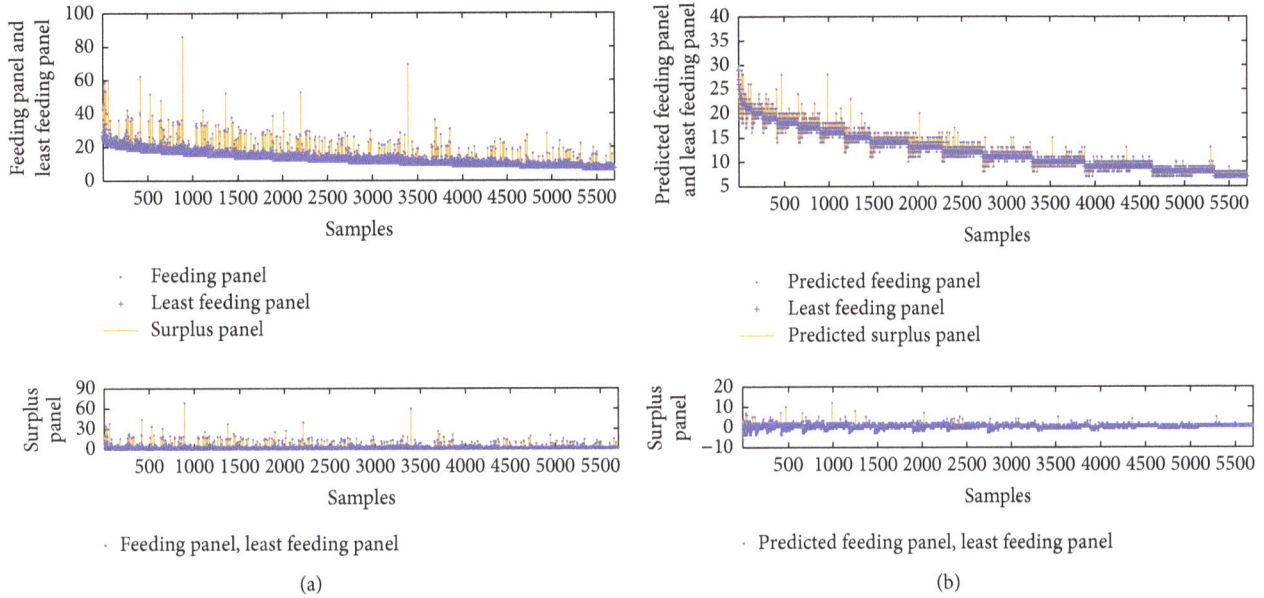

FIGURE 11: (a) Deviation of samples between manual feeding panel and least feeding panel. (b) Deviation between predicted feeding panel and least feeding panel.

MSCM-ANN are more likely to distribute with mean value 0.3 and short tail for training, validation, and test samples, while most of the errors obtained by the manual feeding distributed with mean value 1.725 (Figure 10(a)), and the large positive tail indicates (Figure 10(a)) that the manual feeding can easily lead to high redundancy after delivery of order. The deviations between the manual feeding panel/predicted feeding panel and least feeding panel for sample in group 5 are illustrated in Figure 11. It indicates that the predicted results in Figure 11(b) achieved lower deviation in most of the case compared to the manual feeding results in Figure 11(a), which can bring lower surplus panel, and therefore reduce the cost of material, production, inventory, and disposal.

Figures 12(a) and 12(b) present the regression of manual feeding panel and predicted feeding panel versus least feeding panel, respectively. Results indicate that the predicted feeding panel coincides better with the least feeding panel in Figure 12(b) compared to the manual feeding panel illustrated in Figure 12(a), and therefore the waste of surplus quantity and area can be reduced. The same coefficients and similar regression expressions obtained by MSCM-ANN for training, validation, test, and all samples mutually verify the stability of the proposed approach.

5. Conclusions

Accurate determination of the number of feeding panels for each PCB template order can reduce the cost of material, production, logistics, inventory, disposal, and delivery tardiness compensation. In this paper, a data mining approach (MSCM-ANN) involving the use of multivariate boxplot, MSCM, NCFS, and ANN was developed for establishing the scrap rate prediction model and material feeding optimization for PCB template order considering the structural

change influence for the predicted scrap rate. The various aspects of the approaches have been discussed in detail. Mean squared error, mean absolute error, and mean absolute percentage error, three prediction performance indicators, combined with surplus rate and supplemental feeding rate, two most concerned performances indicators for material optimization in practice, were presented to evaluate the established model. The multivariate boxplot was adopted for scarp rate outlier detection considering the structural changes influence of different input features, while the MSCM was applied to explore the multiple structural changes of the samples and therefore partition the samples into 6 different subgroups. NCFS and ANN were utilized for feature selection and scrap rate prediction model establishment for each group of the samples, respectively. After comparing MSCM-ANN with ANN and the manual feeding, the following conclusions and contributions are highlighted as follows.

(1) The proposed MSCM-ANN shows superior prediction accuracy on training, validation, and test dataset with the lowest MSE, MAE, MAPE, surplus rate, and supplemental feeding rate performance indicators compared to ANN and the manual feeding. MSCM-ANN reduces the surplus rate and supplemental feeding rate from 27.44% and 17.91% obtained by the manual feeding to 11.96% and 11.91%, respectively, but ANN can only reduce them to 15.16% and 12.69%, respectively. The same coefficients and similar regression expressions of the predicted feeding panel versus the least required panel for training, validation, test, and all samples mutually verify the stability of the proposed MSCM-ANN.

(2) The established model provides a new mechanism based on DM for the material feeding optimization of PCB template production that has seldom been studied according to the best of our knowledge. The application of the developed approach can replace the empirical manual feeding and cut

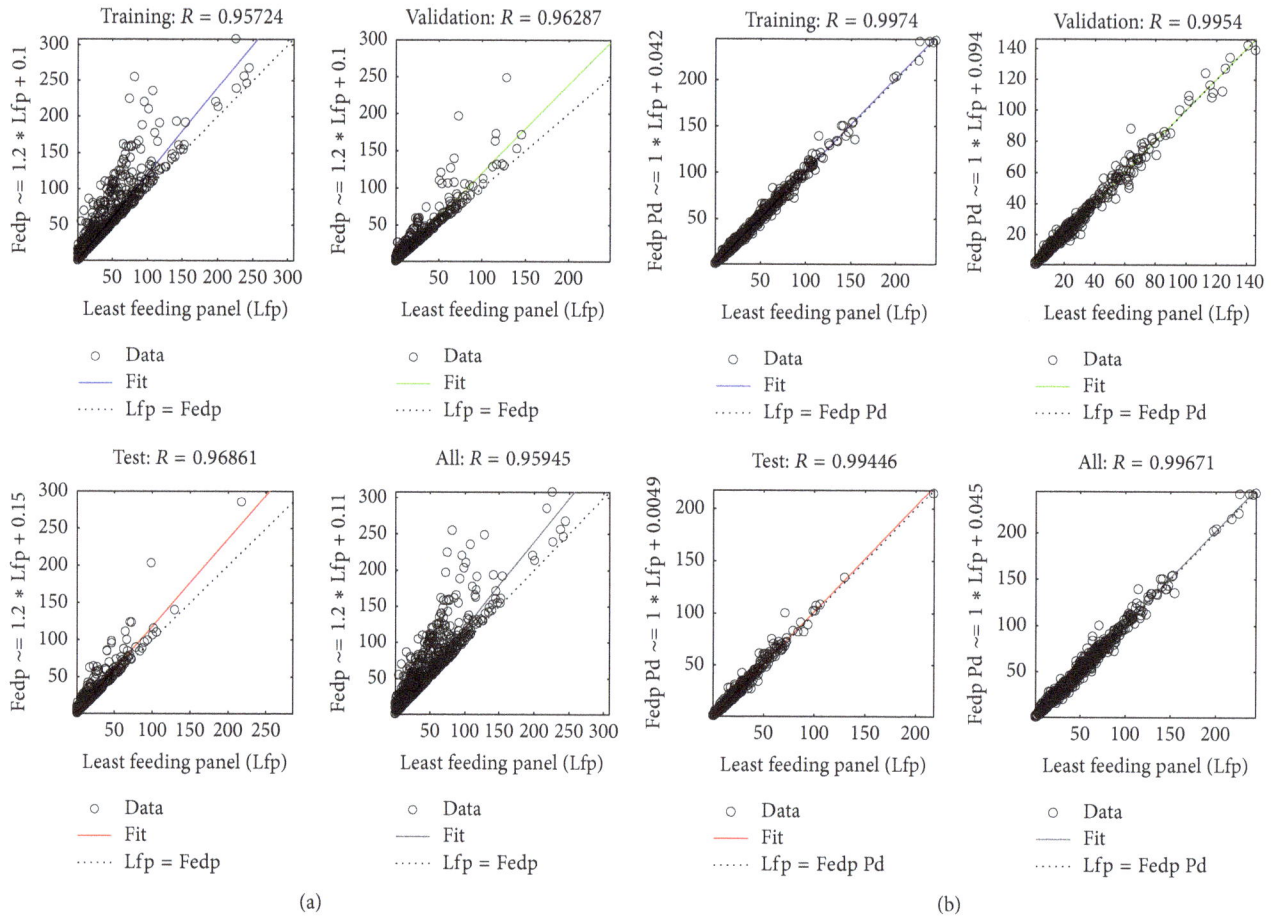

FIGURE 12: (a) Regression of manual feeding panel versus least feeding panel. (b) Regression of predicted feeding panel versus least feeding panel.

down experience dependent workers, but more important it can reduce the comprehensive cost of raw material, production, logistics, inventory, disposal, and delivery tardiness compensation as a whole.

(3) The framework for the material feeding optimization, specified variables, partitioned sample group considering the structural change influence, selected features with mutual verification by different group of the samples, the established ANN-based prediction model, and related transformation can further facilitate other models' development for similar factory.

Tool development based on MSCM-ANN model and practice application should be studied continually. Meanwhile, the structure change of the samples based on other features can be explored and more precise prediction model can also be investigated. The architecture optimization of ANN combined with some heuristic algorithm and the weights updates mechanism considering the balance of surplus and supplemental feeding is being studied by us.

Additional Points

Featured Application. The research was motivated by the requirement of a printed circuit board (PCB) manufacturing

company, and the application of the work is to optimize the material feeding by reducing the surplus rate and supplemental feeding rate simultaneously.

Conflicts of Interest

The authors declare that they have no conflicts of interest.

Authors' Contributions

Shengping Lv proposed and implemented the algorithm and also wrote the paper. Binbin Zheng and Qiangsheng Yue conducted the experiments and analyzed the data. Hoyeol Kim proposed the structure of the paper and helped in proofreading and improving the quality of the article.

Acknowledgments

This paper is supported by the National Natural Science Foundation of China (Grant no. 51605169) and Natural Science Foundation of Guangdong, China (Grant no. 2014A030310345). This study is also supported by the State Scholarship Fund of China (Grant no. 201608440414). The authors wish to thank Guangzhou FastPrint Technology Co., Ltd. to provide data for the study.

References

[1] A. Canal Marques, J.-M. Cabrera, and C. De Fraga Malfatti, "Printed circuit boards: a review on the perspective of sustainability," *Journal of Environmental Management*, vol. 131, pp. 298–306, 2013.

[2] E. Hofmann and M. Rüsch, "Industry 4.0 and the current status as well as future prospects on logistics," *Computers in Industry*, vol. 89, pp. 23–34, 2017.

[3] R. Cupek, A. Ziebinski, L. Huczala, and H. Erdogan, "Agent-based manufacturing execution systems for short-series production scheduling," *Computers in Industry*, vol. 82, pp. 245–258, 2016.

[4] G. Cheng, "Data and knowledge mining with big data towards smart production," *Journal of Industrial Information Integration*, 2017.

[5] G. Köksal, İ. Batmaz, and M. C. Testik, "A review of data mining applications for quality improvement in manufacturing industry," *Expert Systems with Applications*, vol. 38, no. 10, pp. 13448–13467, 2011.

[6] D. Delen and H. Demirkan, "Data, information and analytics as services," *Decision Support Systems*, vol. 55, no. 1, pp. 359–363, 2013.

[7] H. Rostami, J.-Y. Dantan, and L. Homri, "Review of data mining applications for quality assessment in manufacturing industry: support vector machines," *International Journal of Metrology and Quality Engineering*, vol. 6, no. 4, article 401, 2015.

[8] A. K. Choudhary, J. A. Harding, and M. K. Tiwari, "Data mining in manufacturing: a review based on the kind of knowledge," *Journal of Intelligent Manufacturing*, vol. 20, no. 5, pp. 501–521, 2009.

[9] P. O'Donovan, K. Leahy, K. Bruton, and D. T. J. O'Sullivan, "Big data in manufacturing: a systematic mapping study," *Journal of Big Data*, vol. 2, pp. 1–20, 2015.

[10] K. Nagorny, P. Lima-Monteiro, J. Barata, and A. W. Colombo, "Big data analysis in smart manufacturing: a review," *International Journal of Communications, Network and System Sciences*, vol. 10, no. 03, pp. 31–58, 2017.

[11] T. L. Tseng, C. Johnny, H. Lee, and Y. J. Kwon, "Analysis of complex PCB design data using data mining approach to better estimate the work hours," *International Journal of Applied Engineering Research*, vol. 11, pp. 11700–11711, 2016.

[12] T.-N. Tsai and M. Liukkonen, "Robust parameter design for the micro-BGA stencil printing process using a fuzzy logic-based Taguchi method," *Applied Soft Computing*, vol. 48, pp. 124–136, 2016.

[13] T. N. Tsai, "Modeling and optimization of stencil printing operations: a comparison study," *Computers & Industrial Engineering*, vol. 54, pp. 374–389, 2008.

[14] N. Khader, S. W. Yoon, and D. Li, "Stencil printing optimization using a hybrid of support vector regression and mixed-integer linear programming," *Procedia Manufacturing*, vol. 11, pp. 1809–1817, 2017.

[15] H. Haneda, H. Kodama, T. Hirogaki, E. Aoyama, and K. Ogawa, "Investigation of drilling conditions of printed circuit board based on data mining method from tool catalog data-base," *Advanced Materials Research*, vol. 939, pp. 547–554, 2014.

[16] P. K. Srimani and V. Prathiba, "Adaptive data mining approach for PCB defect detection and classification," *Indian Journal of Science and Technology*, vol. 9, no. 44, Article ID 98964, 2016.

[17] H. Sim, D. Choi, and C. O. Kim, "A data mining approach to the causal analysis of product faults in multi-stage PCB manufacturing," *International Journal of Precision Engineering and Manufacturing*, vol. 15, no. 8, pp. 1563–1573, 2014.

[18] E. Hadavandi, H. Shavandi, and A. Ghanbari, "An improved sales forecasting approach by the integration of genetic fuzzy systems and data clustering: case study of printed circuit board," *Expert Systems with Applications*, vol. 38, no. 8, pp. 9392–9399, 2011.

[19] P.-C. Chang, C.-H. Liu, and C.-Y. Fan, "Data clustering and fuzzy neural network for sales forecasting: a case study in printed circuit board industry," *Knowledge-Based Systems*, vol. 22, no. 5, pp. 344–355, 2009.

[20] P.-C. Chang, C.-H. Liu, and R. K. Lai, "A fuzzy case-based reasoning model for sales forecasting in print circuit board industries," *Expert Systems with Applications*, vol. 34, no. 3, pp. 2049–2058, 2008.

[21] P.-C. Chang, Y.-W. Wang, and C.-H. Liu, "The development of a weighted evolving fuzzy neural network for PCB sales forecasting," *Expert Systems with Applications*, vol. 32, no. 1, pp. 86–96, 2007.

[22] A. Tavakkoli, J. Rezaeenour, and E. Hadavandi, "A novel forecasting model based on support vector regression and bat meta-heuristic (Bat-SVR): case study in printed circuit board industry," *International Journal of Information Technology & Decision Making*, vol. 14, no. 1, pp. 195–215, 2015.

[23] V. J. Hodge and J. Austin, "A survey of outlier detection methodologies," *Artificial Intelligence Review*, vol. 22, no. 2, pp. 85–126, 2004.

[24] A. Zimek, E. Schubert, and H.-P. Kriegel, "A survey on unsupervised outlier detection in high-dimensional numerical data," *Statistical Analysis and Data Mining*, vol. 5, no. 5, pp. 363–387, 2012.

[25] J. Bai and P. Perron, "Estimating and testing linear models with multiple structural changes," *Econometrica*, vol. 66, no. 1, pp. 47–78, 1998.

[26] J. Bai and P. Perron, "Computation and analysis of multiple structural change models," *Journal of Applied Econometrics*, vol. 18, no. 1, pp. 1–22, 2003.

[27] Z. Qu, "Estimating and testing structural changes in multivariate regressions," 2017, http://people.bu.edu/perron/code.html.

[28] W. Yang, K. Wang, and W. Zuo, "Neighborhood component feature selection for high-dimensional data," *Journal of Computers*, vol. 7, no. 1, pp. 162–168, 2012.

[29] V. Bolón-Canedo, N. Sánchez-Maroño, and A. Alonso-Betanzos, "A review of feature selection methods on synthetic data," *Knowledge and Information Systems*, vol. 34, no. 3, pp. 483–519, 2013.

[30] G. Chandrashekar and F. Sahin, "A survey on feature selection methods," *Computers & Electrical Engineering*, vol. 40, no. 1, pp. 16–28, 2014.

[31] G. Aldehim and W. Wang, "Determining appropriate approaches for using data in feature selection," *International Journal of Machine Learning and Cybernetics*, vol. 8, no. 3, pp. 915–928, 2017.

[32] "Mathworks," 2017, https://www.mathworks.com/help/stats/neighborhood-component-analysis.html.

[33] "Artificial neural network," 2017, https://en.wikipedia.org/wiki/ Artificial neural_network&gt.

[34] S. T. Hashemi, O. M. Ebadati E., and H. Kaur, "A hybrid conceptual cost estimating model using ANN and GA for power plant projects," *Neural Computing and Applications*, 2017.

[35] G. Wang, T. Xu, T. Tang, T. Yuan, and H. Wang, "A Bayesian network model for prediction of weather-related failures in railway turnout systems," *Expert Systems with Applications*, vol. 69, pp. 247–256, 2017.

11

Impact of Interface Traps on Direct and Alternating Current in Tunneling Field-Effect Transistors

Zhi Jiang, Yiqi Zhuang, Cong Li, Ping Wang, and Yuqi Liu

School of Microelectronics, Xidian University, Xi'an 710071, China

Correspondence should be addressed to Zhi Jiang; zjiang@xidian.edu.cn

Academic Editor: Muhammad Taher Abuelma'atti

We demonstrate the impact of semiconductor/oxide interface traps (ITs) on the DC and AC characteristics of tunnel field-effect transistors (TFETs). Using the Sentaurus simulation tools, we show the impacts of trap density distribution and trap type on the n-type double gate- (DG-) TFET. The results show that the donor-type and acceptor-type ITs have the great influence on DC characteristic at midgap. Donor-like and acceptor-like ITs have different mechanism of the turn-on characteristics. The flat band shift changes obviously and differently in the AC analysis, which results in contrast of peak shift of Miller capacitor C_{gd} for n-type TFETs with donor-like and acceptor-like ITs.

1. Introduction

Tunneling field-effect transistor (TFET) is one of low-power electronics due to lower off-current and steeper slope. The mechanism of tunneling current was produced by band-to-band tunneling (BTBT) in a TFET, so TFET device can break the fundamental subthreshold swing (SS) limit of MOSFET [1–3]. Owing to its extremely low off-state current, the turn-on characteristic of TFET would be superior to MOSFET. Therefore, TFET devices can be recognized as one of the most possible candidates of MOSFETs [4–9]. However, TFET has a drawback of low on-state current (I_{on}). To solve this issue, high-κ dielectric was proposed to enhance I_{on} [10]. Unfortunately, the semiconductor/oxide interface quality is severely tested, and the existence of ITs could introduce instability. Besides, it was not clear how interface traps (ITs) can influence TFET performance [11–15]. What is more, they did not explain influence machine of Miller capacitance and power dissipation. Resolving this issue is important not only to better understand the device operation but also to further research the impacts of interface traps on turn-on and capacitance characteristics of TFETs. In this paper, we address a detailed investigation of the role of trap type, trap density, and trap energy levels on dependence of DG-TFET characteristics with HfO$_2$ high-κ gate insulator.

2. Device Model and TCAD Simulation

In this paper, the investigated device structure for the DG n-channel tunnel field-effect transistor (n-TFET) is shown in Figure 1. The device structure consists of a highly doped p-region (10^{20} atoms·cm^{-3}), a lightly doped intrinsic region (10^{16} atoms·cm^{-3}), and a highly doped n-region (10^{20} atoms·cm^{-3}). The intrinsic region acts as the channel, p-region acts as the source, and n-region acts as the drain and all lengths are 50 nm. The bulk Si thickness (T_{Si}) is 10 nm, the high-κ gate insulator thickness (T_{ox}) is 2 nm, and gate work function Φ is 4.0 eV. According to the uniform electric field limit and Kane's model, the band-to-band tunneling (BTBT) generation rate G is $G = A(F/F_0)^P \exp(-B/F)$, $A = g(m_c m_v)^{3/2}(1 + 2N_{TA})D_{TA}^2(qF_0)^{5/2}/2^{21/4}h^{5/2}m_r^{5/4}\rho\varepsilon_{TA}E_g^{7/4}$, $B = 2^{7/4}\pi m_r^{1/2}E_g^{3/2}/3qh$, and $P = 2.5$ for the indirect tunneling [16]. Specifying $\varepsilon_{TA} > 0$ selects the phonon-assisted tunneling process for Si. The results A and B are 1.4×10^{20} cm^{-3} s^{-1} and 1.12×10^8 V/cm, respectively. For the phonon-assisted tunneling process, the prefactor A and the exponential factor B take into account the material characteristics and external condition (such as optical phonon scattering (OP) and acoustic phonon scattering (AP)). Obviously, the factor B has more impact than A.

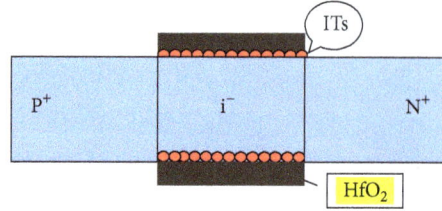

FIGURE 1: Device structures of n DG-TFET with steep doping profiles.

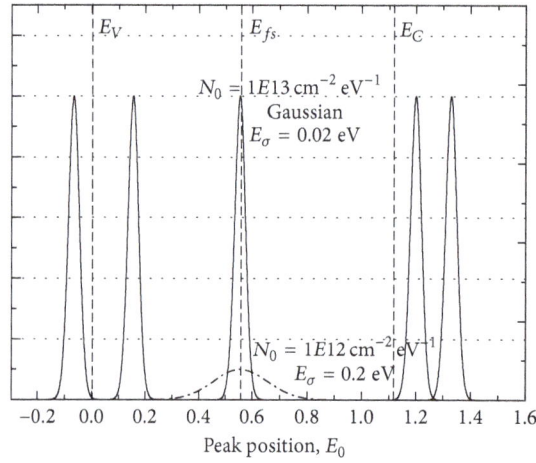

FIGURE 2: High and low Gaussian distributions are E_σ = 0.02 and 0.2, which have 1×10^{13} cm^{-2} eV^{-1} and 1×10^{12} cm^{-2} eV^{-1} interface traps concentration, respectively. The peak position of Gaussian distribution ranges from E_v to E_c and extends beyond the forbidden band.

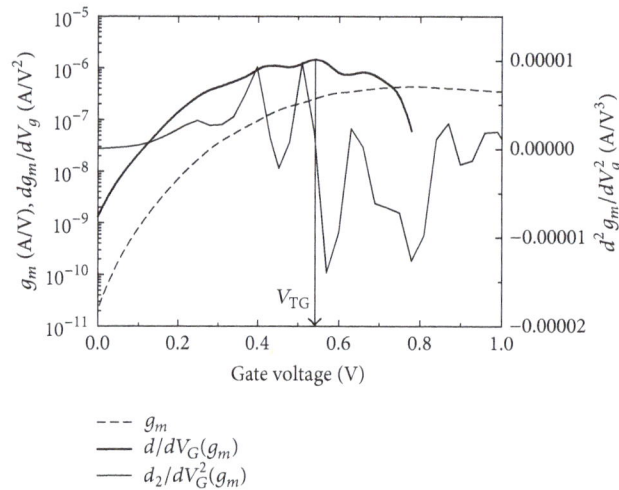

FIGURE 3: First derivative of drain current is g_m, its second derivative is dg_m/dV_g, and its third derivative is dg_m^2/dV_g^2; V_{DS} = 0.5 V; the channel length is 50 nm.

In order to make simulation results more reliable, the doping-dependent mobility model, the dynamical nonlocal-path band-to-band tunneling (BTBT) model, the modified local-density approximation (MLDA) model, the surface SRH recombination model, and the Schenk trap-assisted tunneling (TAT) model are included.

Because high electric fields and silicon process can cause hot-carrier injection (HCI) effects and traps in this semiconductor/oxide interface, we assume that these localized ITs were just located at Si/HfO$_2$ interface and the capture cross section σ ($\sigma_n = \sigma_p$) is 10^{-14} cm^{-2}, as shown in Figure 1. The trap energy and trap distribution consist of the high and low Gaussian distributions, and the peak position (E_0) could be moved in the forbidden band. Hereafter, we study the impact of ITs type, ITs energy level position, and ITs distribution on the turn-on DC characteristics. Besides, AC characteristics were also studied, including the impact of concentrations and type of ITs on Miller capacitance (C_{gd}).

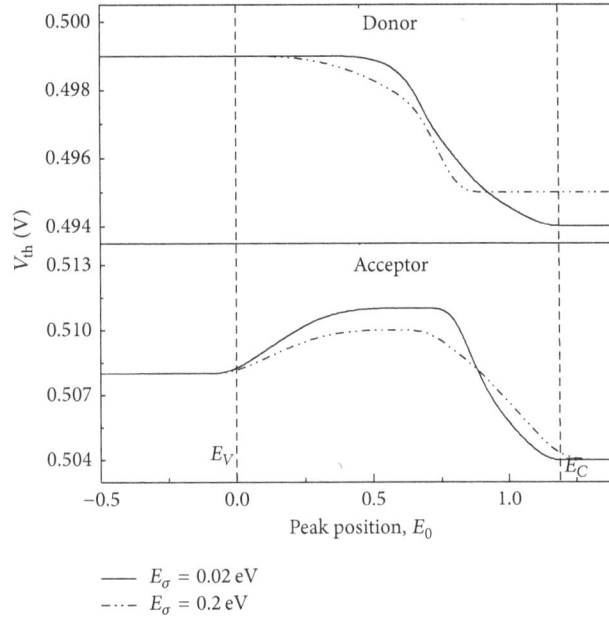

FIGURE 4: Threshold voltage V_{th} versus the peak position E_0 of high distribution and low distribution with donor-type ITs and acceptor-type ITs; $V_{GS} = V_{DS} = 0.5$ V.

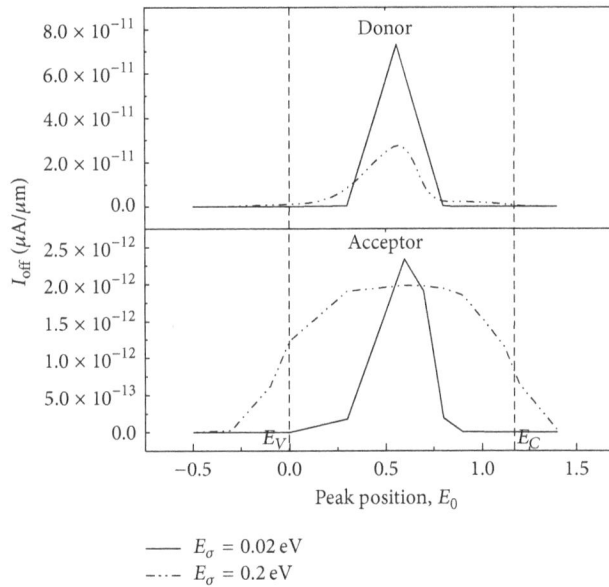

FIGURE 5: The extracted off-state current I_{off} as a function of the peak position E_0 of high distribution and low distribution with donor-type ITs and acceptor-type ITs; $V_{GS} = V_{DS} = 0.5$ V.

3. Results and Discussion

3.1. The Impact of ITs on DC Characteristics of DG-TFET. The high-κ materials have great advantages such as improving the on-state current and reducing the gate leakage current. However, because of the lattice mismatch between HfO_2 and Si, they would introduce many interface state defects by depositing with HfO_2 on nanocrystalline silicon film. It is necessary to discuss issues of the impact of interface traps on the performances of TFETs.

Figure 2 shows two typical Gaussian distributions of ITs energy and peak position. The shape of the Gaussian distribution can be decided by the trap basic vacancy and antisite states. Due to the different proportion of vacancy and antisite states, the thin and tall or fat and short cases are the basic cases. The threshold voltage (V_{th}), the off-state current (I_{off}), the minimum subthreshold swing (miniSS), the on-state current (I_{on}), and I_{on}/I_{off} ration are studied by moving peak position and changing value E_σ of Gaussian distribution. A maximum density $D_{IT}(N_0) = 1 \times 10^{13}\,cm^{-2}\,eV^{-1}$ and

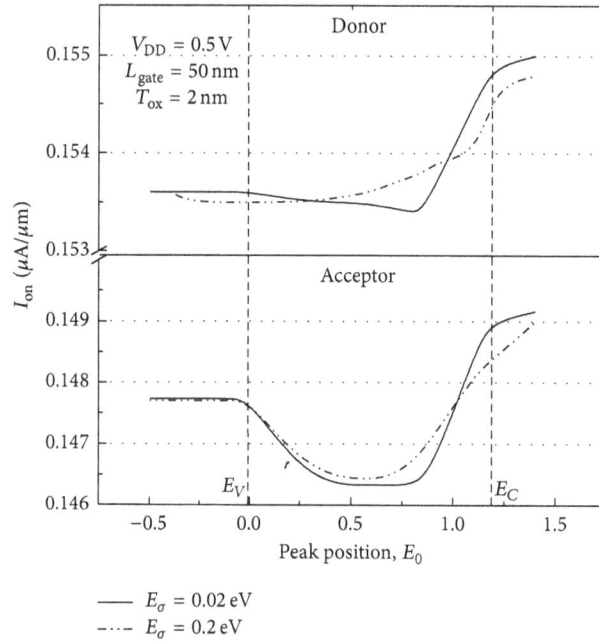

FIGURE 6: The extracted on-state current I_{off} as a function of the peak position E_0 of high distribution and low distribution with donor-type ITs and acceptor-type ITs; $V_{GS} = V_{DS} = 0.5\,V$.

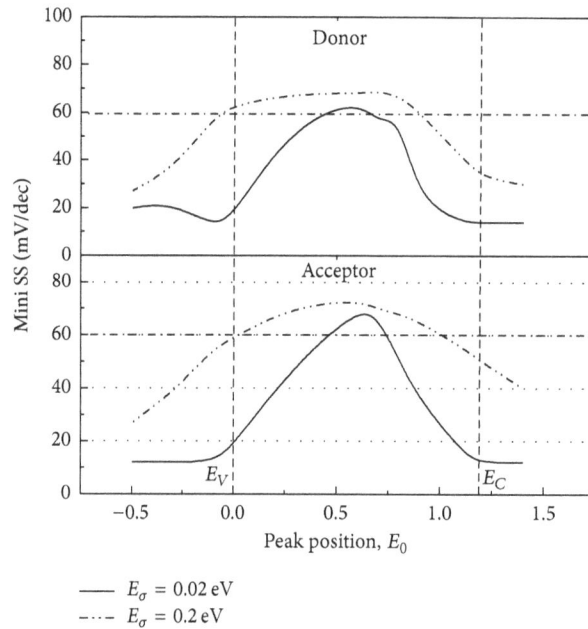

FIGURE 7: Mini SS is extracted as the minimum SS value in the interval between off-state and the threshold voltage V_{th}. Mini SS versus the peak position for DG-TFET at $V_{GS} = V_{DS} = 0.5\,V$.

a minimum $D_{IT}(N_0) = 1 \times 10^{12}\,cm^{-2}\,eV^{-1}$ are employed. Different trends of two trap types were compared in the following simulation. It is worth noting that V_{th} is extracted with the transconductance change method [17]. The method has definitely physical meaning in Figure 3.

Figure 4 shows the V_{th} shifts in acceptor-type trap and donor-type trap DG n-TFET. The impact of acceptor-type trap on V_{th} is greater than donor-type trap. The donor-type

interface traps can make V_{th} smaller from midgap to conduction band (E_c). When the donor-type interface trap level is under the Fermi level, the trap has no effect on V_{th}. Donor-type ITs having lost electrons will be positively charged, which resulted in a small threshold voltage. However, V_{th} will be increased from valence band (E_v) to the Fermi level. This is because acceptor-type ITs capture electron, and then the traps become negatively charged which lead to higher threshold

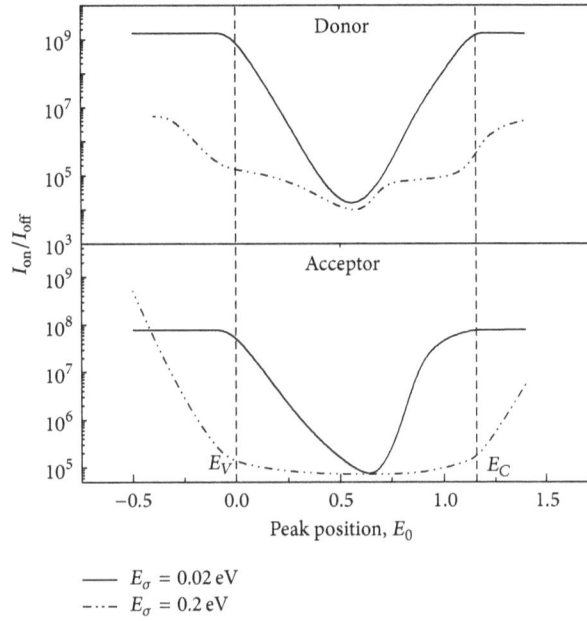

FIGURE 8: Comparison of I_{on}/I_{off} versus the peak position of high and low distribution ITs for DG-TFET.

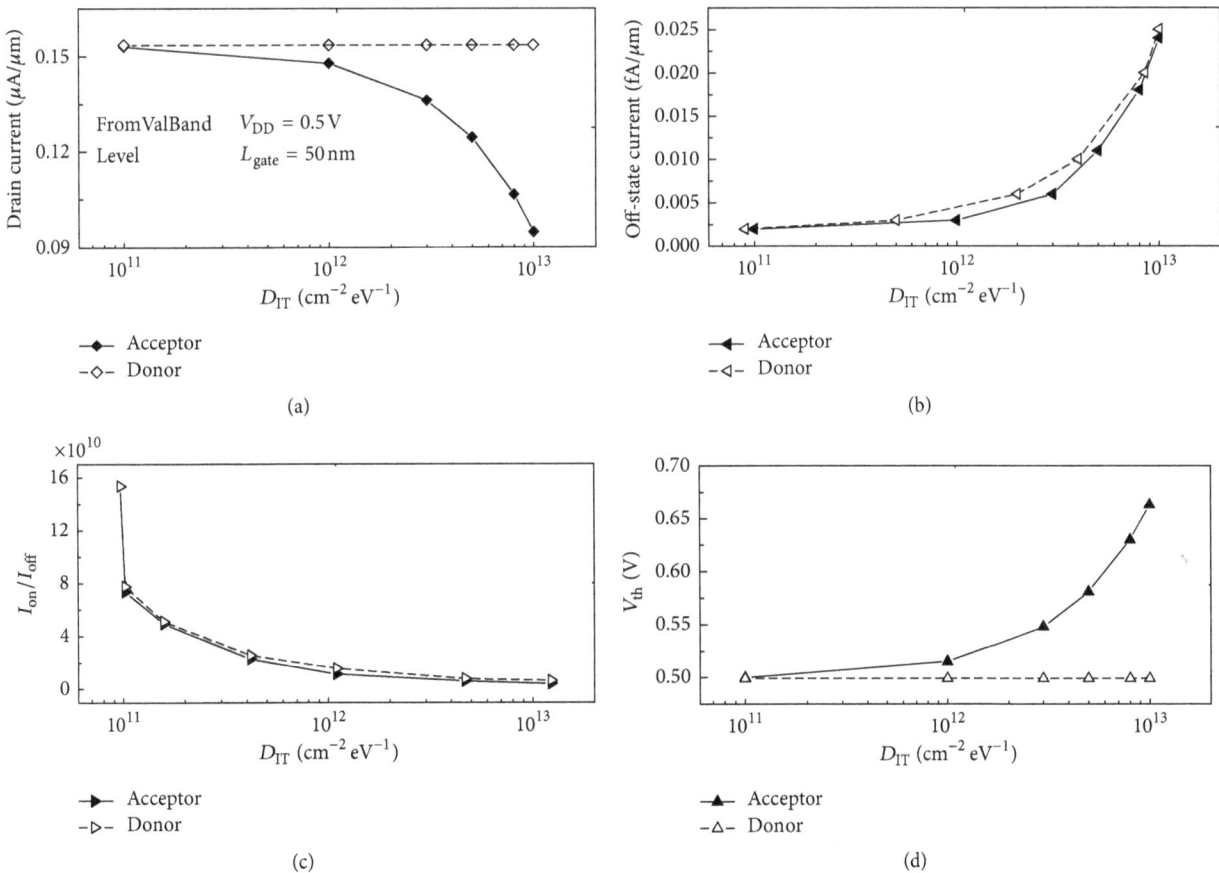

FIGURE 9: (a) Drain current I_{on} at $V_{GS} = V_{DS} = 0.5\,\text{V}$, (b) off-state I_{off} at $V_{GS} = 0\,\text{V}$, $V_{DS} = V_{DD} = 0.5\,\text{V}$, (c) calculated I_{on}/I_{off}, and (d) V_{th} versus ITs density at valence band.

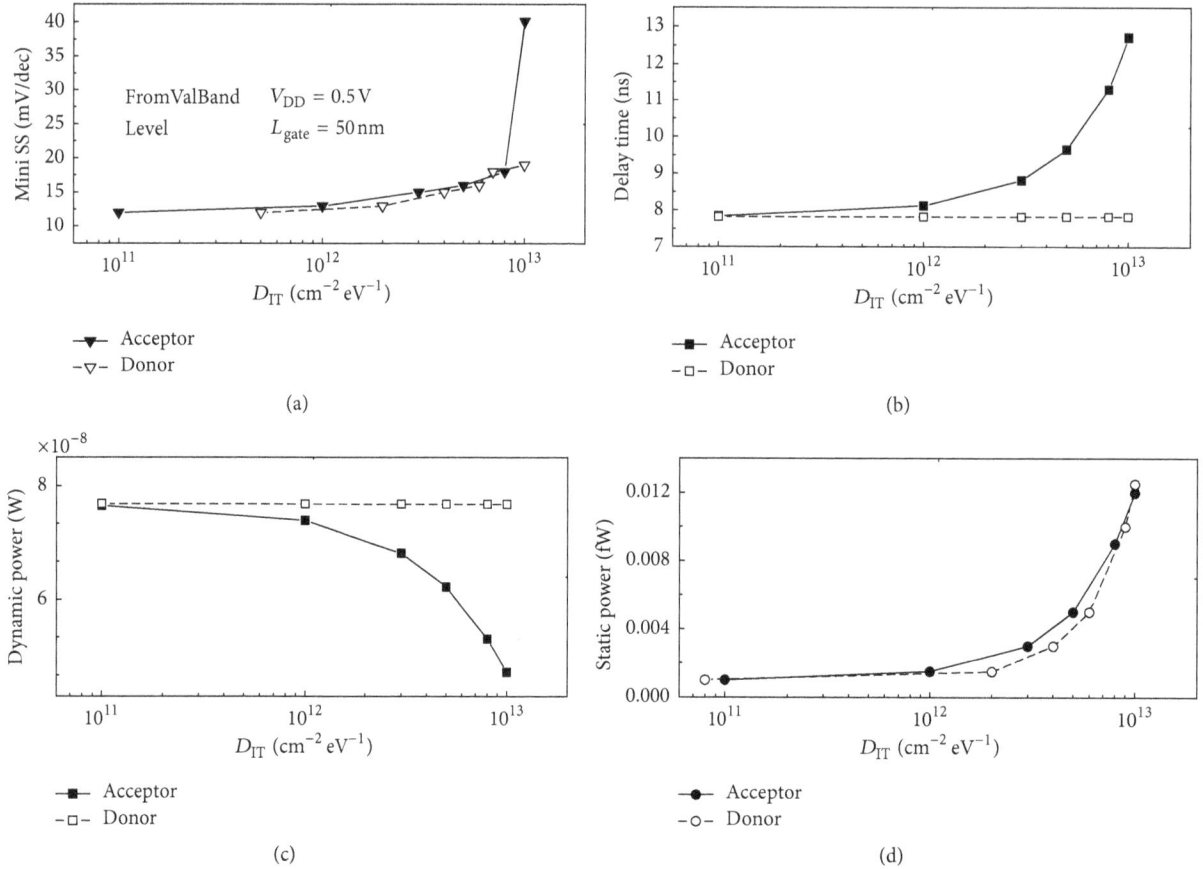

FIGURE 10: (a) The calculated mini SS, (b) delay time, (c) dynamic power, and (d) static power versus ITs density at valence band. Delay time τ is given by $C_L V_{DD}/I_{on}$. The donor-type ITs need more energy in static state than the acceptor-type ITs.

voltage. When acceptor-type trap level goes beyond the Fermi level, the traps having release electrons will be positively charged, which lead to lower threshold voltage again. Besides, it is clearly shown in Figure 4 that small E_σ has more influence than big E_σ.

The extracted off-state current will be increased when traps level is near the Fermi level in Figure 5. The acceptor-type ITs still have greater impact than donor-type. When the trap level is near the Fermi level, the drain-channel junction electric field will be increased (such ambipolar current is not shown under the negative-bias), and this position of trap level would influence electric field gravely between E_v and E_c. It can be observed that donor-type ITs have greater influence than acceptor-type ITs, and the peak position of channel-drain (c/d) tunneling junction field can be determined when traps level was located at midgap, if the electric field appears near the drain end, which results in greater device ambipolar current and off-current. In addition, the low Gaussian distributions of interface trap density E_σ induce smaller peak electric field than high E_σ. It can be seen in Figure 6 that the interface traps can make on-state current degradation between valence band and conduction band. In particular, when the acceptor-like and donor-like traps are located at the energy level 0.3 eV above the Si midgap, the on-state current deteriorates extremely. When ITs are

near the channel-source (c/s) junction, they can change the junction electric field. When traps level is below Fermi level, the donor-type ITs cannot release electrons. Thus, I_{on} could hardly be affected.

Meanwhile, because the acceptor-type ITs capture electrons and c/s junction electric field decreases, I_{on} decrease between the valence band and Fermi level. But when acceptor-type ITs level beyond Fermi level can lose electrons, tunneling field would be increased. After donor-type ITs level is higher than the Fermi level and releases electrons, as a result, the tunneling field increases and I_{on} also rise up rapidly. According to the BTBT (Kane's) model, a small change may increase or decrease abruptly the tunneling rate in the electric field.

The minimum (mini) point SS is defined as SS = $1000/(d/dV_g) \log I_d$ [16]. Figure 7 shows the extracted mini SS. Through the above analysis, the on-state current decreases since the effective source tunneling barrier width increases. The results indicate that the degradation of mini SS is subject to the position of traps level. The source tunneling width attains its maximum value when the traps level is located at Si midgap. It can be seen in Figure 8 that I_{on}/I_{off} rations have reduced between E_c and E_v. On-state current worsens and bipolarity current is produced, which results in smaller value of I_{on}/I_{off} ration for the DG-TFET.

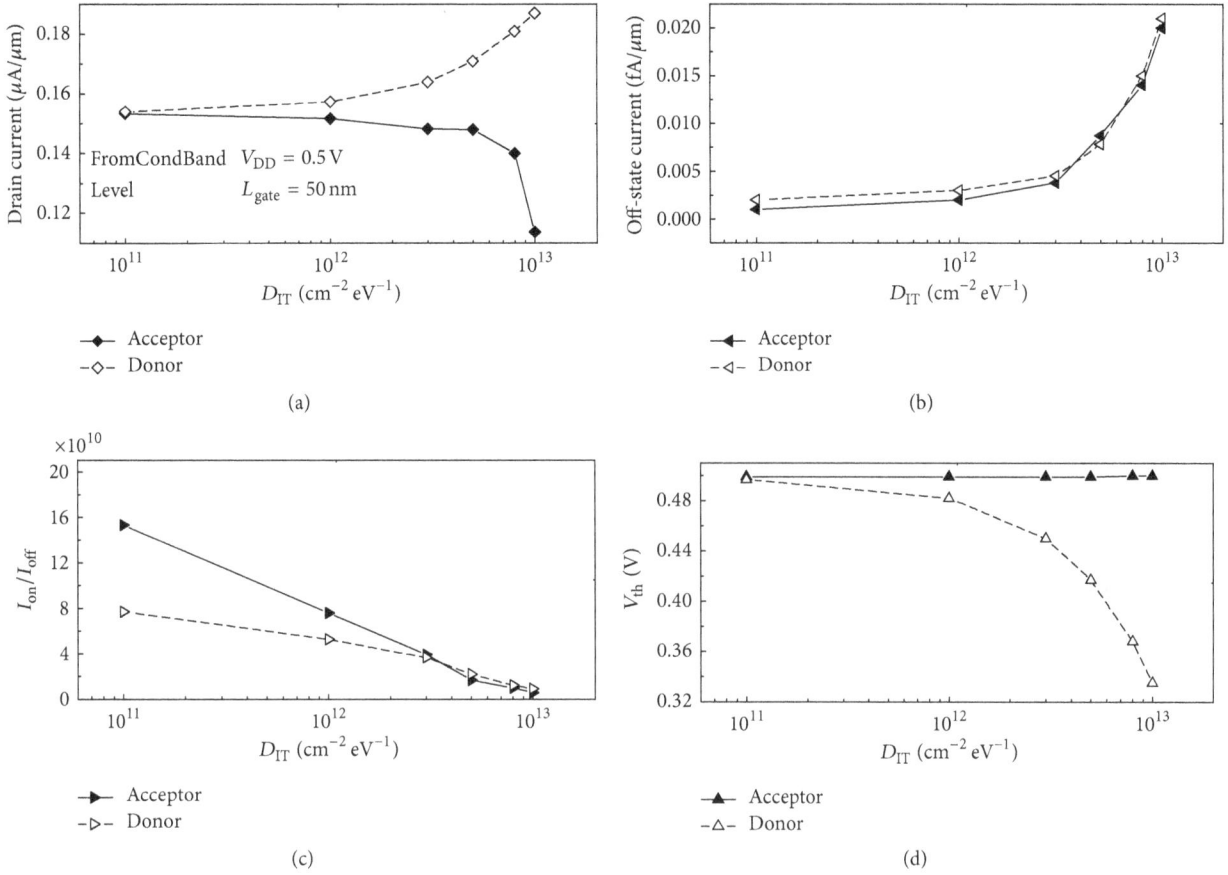

FIGURE 11: (a) Drain current I_{on} at $V_{GS} = V_{DS} = 0.5$ V, (b) off-state I_{off} at $V_{GS} = 0$ V, $V_{DS} = V_{DD} = 0.5$ V, (c) calculated I_{on}/I_{off}, and (d) V_{th} versus ITs density at conduction band. The donor-type ITs contribute to the on-state current which is different from acceptor-type ITs.

In order to get an insight, the impacts of donor-type and acceptor-type ITs density (D_{IT}) located at valence band (FromValBand), middle band (FromMidBandGap), and conduction band (FromCondBand) on drive current were examined. Off-state current, I_{on}/I_{off} ration, threshold voltage (V_{th}), minimum point SS, transistor delay time (τ), dynamic power, and static power were also investigated in Figures 9–14, respectively. For n-type TFETs, the capacitance magnitude is about a few fF/μm. For a DG-TFET device (gate channel length L_g = 50 nm, gate width W = 50 nm), the TFET capacitance (C_L) is about 9 fF, which is shown in Figure 15 where the maximum capacitance value is obtained in most cases.

We can see in Figures 9–14 that donor-type ITs D_{IT} will not have any effect on the drain current, V_{th}, delay time, and dynamic power. The rough delay time is given by $C_L V_{DD}/I_{on}$ ($V_{DD} = V_{DS} = V_{GS}$), and the dynamic power is roughly obtained by $C_L V_{DD}^2/\tau$. The DG-TFET is more immune to donor-type ITs but more susceptible to acceptor-type ITs. It can be seen that the BTBT rate at c/s tunneling junction is not affected obviously by donor-type ITs and I_{on} degradation due to ionized acceptor-type ITs, as shown in Figure 9(a). On the other hand, it is worth noticing that donor-like ITs level is below the Fermi level, and donor-type ITs would not be ionized at the Si midgap (see Figures 13 and 14). Results shown

in Figures 9(a) and 13(a) indicate that donor-type ITs D_{IT} slightly increases I_{on}, which confirm the results previously drawn in Figure 6. However, the acceptor-like ITs will capture electrons under the Fermi level and then reduce the c/s tunneling junction field, so the tunneling current decreases with increasing D_{IT}, as shown in Figure 9(a). For DG-TFET, ambipolarity current was increased by increasing donor-like or acceptor-like D_{IT}. However, traps level is in the middle band which has a larger impact than in the valence band. The off-state current can achieve 0.025 fA/μm in the middle band level and 2.75 pA/μm in the valence band level, as observed in Figures 9(b) and 12(b). According to the above study, the on/off ratio can be drawn from Figures 9(a) and 13(a). Figure 9(c) shows that it has a steeper curve than Figure 13(c). It can be explained that the electron probability occupancy is higher in the valence band than in the middle band. Besides, the acceptor-type ITs can influence V_{th} in Si midgap and the donor-type ITs can change V_{th} in both valence band and conduction band, as evident in Figures 9(d), 11(d), and 13(d).

According to the above formula, the donor-like traps would not affect drain current, so τ and dynamic power are nearly invariable. But the acceptor-like traps increase the delay time and reduce the dynamic power, as shown in Figures 10(b), 10(c), 14(b), and 14(c). The donor-type ITs and acceptor-type ITs have the same properties in Figures 10(a),

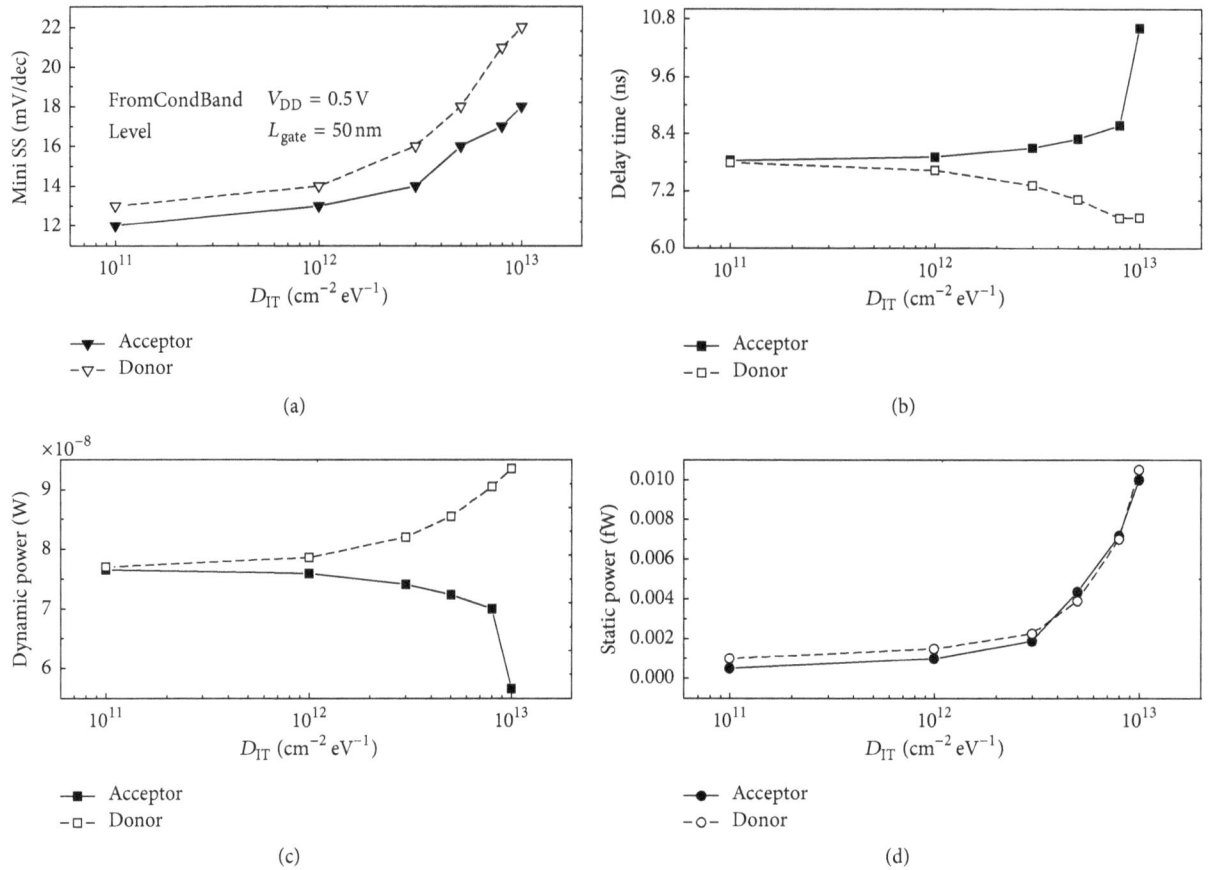

FIGURE 12: (a) The calculated mini SS, (b) delay time, (c) dynamic power, and (d) static power versus ITs density at valence band. Delay time τ is given by $C_L V_{DD}/I_{on}$. The acceptor-type ITs can reduce the dynamic power and the mini SS is more immune to acceptor-type ITs.

10(d), 12(a), 12(d), 14(a), and 14(d). The static power and mini SS would be increased no matter where the ITs level is. Divergent trends in drain current can be seen in Figure 11(a). When traps level is located at the conduction band, drain current would be reduced with increasing acceptor-type D_{IT}. However, drain current increases with increasing donor-type D_{IT}. The simulation results show that the electrons accelerated due to greater tunneling electric field, which was induced through impact ionization. The traps will capture or lose electrons and then weaken or enhance the c/s tunneling junction electric field. The drain current shifts right with increasing the acceptor ITs density. The electrical intensity gradually becomes weak, and then the tunneling carriers decrease. Under the same gate voltage, the tunneling width would not change, so the subthreshold swing would not change obviously. Donor-type ITs inside conduction band can reduce V_{th}. Delay time, dynamic power, and static power have the same changing trend (Figures 12(b), 12(c), and 12(d)).

3.2. The Impact of ITs on Miller Capacitance of DG-TFET. It may be indicated in TFETs that high-κ gate insulator would result in higher fringe capacitance due to the enhanced Miller effects. For the TFET, the gate capacitance is completely controlled by the gate-to-drain capacitance (C_{gd}), C_{gd} makes

up a majority of gate capacitance (C_{gg}) [18–20]. For high-κ gate insulator, traps may exist in Si/high-κ dielectric material interface or high-κ dielectric material. In this case, interface traps affect not only tunneling junction electric field but also capacitive characteristics. Next, in order to obtain further insight, we investigate the impact of ITs density (N_{it}), traps type, and traps level on capacitance characteristics of DG-TFET (see Figures 15–17).

This analysis assumes that all trap capture cross sections are 1×10^{-14} cm^{-2}. Small-signal AC analysis is used to analyze the Miller capacitive characteristics (C_{gd}) of DG-TFET, and the scanning frequency is 100 MHz.

Figure 15 shows the simulated C_{gd}-V_{GS} curves with the acceptor-like ITs. Traps are distributed at the energy levels 0.4 eV and 0.6 eV above/below the Si midgap and the Si midgap. When V_{GS} scans to −0.5 V, electrically neutral acceptor-type ITs are in a releasable state and can capture electrons. ITs can contribute to distribution capacitance. The contribution is proportional to ITs density, as shown in Figures 15(b) and 15(c). Later, surface of channel is in strong inversion state and AC small-signal frequency is very high, which results in time not enough for acceptor-type traps to capture electrons. In this case, the traps reduce the contribution of capacitance value. When traps level is located at the Si midgap, Figure 15(c) shows that gate voltage moves

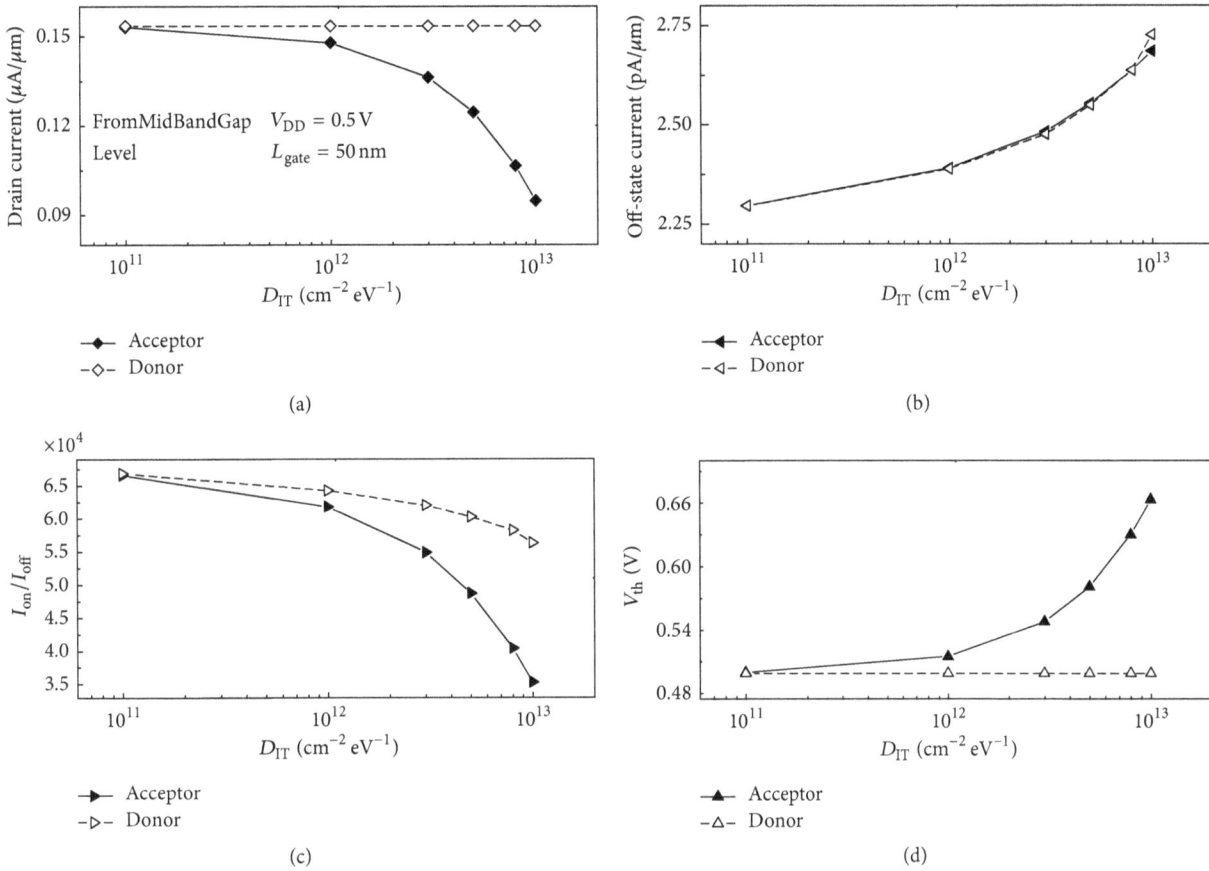

FIGURE 13: (a) Drain current I_{on} at $V_{GS} = V_{DS} = 0.5$ V, (b) off-state I_{off} at $V_{GS} = 0$ V, $V_{DS} = V_{DD} = 0.5$ V, (c) calculated I_{on}/I_{off}, and (d) V_{th} versus ITs density at midgap band. The acceptor-type ITs deteriorates the on-state current which is the same effect at E_c and E_v.

left corresponding to the maximum capacitance contribution value. It is, however, necessary to note that, in Figure 15(c), the maximum capacitance contribution value is also down when traps distribute from E_c to midgap. In addition, the change trend is obvious when $N_{it} = 1 \times 10^{13}$ cm^{-2} eV^{-1}. Gate voltage changes from -1 V to 1 V, and the Fermi level moves from E_v to E_c. Because the Fermi level is below Si midgap, the acceptor-type traps will not capture electrons, which results in having no effect on C_{gd}.

When the Fermi level reaches the Si midgap, the acceptor-type traps begin to capture electrons and make a significant contribution to C_{gd}. With the raising of Fermi level, it enhances capacitance contribution. The position of the Fermi level moves down and improvement of the surface potential is due to negatively charged acceptor-type traps, which results in reduction of capacitance contribution.

The peak point shift of distribution capacitance between donor- and acceptor-type trap is different. The formation energies (E_{form}) $E_{form} = E_0 - qE_F$, and q is the charged defects of charge. Capturing or releasing electrons can result in positive and negative E_{form}, so the Fermi energy can reach firstly the formation energies of the donor-like trap. At the same time, $N_{it} \propto \exp(-E_{form}/kt)$. δE_{form} increase with increasing N_{it}, and the greater the density, the greater the contribution to distribution capacitance. When V_{GS} is less

than 0.5 V, distribution capacitance attains its peak value for higher density.

For the acceptor-like trap, the more the negative E_{form} is, the later it reaches the maximum distribution capacitor. It is the same for the donor-like trap; the greater the density, the greater the contribution of acceptor-like to distribution capacitance.

Figure 15(e) shows an extreme case where traps level is located at energy 0.6 eV below the Si midgap, as shown in Figure 15(d). ITs level has been completely shifted in E_v, which means that Fermi level is always higher than ITs level. Traps can be fully filled by electrons, and then the flat voltage (V_{FB}) will turn right, which indicates that ITs of Si/HfO$_2$ have the same effect with the fixed charges. On the other hand, C_{gd} shift right with increasing traps concentration.

For oxide bulk trap, there are usually a lot of positive fixed charge hydrogen ions (H$^+$) in insulation, and C-V curve moves in the direction of the negative axis. In contrast to C_{gd} curves in Figure 15(e), the acceptor-like trap has the same effect as the negative interface fixed charge trap, and C-V curve moves to the opposite direction. The final effect is the flat band shift. The only difference is the drift direction.

The plots of gate-drain capacitance as a function of V_{GS} for five different level positions of donor-type ITs are shown in Figure 16. Donor-type ITs energy levels are occupied totally

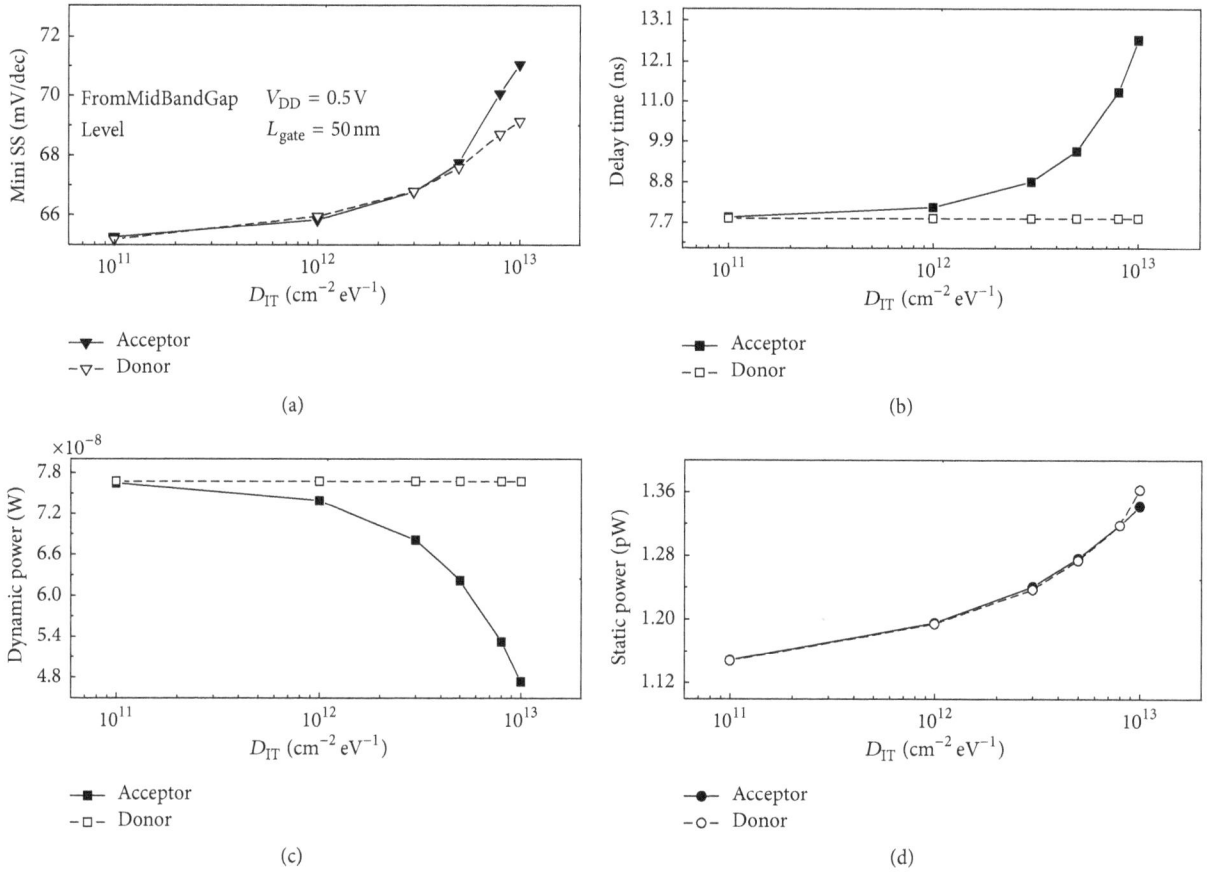

FIGURE 14: (a) The calculated mini SS, (b) delay time, (c) dynamic power, and (d) static power versus ITs density at valence band. Delay time τ is given by $C_L V_{DD}/I_{on}$. The acceptor and acceptor-type ITs are essentially the same effect at midgap band and valence band.

by electrons, so that ITs are electrically neutral. After liberating electrons, the ITs are positive. Figure 16(a) shows that the ITs levels are distributed at the energy level 0.6 eV above the midgap; the Fermi level is under trap level. The result indicates that ITs exert an influence on Miller capacitance. When the gate voltage V_{GS} changes from -1.0 V to 0.2 V, then the Fermi level keeps rising relative to traps level. The influence of traps level on C_{gd} would be shifted left with lowering of the trap level position, as shown in Figures 16(a), 16(b), and 16(c). V_{GS} reaches to -0.6 V, and the Fermi level is near the trap level. The donor-type ITs begin to exchange electrons with channel in Figure 16(d). When the traps level is distributed at the energy level 0.3 eV under the midgap, it can be seen clearly in Figure 16(e) that C_{gd} is hardly affected. C_{gd} fluctuated by donor-type ITs is smaller than acceptor-type ITs, which implies that DG-TFET is more immune to donor-type ITs. Besides, it is found that the peak position shifts left for donor-type ITs and shifts right for acceptor-type ITs.

It is worth noticing that the impact of the different energy distribution of charged traps on Miller capacitance is also necessary to be studied. We assume that the peak concentration of interface traps (donor-type and acceptor-type) is 5×10^{12} cm^{-2} eV^{-1}, and four types of energetic distribution (level, uniform, exponential, and Gaussian) are located at $E_i + 0.4$ eV, E_i, and $E_i - 0.4$ eV, respectively.

High Gaussian distributions ($E_\sigma 0.2$) are adopted, as shown in Figure 17 and Figure 18. First, it was found that the signal level of acceptor or donor traps has the most effect on the C-V curve in Figures 17(a) and 18(a). The shape of ITs energy density distribution has a great influence on capacitance contribution. The smoother the curve is, the smaller the capacitance contribution value is. Due to variations in the positions of traps level and the Fermi level, the electron occupation rate of ITs is different. The greater the occupation chance of ITs is, the more obvious the capacitance effect is. For the uniform, exponential, and Gaussian distribution of ITs, the capacitance effects are almost alike, as shown in Figures 17(b) and 18(b). However, the ITs level is located at the energy level 0.4 eV under the midgap, and the impact of the exponential and Gaussian distribution of ITs is obviously different, as shown clearly in Figure 17(c) and Figure 18(c). In addition, it is clearly shown in Figures 17 and 18 that the effect of acceptor-type ITs on C_{gd} is still more obvious than that of donor-type ITs:

$$C_{ITs} \propto D_{IT} f'_{it}, \tag{1}$$

where C_{ITs} is ITs capacitance contribution and D_{IT} and f'_{it} are ITs density and derivative of occupation rate of ITs,

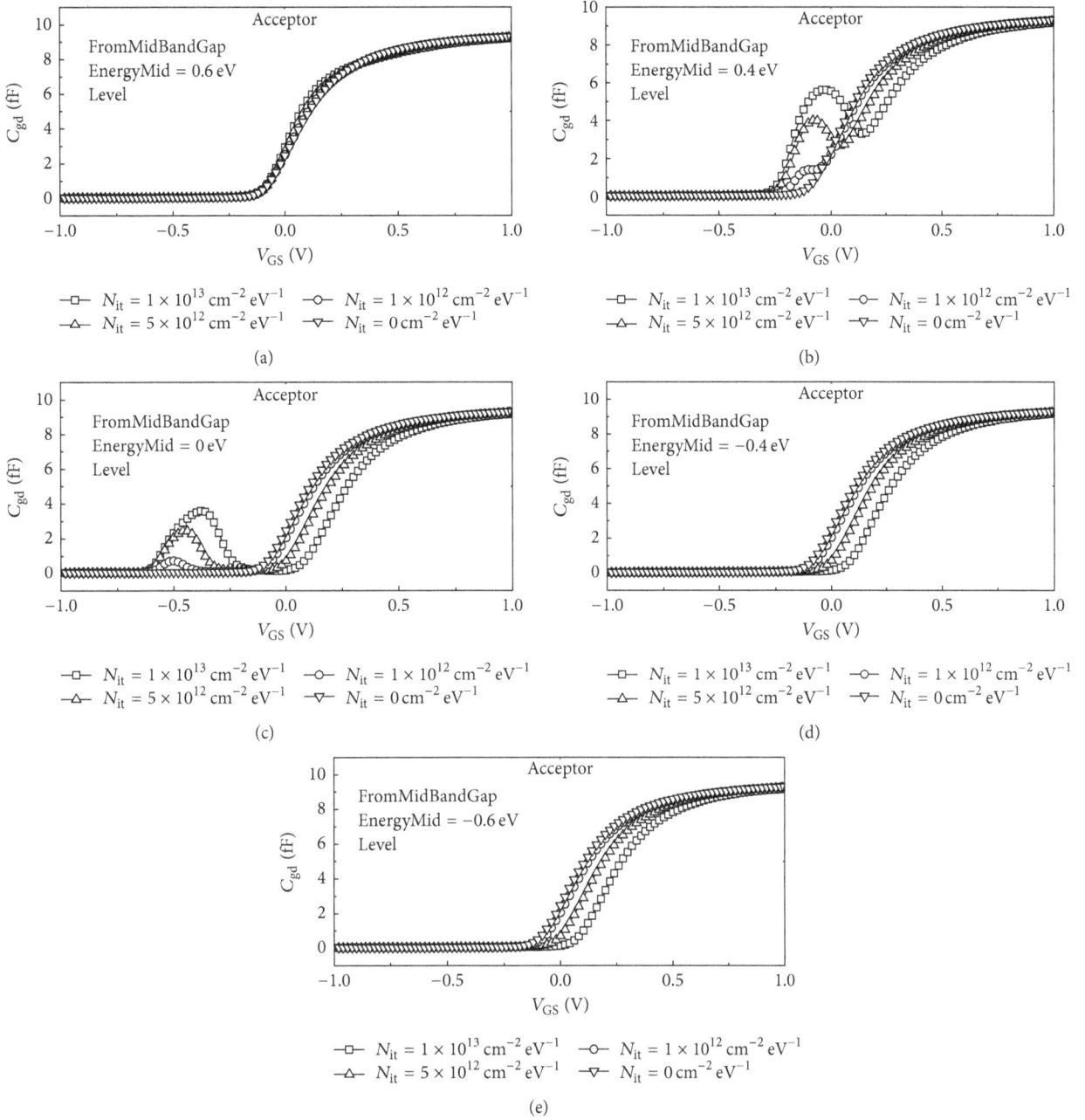

Figure 15: Five C-V curves of a single level ITs are plotted. Miller capacitance C_{gd} of DG-TFET for acceptor-type ITs is studied at $V_{DS} = 0.5\,V$ with variable ITs density N_{it}. (a) The trap is distributed at the energy level 0.6 eV above the midgap, (b) at the energy level 0.4 eV above the midgap, (c) at the midgap, (d) at the energy level 0.4 eV under the midgap, and (e) at the energy level 0.6 eV under the midgap.

respectively. The electron occupancy probability of donor-type or acceptor-type ITs can be expressed as

$$f_{it}(E) = \frac{1}{1 + (1/g)\exp\left((E_{it} - E_F)/kT\right)}, \quad (2)$$

where E_{it} is ITs energy; g is a degeneracy factor; k is Boltzmann's constant; and T is temperature. As mentioned

above, the derivative of electron occupation of ITs can be given as follows:

$$f_{it}' = \frac{1}{1 + (1/g)\exp\left((E_{it} - E_F)/kT\right)^2}\frac{1}{g} \cdot \exp\left(\frac{E_{it} - E_F}{kT}\right)E_{it}'. \quad (3)$$

According to formula (3), it can be found that ITs contribute a lot to C_{gd} for fixed relative positions between ITs level and the Fermi level, where E_{it}' is relatively large.

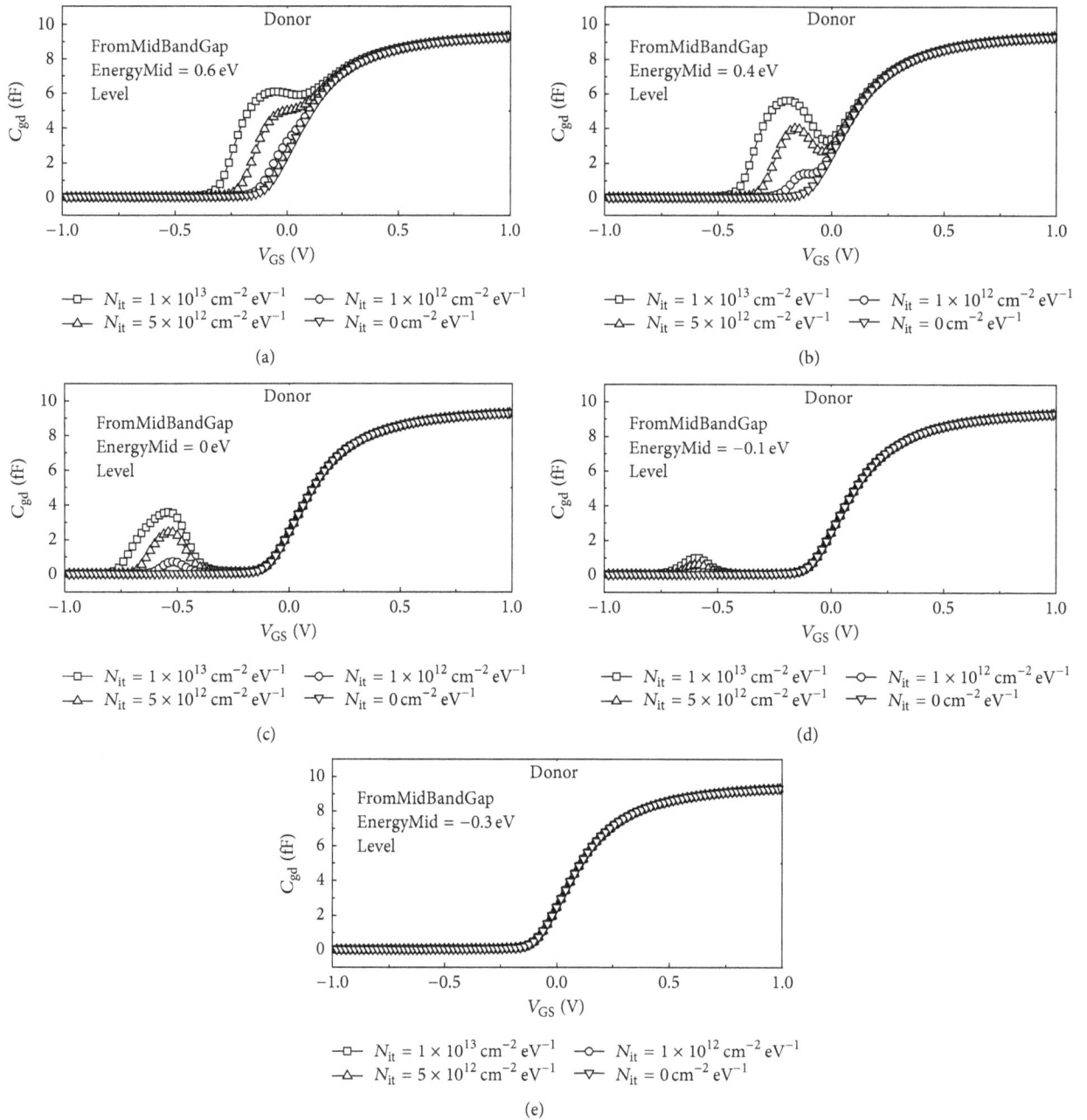

FIGURE 16: Miller capacitance C_{gd} of DG-TFET for donor-type ITs is studied at V_{DS} = 0.5 V with variable ITs density N_{it}. (a) The trap is distributed at the energy level 0.6 eV above the midgap, (b) at the energy level 0.4 eV above the midgap, (c) at the midgap, (d) at the energy level 0.4 eV under the midgap, and (e) at the energy level 0.6 eV under the midgap.

4. Conclusion

The impact of donor-type and acceptor-type ITs density with different levels and distributions on DC and AC characteristics has been investigated. Peak position of traps is located between E_v and E_c which results in degradation of I_{on}/I_{off} ratio. In particular, the attenuation of tunneling current is fierce when the ITs are distributed at the Si midgap. The donor-type ITs are with the valence band and the Si midgap, which would not affect the drain current, the threshold voltage, delay time, and dynamic power. However,

the donor-type ITs and acceptor-type ITs in the conduction band exhibited an opposite trend, and the donor-type ITs have contributed to the drain current. In addition, the impacts of the different types and energy level positions of ITs on the C-V characteristics are qualitatively investigated. A single energy distribution has the most impact on Miller capacitance. For ITs level that is below the Fermi level, ITs have a very small impact on C-V curve, but the exponential and Gaussian distribution of trap now start playing a role in determining the C-V characteristics at V_{GS} = −0.5 V.

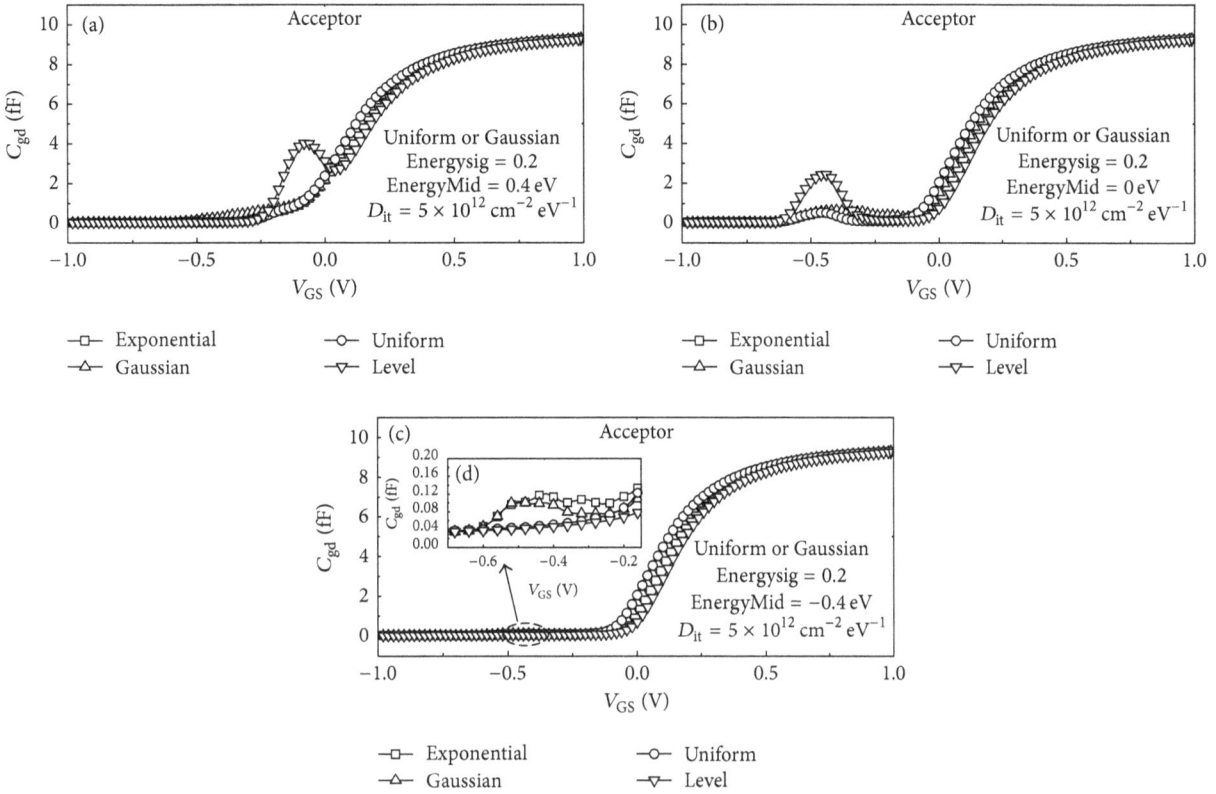

FIGURE 17: Four types of energetic distribution for acceptor-type are located at three representative level positions which are (a) EnergyMid = 0.4 eV (E_i + 0.4 eV), (b) EnergyMid = 0.0 eV, and (c) EnergyMid = 0.4 eV, respectively.

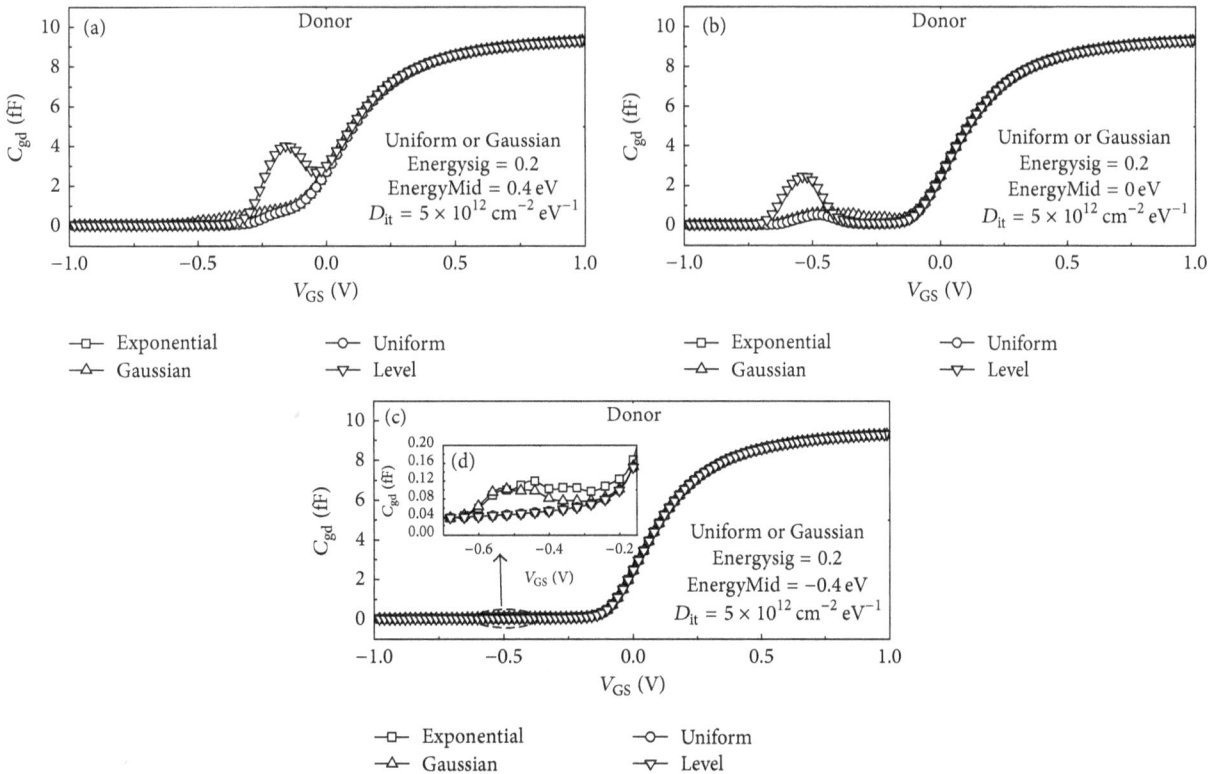

FIGURE 18: Four types of energetic distribution for donor-type are located at three representative level positions which are (a) EnergyMid = 0.4 eV (E_i + 0.4 eV), (b) EnergyMid = 0.0 eV, and (c) EnergyMid = 0.4 eV, respectively.

Conflict of Interests

The authors declare that there is no conflict of interests regarding the publication of this paper.

Acknowledgment

The work at Xidian University has been supported by National Natural Science Foundation of China (Award nos. 61574109 and 61204092).

References

[1] A. M. Ionescu and H. Riel, "Tunnel field-effect transistors as energy-efficient electronic switches," *Nature*, vol. 479, no. 7373, pp. 329–337, 2011.

[2] E. O. Kane, "Theory of tunneling," *Journal of Applied Physics*, vol. 31, no. 1, pp. 83–89, 1961.

[3] A. Mallik and A. Chattopadhyay, "Tunnel field-effect transistors for analog/mixed-signal system-on-chip applications," *IEEE Transactions on Electron Devices*, vol. 59, no. 4, pp. 888–894, 2012.

[4] C. Li, Y. Zhuang, and R. Han, "Cylindrical surrounding-gate MOSFETs with electrically induced source/drain extension," *Microelectronics Journal*, vol. 42, no. 2, pp. 341–346, 2011.

[5] C. Li, Y. Zhuang, R. Han, and G. Jin, "Subthreshold behavior models for short-channel junctionless tri-material cylindrical surrounding-gate MOSFET," *Microelectronics Reliability*, vol. 54, no. 6-7, pp. 1274–1281, 2014.

[6] T. S. A. Samuel, N. B. Balamurugan, S. Bhuvaneswari, D. Sharmila, and K. Padmapriya, "Analytical modelling and simulation of single-gate SOI TFET for low-power applications," *International Journal of Electronics*, vol. 101, no. 6, pp. 779–788, 2014.

[7] E.-H. Toh, G. H. Wang, G. Samudra, and Y.-C. Yeo, "Device physics and design of double-gate tunneling field-effect transistor by silicon film thickness optimization," *Applied Physics Letters*, vol. 90, no. 26, Article ID 263507, 2007.

[8] P. Wang, Y. Zhuang, C. Li, Y. Li, and Z. Jiang, "Subthreshold behavior models for nanoscale junctionless double-gate MOSFETs with dual-material gate stack," *Japanese Journal of Applied Physics*, vol. 53, no. 8, Article ID 084201, 7 pages, 2014.

[9] L. Shi, Y. Zhuang, C. Li, and D. Li, "Analytical modeling of the direct tunneling current through high-k gate stacks for long-channel cylindrical surrounding-gate MOSFETs," *Journal of Semiconductors*, vol. 35, no. 3, Article ID 034009, 2014.

[10] A. Chattopadhyay and A. Mallik, "The impact of a high-κ gate dielectric on a p-channel tunnel field-effect transistor," in *16th International Workshop on Physics of Semiconductor Devices*, vol. 8549 of *Proceedings of SPIE*, p. 5, The International Society for Optical Engineering, Kanpur, India, 2011.

[11] G. B. Beneventi, E. Gnani, A. Gnudi, S. Reggiani, and G. Baccarani, "Can interface traps suppress TFET ambipolarity?" *IEEE Electron Device Letters*, vol. 34, no. 12, pp. 1557–1559, 2013.

[12] Y. Qiu, R. Wang, Q. Huang, and R. Huang, "A comparative study on the impacts of interface traps on tunneling FET and MOSFET," *IEEE Transactions on Electron Devices*, vol. 61, no. 5, pp. 1284–1291, 2014.

[13] X. Y. Huang, G. F. Jiao, W. Cao et al., "Effect of interface traps and oxide charge on drain current degradation in tunneling field-effect transistors," *IEEE Electron Device Letters*, vol. 31, no. 8, pp. 779–781, 2010.

[14] M. G. Pala, D. Esseni, and F. Conzatti, "Impact of interface traps on the IV curves of InAs tunnel-FETs and MOSFETs: a full quantum study," in *Proceedings of the IEEE Electron Devices Meeting (IEDM '12)*, pp. 135–138, December 2012.

[15] S. Hanson, B. Zhai, K. Bernstein et al., "Ultralow-voltage, minimum-energy CMOS," *IBM Journal of Research and Development*, vol. 50, no. 4-5, pp. 469–490, 2006.

[16] *TCAD Sentaurus Device Manual*, Synopsys, Mountain View, Calif, USA, 2012.

[17] K. Boucart and A. M. Ionescu, "A new definition of threshold voltage in Tunnel FETs," *Solid-State Electronics*, vol. 52, no. 9, pp. 1318–1323, 2008.

[18] B. Laikhtman and E. L. Wolf, "Tunneling time and effective capacitance for single electron tunneling," *Physics Letters A*, vol. 139, no. 5-6, pp. 257–260, 1989.

[19] J. Boehmer, J. Schumann, and H. Eckel, "Effect of the miller-capacitance during switching transients of IGBT and MOSFET," in *Proceedings of the 15th International Power Electronics and Motion Control Conference (EPE/PEMC '12)*, pp. LS6d.3-1–LS6d.3-5, IEEE, Novi Sad, Serbia, September 2012.

[20] Y. Yang, X. Tong, L.-T. Yang, P.-F. Guo, L. Fan, and Y.-C. Yeo, "Tunneling field-effect transistor: capacitance components and modeling," *IEEE Electron Device Letters*, vol. 31, no. 7, pp. 752–754, 2010.

12

Circuit Implementation, Synchronization of Multistability, and Image Encryption of a Four-Wing Memristive Chaotic System

Guangya Peng ⓘ, **Fuhong Min** ⓘ, **and Enrong Wang**

School of Electrical and Automatic Engineering, Nanjing Normal University, Nanjing 210042, China

Correspondence should be addressed to Fuhong Min; minfuhong@njnu.edu.cn

Academic Editor: Ephraim Suhir

The four-wing memristive chaotic system used in synchronization is applied to secure communication which can increase the difficulty of deciphering effectively and enhance the security of information. In this paper, a novel four-wing memristive chaotic system with an active cubic flux-controlled memristor is proposed based on a Lorenz-like circuit. Dynamical behaviors of the memristive system are illustrated in terms of Lyapunov exponents, bifurcation diagrams, coexistence Poincaré maps, coexistence phase diagrams, and attraction basins. Besides, the modular equivalent circuit of four-wing memristive system is designed and the corresponding results are observed to verify its accuracy and rationality. A nonlinear synchronization controller with exponential function is devised to realize synchronization of the coexistence of multiple attractors, and the synchronization control scheme is applied to image encryption to improve secret key space. More interestingly, considering different influence of multistability on encryption, the appropriate key is achieved to enhance the antideciphering ability.

1. Introduction

The memristor has been thought to be the fourth basic element, and the concept was proposed by Professor Chua in 1971 [1]. The first memristor physical model was implemented by HP Labs in 2008 [2], which has lots of advantages [3], such as low energy consumption, small size, and high integration, so that it has been investigated in memory storage [4, 5], artificial neural networks [6, 7], chaotic circuits [8, 9], secure communications [10–12], and so forth.

Nowadays, an in-depth study of connecting the memristors with chaotic systems has been conducted by many scholars. Some researchers added the memristor model to Chua's circuit [13], and others combined the Lorenz system with memristors [14]. In recent years, a lot of attention has been paid to the research on the memristive chaotic circuits applied to synchronization, and the synchronization scheme of memristive chaos based on Lorenz system has been developed initially [15, 16]. Master system and slave system can achieve chaotic synchronization under controlled constraints, and it has been widely used in secure communication

and other fields [17, 18]. However, a lot of literature with respect to the memristive chaotic system applied to secure communication only extracted its own chaotic sequence, without involving the memristive synchronization scheme [12, 19, 20]. In addition, the multiwings [21, 22] and multistability [23, 24] of memristive chaotic systems have been a hot topic. Nevertheless, there were few studies on the multiwing memristive circuit based on Lorenz system, and less literature took the fact that the multistability may produce different encryption effects into consideration. In contrast with the aforementioned literature, the multiwing memristive circuit, the synchronization control used in image encryption, and different influences of multistability will be carried out in this work.

In this paper, a novel four-wing memristive chaotic system is proposed by introducing a smooth flux-controlled memristor to Lorenz-like system. First of all, theoretical analyses are investigated by means of calculating dissipation and equilibrium point set. Secondly, the conventional dynamic analyses are carried out by Lyapunov exponents, bifurcation diagrams, Poincaré maps, phase diagrams, and so forth. The circuit of new system is devised by utilizing resistors, capacitors, op amps, and multipliers, and the results of circuit simulation are consistent well with those of numerical simulations. Then, a nonlinear feedback controller for synchronization of multistability is designed according to the system equations. The error curves and timing diagrams are used to illustrate the synchronization of different coexistence states indicating that the synchronization effect is effective. Finally, the synchronization scheme is applied to image encryption, so that the key space is more complicated. Taking the influence of different synchronization states on encryption into account, it is beneficial to select suitable keys and improve the antideciphering ability.

2. Four-Wing Memristive Chaotic System

Memristor is a basic circuit element describing the constitutive relation between charge and flux as $d\varphi(q) = M(q)dq$ or $dq(\varphi) = W(\varphi)d\varphi$, where $q, \varphi, M(q), W(\varphi)$ are charge, flux, memristance, and memductance [1]. A cubic flux-controlled memristor [25] described as $q(\varphi) = a\varphi + b\varphi^3$, $W(\varphi) = dq(\varphi)/d\varphi$ will be carried out in this paper, where a and b are memristor parameters.

A new four-dimensional four-wing memristive chaotic system is obtained by adding a memristor with the constitutive relation of $W(\varphi) = dq(\varphi)/d\varphi$ to a Lorenz-like system deformed by classical Lorenz system [26]. The mathematical model can be described as

$$\begin{aligned}
\dot{x}_1 &= 36y_1z_1 - \alpha x_1 - z_1 W(w_1), \\
\dot{y}_1 &= \xi y_1 - \beta x_1 z_1, \\
\dot{z}_1 &= 8x_1y_1 - \gamma z_1, \\
\dot{w}_1 &= x_1,
\end{aligned} \tag{1}$$

where $w_1 = \varphi$ is magnetic flux; besides, $W(w_1) = a + bw_1{}^2$ is memductance. Both a and b are memristor parameters determining its characteristics, and $\alpha, \beta, \xi, \gamma$ are system parameters which determine the motion states. All the parameters are chosen as $a = -0.5$, $b = 2.4$, $\alpha = 15$, $\beta = 8$, $\xi = 1.68$, and $\gamma = 15.15$. When initial conditions are set as $(0.1, 0.1, 0, 0)$, the system in (1) produces four-wing chaotic attractors, as shown in Figure 1. The four Lyapunov exponents (LE_i) of the system in (1) are obtained by singular value decomposition as $LE_1 = 0.1354$, $LE_2 = 0.0076$, $LE_3 = -0.2936$, and $LE_4 = -25.6091$. It is clear that $LE_1 > 0$, $LE_2 \approx 0$, $LE_3 < 0$, and $LE_4 < 0$ indicating that the attractor is a strange attractor with chaotic characteristics. The Lyapunov dimension [27] of the new 4D four-wing memristive system is

$$\begin{aligned}
D_L &= j + \frac{1}{\left|LE_{j+1}\right|}\sum_{i=1}^{j}LE_i = 2 + \frac{(LE_1 + LE_2)}{|LE_3|} \\
&= 2 + \frac{(0.1354 + 0.0076)}{|-0.2936|} = 2.487.
\end{aligned} \tag{2}$$

It can be seen that the Lyapunov dimension is fractional which means it is chaotic system. The phase diagrams listed

in Figure 1 show that the chaotic attractor trajectory has ergodicity and boundedness in a specific attraction domain.

Equation (5) shows the dissipation of the system in (4):

$$\nabla V = \frac{\partial \dot{x}_1}{\partial x_1} + \frac{\partial \dot{y}_1}{\partial y_1} + \frac{\partial \dot{z}_1}{\partial z_1} + \frac{\partial \dot{w}_1}{\partial w_1} = -\alpha + \xi - \gamma \tag{3}$$

when these parameters involved in (3) are selected as $\alpha = 15$, $\xi = 1.68$, and $\gamma = 15.15$; that is, $\nabla V = -\alpha + \xi - \gamma = -15 + 1.68 - 15.15 = -28.47 < 0$. It means that system in (1) is dissipative, and the system trajectories starting from the region converge exponentially with time and form chaos asymptotically.

Let $\dot{x}_1 = \dot{y}_1 = \dot{z}_1 = \dot{w}_1 = 0$; the equilibrium points in (1) are obtained as a set $A = \{(x_1, y_1, z_1, w_1)|x_1 = y_1 = z_1 = 0, w_1 = l_1\}$, where l_1 is a real constant; thus all points on the w_1-axis are equilibrium points. The Jacobi matrix J_A of the new four-wing memristive chaotic system at the equilibrium point set is got:

$$J_A = \begin{bmatrix} -\alpha & 36z_1 & 36y_1 - W(w_1) & -z_1(2bw_1) \\ -\beta z_1 & \xi & -\beta x_1 & 0 \\ 8y_1 & 8x_1 & -\gamma & 0 \\ 1 & 0 & 0 & 0 \end{bmatrix}. \tag{4}$$

The eigenvalue equation of the equilibrium point set A is

$$\begin{aligned}
&\lambda(\lambda + \alpha)(\lambda - \xi)(\lambda + \gamma) \\
&= \lambda\left[\lambda^3 + (\alpha + \gamma - \xi)\lambda^2 + (\alpha\gamma - \alpha\xi - \gamma\xi)\lambda - \alpha\gamma\xi\right] \tag{5} \\
&= 0.
\end{aligned}$$

Simplify (5) as

$$\lambda\left[\lambda^3 + a_1\lambda^2 + a_2\lambda + a_3\right] = 0. \tag{6}$$

Equation (6) reveals that system in (1) has one zero eigenvalue and three nonzero eigenvalues, where

$$\begin{aligned}
a_1 &= \alpha + \gamma - \xi = 28.47 > 0, \\
a_2 &= \alpha\gamma - \alpha\xi - \gamma\xi = 176.598 > 0, \tag{7} \\
a_3 &= -\alpha\gamma\xi = -381.78 < 0.
\end{aligned}$$

According to the stability condition, the set of the unstable range of system in (1) can be obtained as the whole real field due to $a_3 < 0$. No matter the value of l_1, the eigenvalues of equilibrium set are calculated as $\lambda_1 = 0$, $\lambda_2 = -15$, $\lambda_3 = 1.68$, $\lambda_4 = -15.15$. λ_2 and λ_4 correspond to two stable solutions, and λ_3 is consistent with an unstable solution. Therefore, all the points on equilibrium point set are unstable. Self-excited attractors are connected with unstable equilibriums, but a hidden attractor is not connected with equilibria, for instance, those in the systems without equilibria or only one stable equilibrium [28, 29]. It is clear that the attractors in (1) are closely related to unstable equilibriums, so those are self-excited attractors.

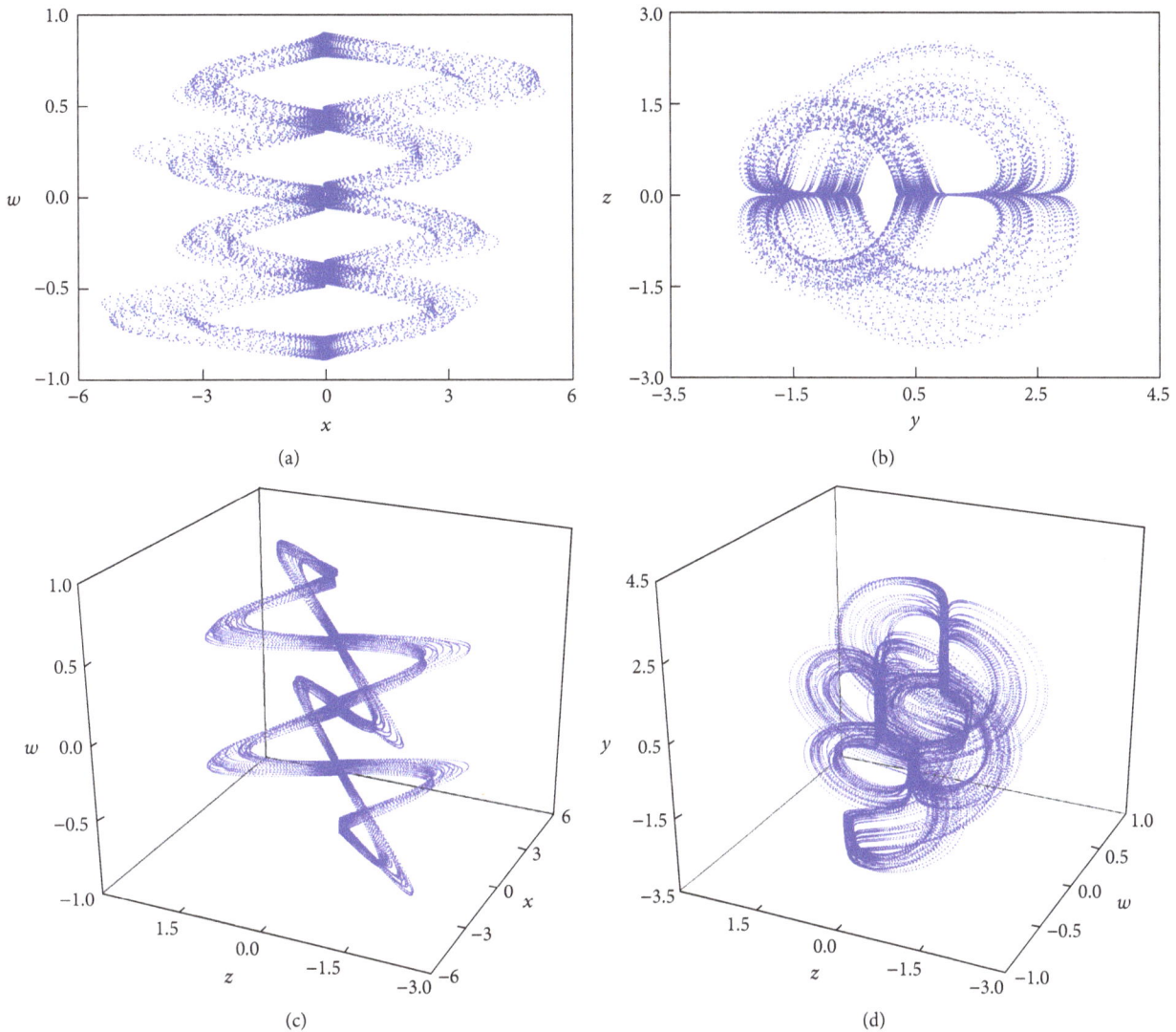

FIGURE 1: Phase diagrams: (a) x-w, (b) y-z, (c) x-z-w, and (d) w-z-y.

3. Dynamic Analyses and Circuit Implementation of the New Chaotic System

3.1. Dynamic Analyses

3.1.1. The Effect of System Parameter Changing. The chaotic system will present different dynamic behaviors with varying system parameters. The bifurcation diagram for the state variable $y(t)$ and the Lyapunov exponents are presented in Figure 2 with the change of the system parameter ξ. The initial values $(0.1, 0.1, 0, 0)$ in (1) are taken as initial conditions, and the range of ξ is $0 < \xi < 4.8$. The dynamics of system in (1) can be intuitively observed from Figure 2. It is clear that this system is periodic with ξ varying in interval of $(1.43, 1.46)$, $(1.77, 3.41)$, and $(3.50, 3.95)$. The partial bifurcation diagram in Figure 2(b) shows that it is period-3 motion when $\xi \in (1.43, 1.46)$ and generates saddle bifurcation (SNB) at $\xi = 1.46$. The phase diagram of $\xi = 1.44$ is shown as Figure 3(a). With changing the parameter ξ in

$(0, 1.42)$, $(1.47, 1.76)$, $(3.42, 3.49)$, and $(3.96, 4.65)$, the four-wing memristive system is chaotic and the corresponding largest Lyapunov exponent is positive. Figure 3(b) is the phase diagram of $\xi = 1.68$, and this system diverges in $\xi \in (4.66, 4.8)$.

3.1.2. The Effect of Initial Condition Changing. The memristor-based chaotic system is extremely sensitive to initial values, and its dynamic characteristics directly transform with the variations of initial conditions. Fix the circuit parameters as $a = -0.5$, $b = 2.4$, $\alpha = 15$, $\beta = 8$, $\xi = 1.68$, and $\gamma = 15.15$. The initial values $(0.1, 0.1, 0, w_{1o})$ are selected as initial conditions, where w_{1o} is initial value of the state variable w_1. The Lyapunov exponent spectrum of system in (1) varying with initial condition $l_1 = w_{1o}$ is shown in Figure 4(a), and the corresponding bifurcation diagram for $y(t)$ is illustrated as Figure 4(b) with $l_1 \in [-5, 5]$.

The multistability refers to a variety of stable states generating with changing initial values under the same system

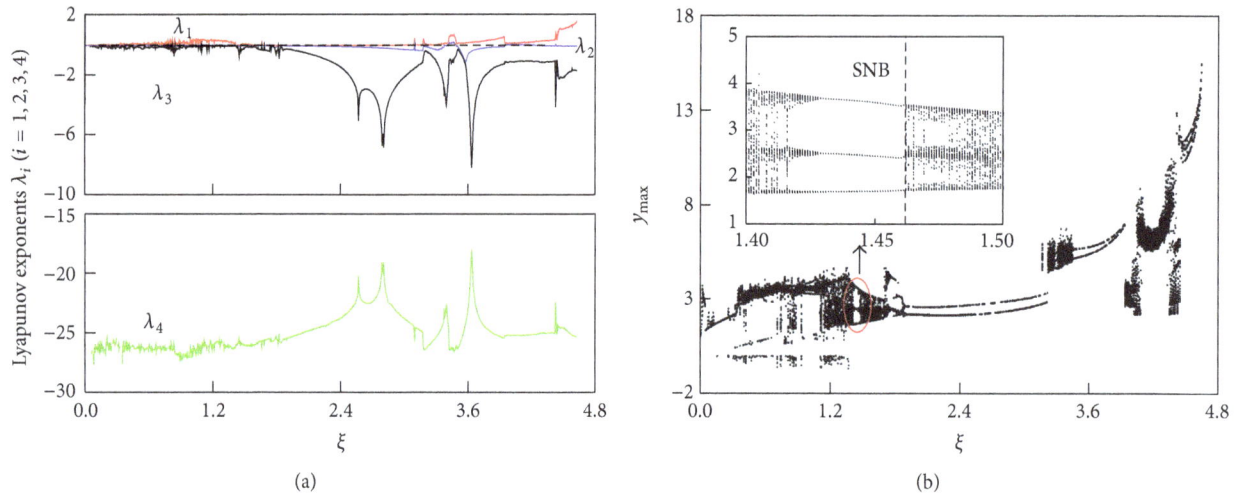

FIGURE 2: Lyapunov exponents and bifurcation diagram with parameter ξ changing (a) $l_1 = 0$, Lyapunov exponents, and (b) $l_1 = 0$, bifurcation diagram.

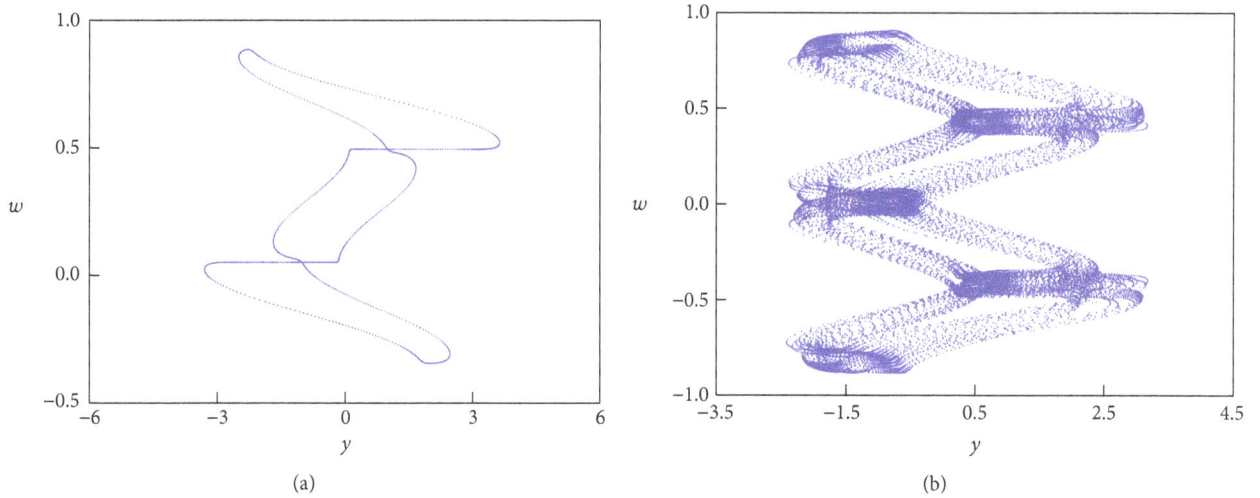

FIGURE 3: Phase diagrams of different system parameters ξ: (a) $\xi = 1.44$ and (b) $\xi = 1.68$.

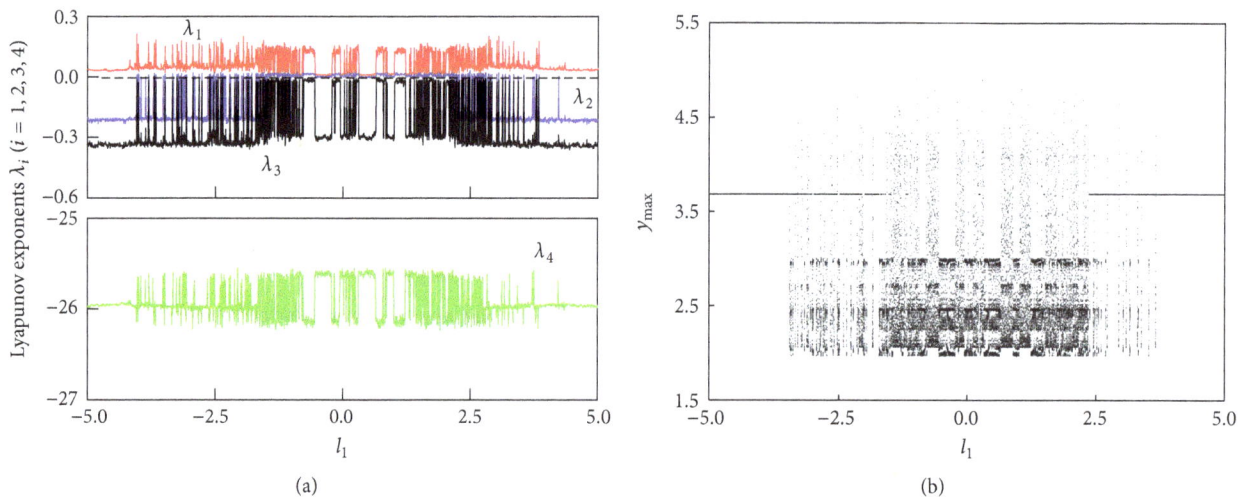

FIGURE 4: Lyapunov exponents and bifurcation diagram with parameter l_1 changing (a) $\xi = 1.68$, Lyapunov exponents, and (b) $\xi = 1.68$, bifurcation diagram.

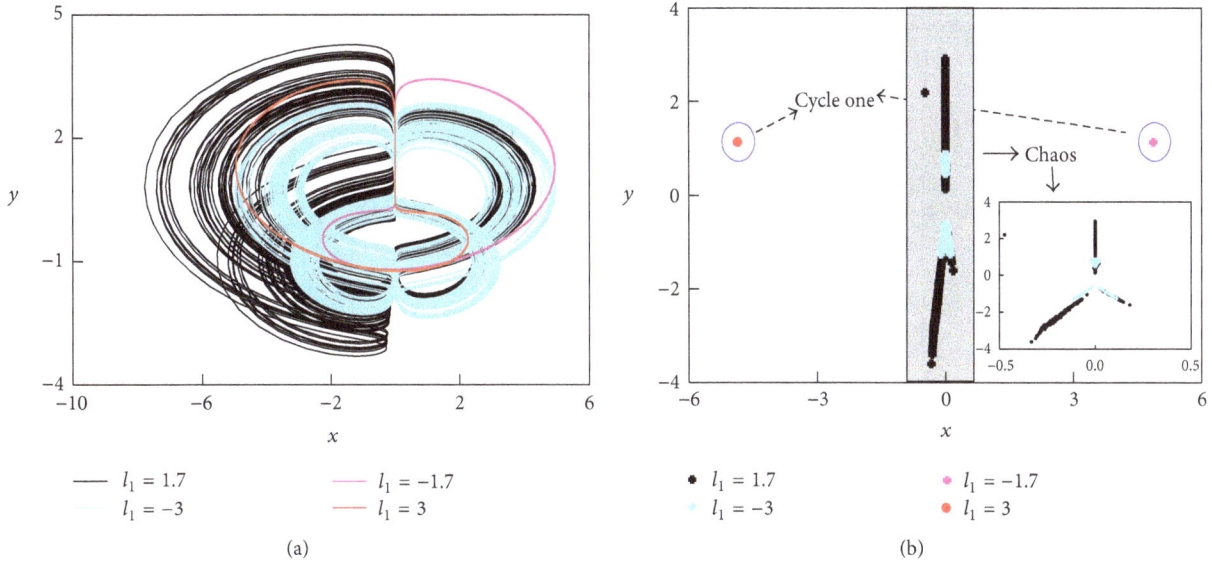

FIGURE 5: Coexistence of initial value l_1 changing (a) coexistence phase diagram and (b) coexistence of Poincaré section.

parameters. Figure 4 reveals that the chaotic system in (1) has a strong sensitivity to the transformation of l_1. The dynamic behaviors of system in (1) appear as the switch of periodic and chaotic motions with the variable l_1 varying. To observe the multistability clearly, select partial values for analyses randomly. Figure 5(a) shows the coexistence of four phase diagrams with different initial conditions. The black curves correspond to asymmetric double scrolls for $l_1 = 1.7$, and the light blue curves are associated with a butterfly attractor for $l_1 = -3$. In addition, the pink and red curves are relative to period-1 with $l_1 = -1.7$ and $l_1 = 3$, respectively. Both periodic states are symmetrical about the x-axis. The corresponding coexistence of Poincaré sections is depicted in Figure 5(b). The gray domain means the chaotic region as $l_1 = 1.7$ (black points) and $l_1 = -3$ (light blue points). The local gray area ($x \in [-0.5, 0.5]$) is enlarged to observe the discrete mapping on Poincaré section clearly. The several successive curves can be seen from the partial magnification, proving that system in (1) is indeed in chaotic motion. Each of the two-blue ellipses in Figure 5(b) has a discrete point, corresponding to period-1 as $l_1 = 3$ (red) and $l_1 = -1.7$ (pink). Two discrete points are also symmetrical about the x-axis. Taking the length of the article into consideration, only four coexistence states are listed here.

3.1.3. Attraction Basins.

The attraction basin is the distribution of different states and it has been a hot topic in the research on multiple attractors. In [9, 30], the attraction basins were carried out with respect to initial conditions. Similarly, the domains of initial conditions are analyzed in which each attraction can be found. The system parameters in (1) are set as $a = -0.5$, $b = 2.4$, $\alpha = 15$, $\beta = 8$, $\xi = 1.68$, and $\gamma = 15.15$, and the initial values are assigned as $x_1(0) = 0.1$, $y_1(0) = 0.1$, $z_1(0) = p$, $w_1(0) = l_1$, where p and l_1 are the variables. The attraction basins with respect to p and l_1

are shown in Figure 6(a) which presented with the periods (purple) and chaotic attraction (red).

In addition, the basins of attractions are investigated with respect to initial condition l_1 and system parameter ξ. The parameters are set as $a = -0.5$, $b = 2.4$, $\alpha = 15$, $\beta = 8$, $\gamma = 15.15$ and initial values are given as $x_1(0) = 0.1$, $y_1(0) = 0.1$, $z_1(0) = 0$, $w_1(0) = l_1$. The initial condition l_1 and system parameter ξ are taken together to analyze the attraction basins shown in Figure 6(b), which is also associated with the periods (purple) and chaotic attraction (red). The boundaries are marked by yellow in Figure 6 and they conform with the analysis above; for example, the periodic motion (purple) as $l_1 = 3$ and $p = 0$ in Figure 6(a) corresponds to the red curves in Figure 5(a) and the chaotic motion (red) as $\xi = 1.68$ and $l_1 = 0$ observed in Figure 6(a) matches the phase diagram shown in Figure 3(b), which means that the phenomenon of multistability indeed coexists in the proposed four-wing memristive chaotic system.

3.2. Circuit Implementation.

The dynamic analysis of the four-wing memristive chaotic system shows its rich motion states. The circuit construction of system in (1) has further verified its complex dynamic behaviors by numerical simulations with the software MULTISIM.11. The modular design of the equivalent circuit is made up of basic components, such as resistors, capacitors, op amps, and multipliers, as shown in Figure 8. There are five channels: the first channel aims to achieve the function of memristor and the other four channels output x, y, z, and w, respectively. On the basis of Figure 8 and circuit principle, the circuit realization equation of four-wing memristive chaotic system in (1) is shown in (8):

$$\frac{dx}{dt} = -\frac{1}{10R_6C_1}zM - \frac{1}{10R_7C_1}(-yz) - \frac{1}{R_8C_1}x,$$

$$\frac{dy}{dt} = -\frac{1}{10R_{11}C_2}xz - \frac{1}{R_{12}C_2}(-y),$$

(a)

(b)

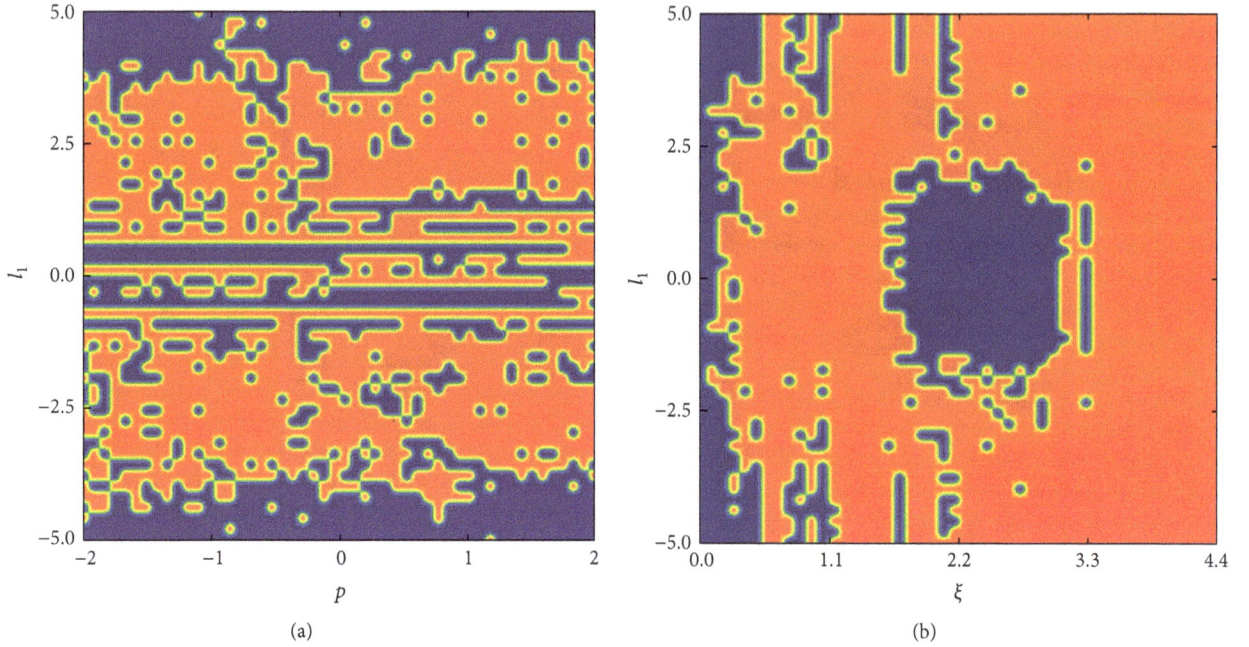

FIGURE 6: Attraction basins: (a) under initial conditions $(0.1, 0.1, p, l_1)$ and (b) under initial condition l_1 and system parameter ξ.

$$\frac{dz}{dt} = -\frac{1}{10R_{15}C_3}\left(-xy\right) - \frac{1}{R_{16}C_3}z,$$

$$\frac{dw}{dt} = -\frac{1}{R_{17}C_4}\left(-x\right).$$

$$(8)$$

Comparing (1) and (8), the parameters of the expression can be obtained:

$$\frac{1}{R_8 C_1} = \alpha,$$

$$\frac{1}{10R_{11}C_2} = \beta,$$

$$\frac{1}{R_{12}C_2} = \xi,$$

$$\frac{1}{R_{16}C_3} = \gamma, \qquad (9)$$

$$\frac{1}{10R_7 C_1} = 36,$$

$$\frac{1}{10R_{15}C_3} = 8,$$

$$\frac{1}{10R_6 C_1} = \frac{1}{R_{17}C_4} = 1.$$

Let $\alpha = 15$, $\beta = 8$, $\xi = 1.68$, and $\gamma = 15.15$; the values of four capacitors in Figure 8 are taken as $C_1 = C_2 = C_3 = C_4 = 1\,\text{nF}$. The resistance values in circuit are calculated as $R_1 = 1\,\text{k}\Omega$, $R_2 = 48\,\text{k}\Omega$, $R_3 = 24\,\text{k}\Omega$, $R_6 = 100\,\text{k}\Omega$, $R_7 = 2.78\,\text{k}\Omega$, $R_8 = 66.7\,\text{k}\Omega$, $R_{11} = 12.5\,\text{k}\Omega$, $R_{12} = 596\,\text{k}\Omega$,

$R_{15} = 16.7\,\text{k}\Omega$, $R_{16} = 67\,\text{k}\Omega$, $R_{17} = 1\,\text{M}\Omega$, $R_4 = 10\,\text{k}\Omega$, $R_5 = R_9 = R_{10} = 10\,\text{k}\Omega$, $R_{13} = R_{14} = 10\,\text{k}\Omega$. The value of system parameter ξ is determined by resistor R_{12}, and the different motions can be got by adjusting the value of R_{12}. The resistance R_{12} can be accessed by a potentiometer in the actual circuit. Take $R_{12} = 596\,\text{k}\Omega$, $C_2 = 1\,\text{nF}$ when $\xi = 1.68$, then the circuit simulation result is illustrated in Figure 7(a). The coexistence phase diagrams are shown as Figure 7(b) with the variation of l_1. Only two cases $l_1 = -3$ (light blue) and $l_1 = -1.7$ (pink) are listed here. It is possible to show that there is a multistability characteristic in system in (1) as well. The result in Figure 7(a) is in accordance with phase diagram plotted in Figure 1(b), and coexistence phase diagrams shown in Figure 7(b) agree with the light blue points ($l_1 = -3$) and pink points ($l_1 = -1.7$) shown in Figure 5(a). Hence the results of circuit and numerical simulation basically correspond under the same conditions.

4. Synchronization of the Memristive System and Its Application in Image Encryption

In 1990, a chaotic synchronization control method was firstly proposed by Pecora and Carroll [31]. Then, lots of results of chaos synchronization have been taken by scholars [32, 33]. With the gradual development of memristor, some scholars have applied the synchronization control to memristive chaotic systems, which has aroused a new upsurge in people's research. The nonlinear controller is designed according to this new four-dimensional four-wing memristive chaotic system for completing the synchronization of multistability. Moreover, the reliability of secret keys has been improved and the security of encryption has been enhanced by applying it to the image encryption.

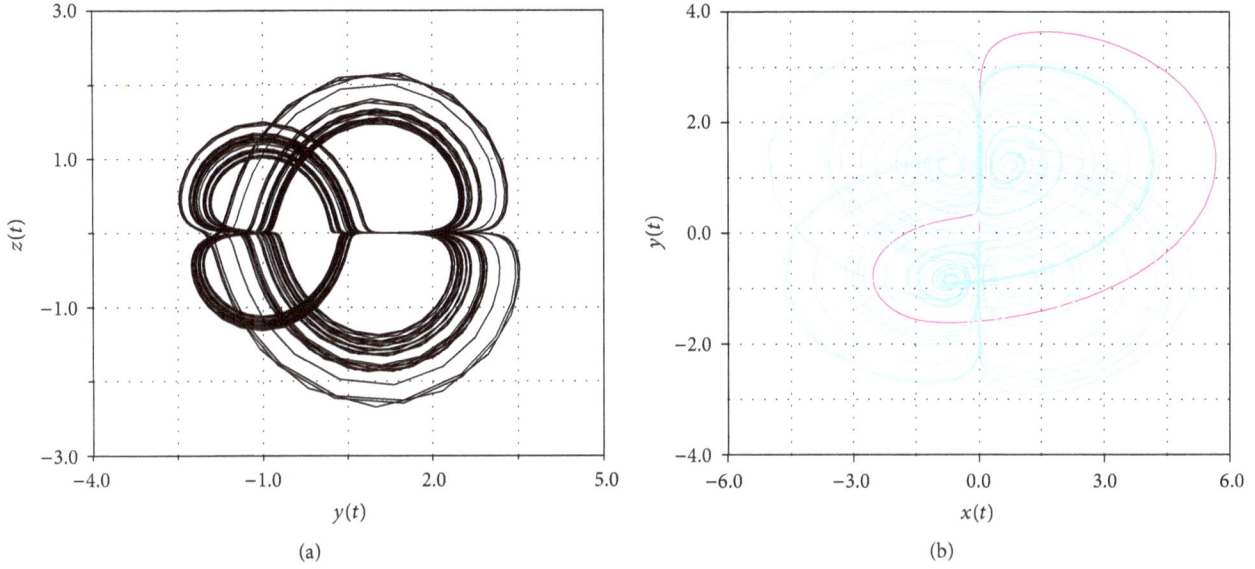

FIGURE 7: Circuit simulation diagrams: (a) y-z phase diagram and (b) x-y coexistence phase diagrams.

4.1. Synchronization of Four-Wing Memristive Chaotic System. According to the dynamic analysis of this four-wing memristive chaotic system, multistability will appear with initial conditions varying. A reasonable controller is devised to achieve the synchronization for coexistence attractors. Equation (1) is set as master system, so slave system is given by

$$\begin{aligned}
\dot{x}_2 &= 36 y_2 z_2 - \alpha x_2 - z_2 W\left(w_2\right) + u_1, \\
\dot{y}_2 &= \xi y_2 - \beta x_2 z_2 + u_2, \\
\dot{z}_2 &= 8 x_2 y_2 - \gamma z_2 + u_3, \\
\dot{w}_2 &= x_2 + u_4,
\end{aligned} \tag{10}$$

where $W(w_2) = a + bw_2^2$, $a = -0.5$, $b = 2.4$, $\alpha = 15$, $\beta = 8$, $\xi = 1.68$, and $\gamma = 15.15$. u_1, u_2, u_3, and u_4 are synchronization controllers. The synchronization error is described as

$$\begin{aligned}
e_1 &= x_2 - x_1, \\
e_2 &= y_2 - y_1, \\
e_3 &= z_2 - z_1, \\
e_4 &= w_2 - w_1.
\end{aligned} \tag{11}$$

The equation of error system in (11) is

$$\begin{aligned}
\dot{e}_1 &= 36\left(y_2 z_2 - y_1 z_1\right) - 15 e_1 \\
&\quad - \left[z_2\left(-0.5 + 2.4 w_2^2\right) - z_1\left(-0.5 + 2.4 w_1^2\right)\right] \\
&\quad + u_1, \\
\dot{e}_2 &= 1.68 e_2 - 8\left(x_2 z_2 - x_1 z_1\right) + u_2, \\
\dot{e}_3 &= 8\left(x_2 y_2 - x_1 y_1\right) - 15.15 e_3 + u_3, \\
\dot{e}_4 &= e_1 + u_4.
\end{aligned} \tag{12}$$

Theorem 1. *Select the following nonlinear feedback synchronization controller functions:*

$$\begin{aligned}
u_1 &= -k_1^{|e_1|} e_1 + 0.5 e_3 - 36\left(y_2 z_2 - y_1 z_1\right) \\
&\quad - 2.4\left(z_2 w_2^2 - z_1 w_1^2\right), \\
u_2 &= -k_2^{|e_2|} e_2 - 2 e_2 + 8\left(x_2 z_2 - x_1 z_1\right), \\
u_3 &= -k_3^{|e_3|} e_3 - 8\left(x_2 y_2 - x_1 y_1\right), \\
u_4 &= -k_4^{|e_4|} e_4 - e_1.
\end{aligned} \tag{13}$$

Master system in (1) is asymptotically synchronized with slave system in (10) when the feedback control gain k_i ($i = 1, 2, 3, 4$) ≥ 0.

Proof. Synchronization controller in (13) is designed with the specific form of error system in (12). The expressions with $-k_i^{|e_i|} e_i$ ($i = 1, 2, 3, 4$) [34] are added to increase the speed of synchronization and reduce the synchronization error in transition process. The error system can be simplified from (12) and (13) as

$$\begin{aligned}
\dot{e}_1 &= -k_1^{|e_1|} e_1 - 15 e_1, \\
\dot{e}_2 &= -k_2^{|e_2|} e_2 - 0.32 e_2, \\
\dot{e}_3 &= -k_3^{|e_3|} e_3 - 15.15 e_3, \\
\dot{e}_4 &= -k_4^{|e_4|} e_4.
\end{aligned} \tag{14}$$

Construct the Lyapunov function as $V = (1/2) e^T e$, and the derivative of V is calculated as

$$\begin{aligned}
\dot{V} &= \dot{e}^T e = e_1 \dot{e}_1 + e_2 \dot{e}_2 + e_3 \dot{e}_3 + e_4 \dot{e}_4 \\
&= -k_1^{|e_1|} e_1^2 - 15 e_1^2 - k_2^{|e_2|} e_2^2 - 0.32 e_2^2 - k_3^{|e_3|} e_3^2 \\
&\quad - 15.15 e_3^2 - k_4^{|e_4|} e_4^2
\end{aligned}$$

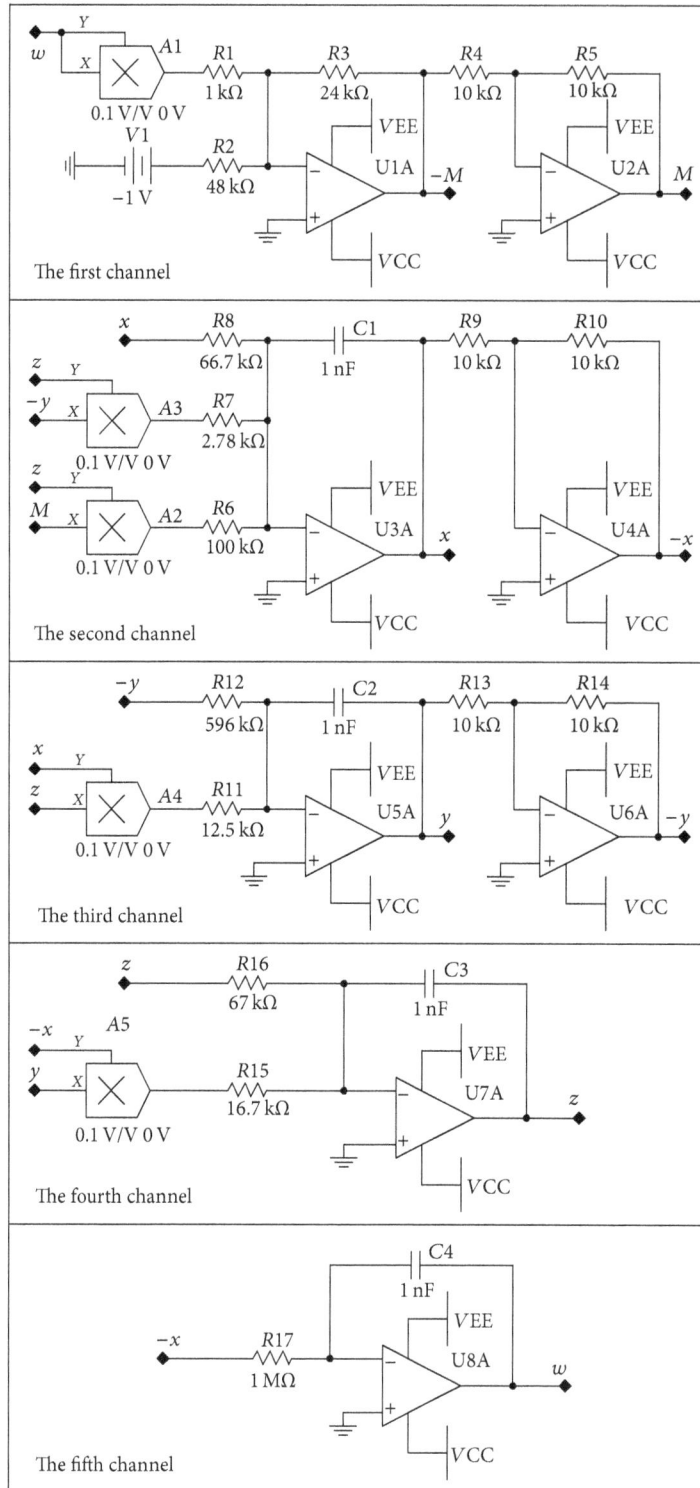

FIGURE 8: Circuit schematic.

$$= - \left(k_1{}^{|e_1|} e_1{}^2 + k_2{}^{|e_2|} e_2{}^2 + k_3{}^{|e_3|} e_3{}^2 + k_4{}^{|e_4|} e_4{}^2 \right)$$
$$- \left(15 e_1{}^2 + 0.32 e_2{}^2 + 15.15 e_3{}^2 \right).$$

(15)

It is obvious that $\dot{V}$ is negative owing to k_i $(i = 1, 2, 3, 4) \geq 0$, and the error system in (12) is large-scale asymptotic stable. Master system (1) and slave system (10) can be synchronized under the nonlinear synchronization controller in (13). $\square$

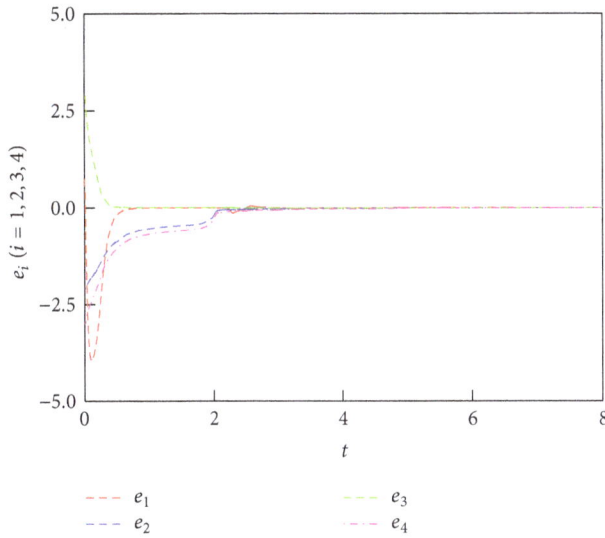

FIGURE 9: Synchronization errors.

According to the proof afore-discussed, the feedback gain is chosen as k_i $(i = 1, 2, 3, 4) = 2$ randomly. $(x_1, y_1, z_1, w_1, x_2, y_2, z_2, w_2) = (0.1, 0.1, 0, 0, 1, -2, 3, -3)$ is set to be the initial values of system (1) and system (10). The response curves of system errors are shown in Figure 9, and the synchronization errors converge to zero at about 4 s.

To realize the synchronization of multiple dynamics in the four-wing memristive chaotic system, initial conditions of master system in (1) are set to $(0.1, 0.1, 0, w_{1o})$ and those of slave system in (10) are taken as $(0.1, 0.1, 0, w_{2o})$. It is clear that the first three terms of two initial conditions remain unified, and the fourth terms w_{1o} and w_{2o} are set as variables. According to the aforementioned dynamics analyses above, the system in (1) will produce a rich multistable phenomenon along with initial condition $l_1 = w_{1o}$ varying. The motion of master system is a chaotic attractor as shown in Figure 1 as $l_1 = 0$, and it is in period-1 motion with the red points in Figure 5(b) as $l_1 = 3$. Similarly, the same coexistence attractors can be observed with the variable $l_2 = w_{2o}$ without regarding the influence of nonlinear synchronization controllers. The following two cases will be used to demonstrate the synchronization of coexistence states in the four-wing memristive chaotic system.

Case 1: setting two variables as $l_1 = 0$ and $l_2 = 3$, slave system in (10) can be synchronized with master system in (1) as a chaotic state. The timing diagrams for variables l_1 and l_2 are depicted in Figure 10(a). Similarly, case 2: for the variables above obtained as $l_1 = 3$ and $l_2 = 0$, the slave and master systems are synchronized as a periodic state. The state evolutions in this case are shown in Figure 10(b). Figure 10 indicates that these two cases are fully synchronized at about 4 s; that is, the multi-steady-state synchronization of the four-wing memristive chaotic system can be realized. Both different synchronization effects lay foundation for the application of memristive chaotic synchronization in image encryption.

4.2. Image Encryption. Chaotic system can provide sequences for information encryption, and most of the current researches on secure communication are for chaotic system without memristor [35, 36]. As we all know, the memristive chaotic circuit has a complex topology which can extend the dimension of chaotic systems, and it is extremely sensitive to the variation of initial condition that the better sequence can be extracted. So far, there were few studies on the application of synchronization of memristive chaotic system in image encryption. The initial conditions of master system in (1) and slave system in (8) are used as keys to improve the complexity of secret keys and anticrack ability. Figure 11 shows the process of encryption and decryption. A chaotic synchronization control signal is generated by the four-wing chaotic system to encrypt the original image, and the discrete sequences are also obtained from it. The encrypted image is got by the encryption transformation of discrete sequences and original image. The encrypted image will be transformed to the decryption terminal through four channels, including all data of the encrypted image and the data in each channel with unequal quantity to enhance the difficulty of deciphering. The decryption terminal receives encrypted signals and generates the decrypted image according to the key and encryption transformation, completing the synchronization of four-wing memristive chaotic system applied in image encryption.

The typical image "cameraman" [37] is selected for image encryption analysis, and the original image and its histogram are presented in Figure 12. According to the above analysis of the synchronization method, for the first case, initial conditions of master system (1) and slave system (10) are taken as $(0.1, 0.1, 0, 0)$ and $(0.1, 0.1, 0, 3)$, respectively. The synchronization result of master and slave system is chaos, and the encrypted image and its histogram are shown in Figure 13. The encrypted image exhibits a fuzzy state, and there is no feature of the original image. The histogram exhibited in Figure 13(b) no longer has the statistical characteristics of original image, so that the effect of encryption is good. The same method is utilized for the second case, and the initial values of system in (1) and system in (10) are, respectively, sited as $(0.1, 0.1, 0, 3)$ and $(0.1, 0.1, 0, 0)$. Thus, the result of synchronization is periodic motion in this case, and the encrypted image and its histogram are plotted in Figure 14. Contrasting Figure 14 with Figures 12 and 13, the encrypted image depicted in Figure 14(a) retains some features of original image exhibited in Figure 12(a). Similarly, the histogram shown in Figure 14(b) still has some statistical features of the original image, and the encryption is flawed. Therefore, the synchronization control of different motions will have different effects on image encryption, playing an important role in the study of secure communication.

On the basis of different effects on image encryption, it is better to apply case 1 to image encryption so that the image encryption with case 1 will be considered to analyze the noise attack. It is a common phenomenon to add noise in the process of information transmission. The Gaussian noise with the percentage of 1% is added to the encrypted image of case 1, and the decrypted image is depicted in Figure 15(a). Similarly, the salt-and-pepper noise with the percentage of 2% is put in

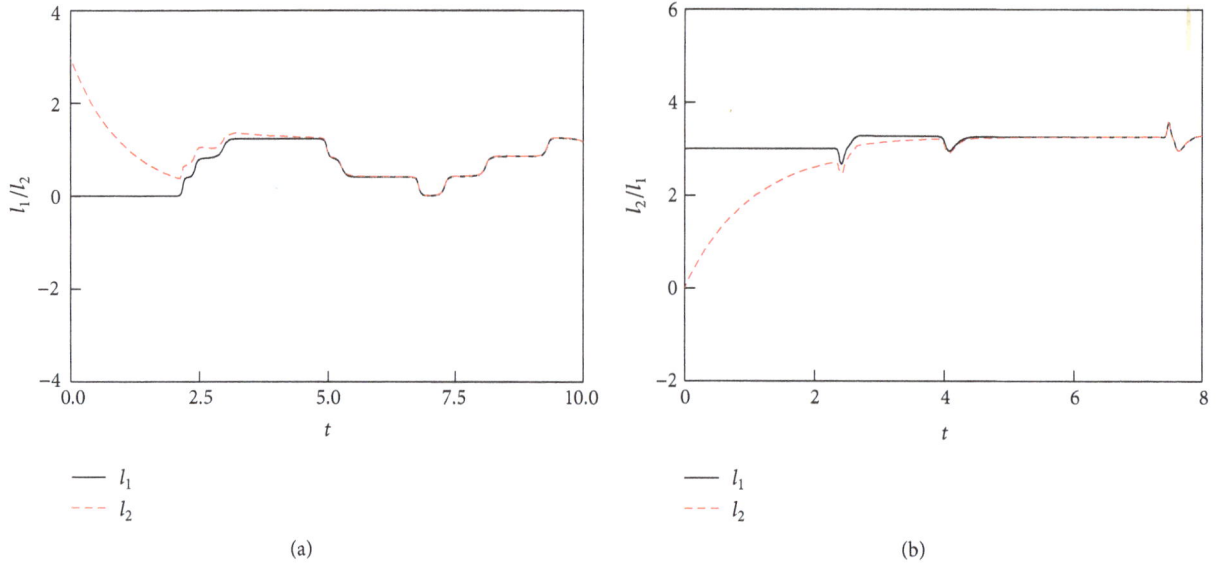

FIGURE 10: Timing diagrams: (a) case 1, $l_1 = 0$ & $l_2 = 3$, and (b) case 2, $l_1 = 3$ & $l_2 = 0$.

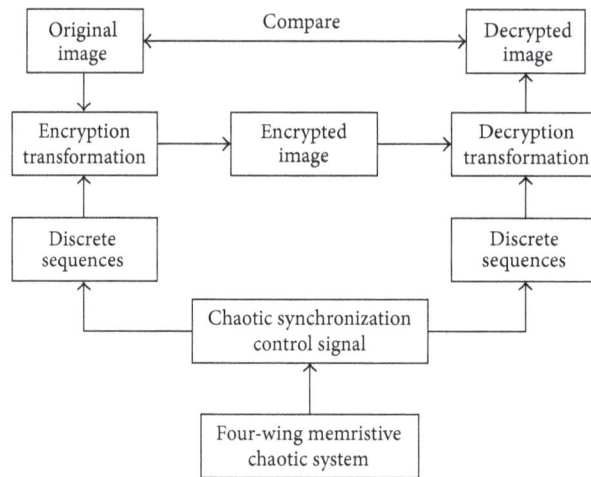

FIGURE 11: Encryption and decryption flow chart.

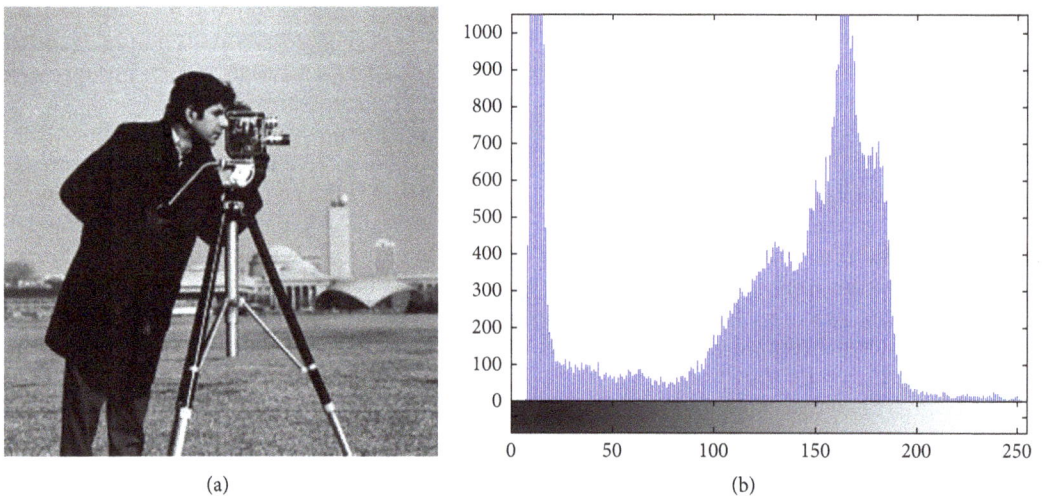

FIGURE 12: Typical image "cameraman": (a) original image and (b) corresponding histogram.

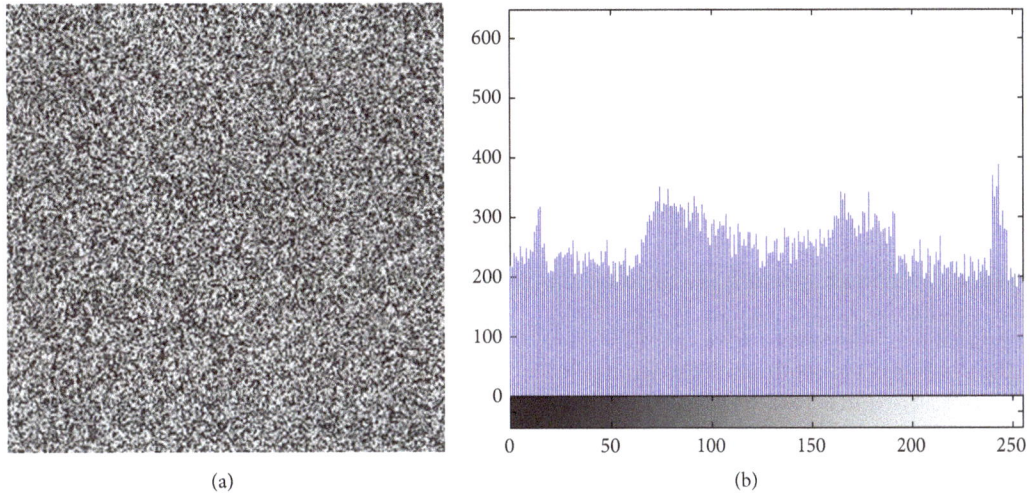

(a)   (b)

FIGURE 13: Case 1: (a) encrypted image and (b) encrypted histogram.

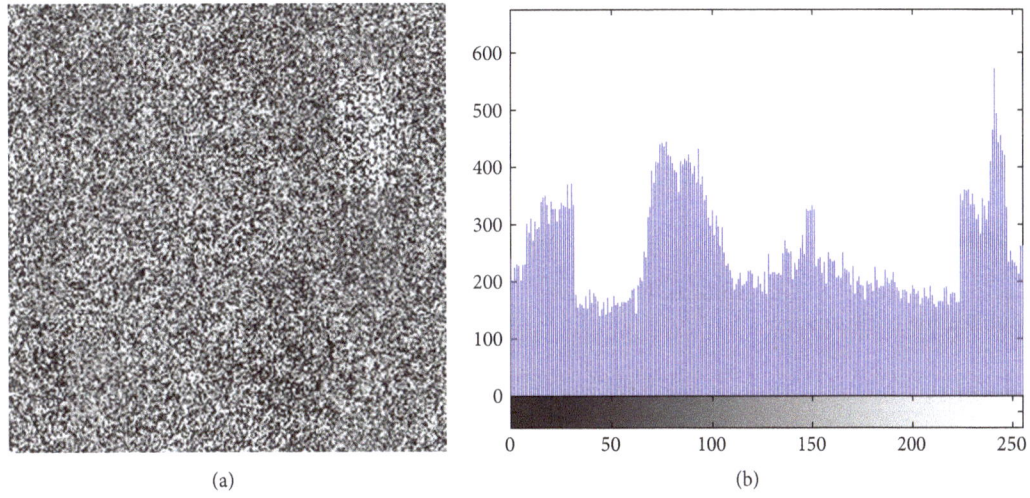

(a)   (b)

FIGURE 14: Case 2: (a) encrypted image and (b) encrypted histogram.

(a)   (b)

FIGURE 15: Decrypted images: (a) attacked by Gaussian noise and (b) attacked by salt-and-pepper noise.

the encrypted image in Figure 13, and the decrypted image is exhibited in Figure 15(b). Although noise interference can be observed in decrypted images, the main information of original image presented in Figure 12(a) remains. The good decryption effects prove that the encryption algorithm can effectively resist different noise attacks.

5. Conclusion

In summary, a modified Lorenz system is obtained on the basis of classical continuous Lorenz chaotic system, and a new four-dimensional four-wing memristive chaotic system is proposed by the addition of a smooth active flux-controlled memristor. The rich dynamical behaviors are exhibited through Lyapunov exponents, bifurcation diagrams, coexistence Poincaré sections, coexistence phase diagrams, and attraction basins. The multistability can be observed with respect to the sensitivity of initial conditions diametrically. Moreover, a new nonlinear synchronization controller with exponential term is designed to realize the synchronization control of multiple attractors and applied to image encryption. The chaotic sequence based on the synchronization scheme is extracted to increase the complexity of keys. At the same time, it is analyzed that the keys generated in different states have an important impact on encryption, and the choice of appropriate key is valuable to improve security.

Conflicts of Interest

The authors declare that there are no conflicts of interest regarding the publication of this paper.

Acknowledgments

The work is supported by the National Natural Science Foundation of China (no. 51475246), the Natural Science Foundation of Jiangsu Province under Grant no. Bk20131402, and the Postgraduate Research & Practice Innovation Program of Jiangsu Province of China under Grant no. KYCX17_1082.

References

[1] L. O. Chua, "Memristor—the missing circuit element," *IEEE Transactions on Circuit Theory*, vol. 18, no. 5, pp. 507–519, 1971.

[2] D. B. Strukov, G. S. Snider, D. R. Stewart, and R. S. Williams, "The missing memristor found," *Nature*, vol. 453, pp. 80–83, 2008.

[3] J. J. Yang, D. B. Strukov, and D. R. Stewart, "Memristive devices for computing," *Nature Nanotechnology*, vol. 8, no. 1, pp. 13–24, 2013.

[4] C. Yakopcic, V. Bontupalli, R. Hasan, D. Mountain, and T. M. Taha, "Self-biasing memristor crossbar used for string matching and ternary content-addressable memory implementation," *IEEE Electronics Letters*, vol. 53, no. 7, pp. 463–465, 2017.

[5] Y. Chang, F. Zhou, B. W. Fowler et al., "Memcomputing (Memristor + Computing) in intrinsic SiOx-based resistive switching memory: arithmetic operations for logic applications," *IEEE Transactions on Electron Devices*, vol. 64, no. 7, pp. 2977–2983, 2017.

[6] L. Wang, Q. Song, Y. Liu, Z. Zhao, and F. E. Alsaadi, "Finite-time stability analysis of fractional-order complex-valued memristor-based neural networks with both leakage and time-varying delays," *Neurocomputing*, vol. 245, pp. 86–101, 2017.

[7] J. Feng, Q. Ma, and S. Qin, "Exponential stability of periodic solution for impulsive memristor-based Cohen-Grossberg neural networks with mixed delays," *International Journal of Pattern Recognition and Artificial Intelligence*, vol. 31, no. 7, 1750022, 17 pages, 2017.

[8] R. Rocha, J. Ruthiramoorthy, and T. Kathamuthu, "Memristive oscillator based on Chua's circuit: stability analysis and hidden dynamics," *Nonlinear Dynamics*, vol. 88, no. 4, pp. 2577–2587, 2017.

[9] J. Kengne, A. N. Negou, and D. Tchiotsop, "Antimonotonicity, chaos and multiple attractors in a novel autonomous memristor-based jerk circuit," *Nonlinear Dynamics*, vol. 88, no. 4, pp. 2589–2608, 2017.

[10] H. Abunahla, D. Shehada, C. Y. Yeun, B. Mohammad, and M. A. Jaoude, "Novel secret key generation techniques using memristor devices," *AIP Advances*, vol. 6, no. 2, Article ID 025107, 2016.

[11] S. Kannan, N. Karimi, O. Sinanoglu, and R. Karri, "Security Vulnerabilities of Emerging Nonvolatile Main Memories and Countermeasures," *IEEE Transactions on Computer-Aided Design of Integrated Circuits and Systems*, vol. 34, no. 1, pp. 2–15, 2015.

[12] Z. Wang, F. Min, and E. Wang, "A new hyperchaotic circuit with two memristors and its application in image encryption," *AIP Advances*, vol. 6, no. 9, Article ID 095316, 2016.

[13] Q. Xu, Y. Lin, B. Bao, and M. Chen, "Multiple attractors in a non-ideal active voltage-controlled memristor based Chua's circuit," *Chaos, Solitons & Fractals*, vol. 83, pp. 186–200, 2016.

[14] H. Xi, Y. Li, and X. Huang, "Generation and nonlinear dynamical analyses of fractional-order memristor-based Lorenz systems," *Entropy*, vol. 16, no. 12, pp. 6240–6253, 2014.

[15] B. Zhang and F. Q. Deng, "Double-compound synchronization of six memristor-based Lorenz systems," *Nonlinear Dynamics*, vol. 77, no. 4, pp. 1519–1530, 2014.

[16] S. Wang, X. Wang, and Y. Zhou, "A memristor-based complex lorenz system and its modified projective synchronization," *Entropy*, vol. 17, no. 11, pp. 7628–7644, 2015.

[17] S. Kassim, H. Hamiche, S. Djennoune, and M. Bettayeb, "A novel secure image transmission scheme based on synchronization of fractional-order discrete-time hyperchaotic systems," *Nonlinear Dynamics*, vol. 88, no. 4, pp. 2473–2489, 2017.

[18] S. Wang, J. Zhao, X. Li, and L. Zhang, "Image blocking encryption algorithm based on laser chaos synchronization," *Journal of Electrical and Computer Engineering*, vol. 2016, Article ID 4138654, 14 pages, 2016.

[19] Z. Lin and H. Wang, "Image encryption based on chaos with PWL memristor in Chua's circuit," in *Proceedings of the 2009 International Conference on Communications, Circuits and Systems (ICCCAS)*, pp. 964–968, Milpitas, Ca, USA, July 2009.

[20] C. Yang, Q. Hu, Q. Yu, R. Zhang, Y. Yao, and J. Cai, "Memristor-Based Chaotic Circuit for Text/Image Encryption and Decryption," in *Proceedings of the 8th International Symposium on Computational Intelligence and Design, ISCID 2015*, pp. 447–450, China, December 2015.

[21] J. Ma, Z. Chen, Z. Wang, and Q. Zhang, "A four-wing hyperchaotic attractor generated from a 4D memristive system with a line equilibrium," *Nonlinear Dynamics*, vol. 81, no. 3, pp. 1275–1288, 2015.

[22] L. Zhou, C. Wang, and L. Zhou, "Generating Four-Wing Hyper-chaotic Attractor and Two-Wing, Three-Wing, and Four-Wing Chaotic Attractors in 4D Memristive System," *International Journal of Bifurcation and Chaos*, vol. 27, no. 2, Article ID 1750027, 2017.

[23] N. H. Alombah, H. Fotsin, and K. Romanic, "Coexistence of multiple attractors, metastable chaos and bursting oscillations in a multiscroll memristive chaotic circuit," *International Journal of Bifurcation and Chaos*, vol. 27, no. 5, 1750067, 20 pages, 2017.

[24] Z. T. Njitacke, J. Kengne, H. B. Fotsin, A. N. Negou, and D. Tchiotsop, "Coexistence of multiple attractors and crisis route to chaos in a novel memristive diode bidge-based Jerk circuit," *Chaos, Solitons & Fractals*, vol. 91, pp. 180–197, 2016.

[25] G. Peng and F. Min, "Multistability analysis, circuit implementations and application in image encryption of a novel memristive chaotic circuit," *Nonlinear Dynamics*, vol. 90, no. 3, pp. 1607–1625, 2017.

[26] E. N. Lorenz, "Deterministic nonperiodic flow," *Journal of the Atmospheric Sciences*, vol. 20, no. 2, pp. 130–141, 1963.

[27] S. M. Yv, *Chaotic Systems and Chaotic Circuit: Principle, Design and Its Application in Communications*, Xidian University Publishing House, Xi'an, China, 2011.

[28] G. A. Leonov, N. V. Kuznetsov, and T. N. Mokaev, "Hidden attractor and homoclinic orbit in Lorenz-like system describing convective fluid motion in rotating cavity," *Communications in Nonlinear Science and Numerical Simulation*, vol. 28, no. 1-3, pp. 166–174, 2015.

[29] H. Bao, B. C. Bao, Y. Lin et al., "Hidden attractor and its dynamical characteristic in memristive self-oscillating system," *Acta Physica Sinica*, vol. 65, no. 18, article 180501, 2016.

[30] B. Bao, T. Jiang, G. Wang, P. Jin, H. Bao, and M. Chen, "Two-memristor-based Chua's hyperchaotic circuit with plane equilibrium and its extreme multistability," *Nonlinear Dynamics*, pp. 1–15, 2017.

[31] L. M. Pecora and T. L. Carroll, "Synchronization in chaotic systems," *Physical Review Letters*, vol. 64, no. 8, pp. 821–824, 1990.

[32] F. Min and A. C. Luo, "Complex dynamics of projective synchronization of Chua circuits with different scrolls," *International Journal of Bifurcation and Chaos*, vol. 25, no. 5, 1530016, 28 pages, 2015.

[33] B. S. Reddy and A. Ghosal, "Chaotic Motion in a Flexible Rotating Beam and Synchronization," *Journal of Computational and Nonlinear Dynamics*, vol. 12, no. 4, p. 044505, 2017.

[34] L. Shan, J. Li, F. H. Min, and Z. Q. Wang, "Synchronization control of new piecewise fractional-order chaotic systems," *Systems Engineering and Electronics*, vol. 32, no. 10, pp. 2198–2202, 2010.

[35] C. L. Wang, Z. B. Chen, and G. E. Yong, "Infrared image encryption scheme using Lorenz chaotic system," *Journal of Computer Applications*, vol. 35, no. 8, pp. 2205–2209, 2015.

[36] J. Zhao, S. Wang, L. Zhang, and X. Wang, "Image encryption algorithm based on a novel improper fractional-order attractor and a wavelet function map," *Journal of Electrical and Computer Engineering*, vol. 2017, Article ID 8672716, 10 pages, 2017.

[37] I. F. Elashry, O. S. F. Allah, A. M. Abbas, S. El-Rabaie, and F. E. A. El-Samie, "Homomorphic image encryption," *Journal of Electronic Imaging*, vol. 18, no. 3, Article ID 033002, 2009.

13

Compact Design of Circularly Polarized Antenna with Vertical Slotted Ground for RFID Reader Application

Hesheng Cheng,[1] **Jin Zhang,**[2] **Hexia Cheng,**[3] **and Qunli Zhao**[1]

[1]*College of Computer Science, Hefei Normal University, Hefei, China*
[2]*School of Electronic and Information Engineering, Jinling Institute of Technology, Nanjing, China*
[3]*College of Computer & Information, Anqing Normal University, Anqing, China*

Correspondence should be addressed to Hesheng Cheng; chenghesheng238@126.com

Academic Editor: Jit S. Mandeep

A novel compact circular polarization (CP) microstrip antenna is proposed for UHF ultrahigh frequency (UHF) radio frequency identification (RFID) reader applications. The proposed antenna is composed of a corner truncated square-ring radiating patch on a substrate and a vertical slotted ground surrounding four sides of the antenna. A new feeding scheme is designed from flexible impedance matching techniques. The impedance bandwidths for $S_{11} < -10$ dB and 3 dB axial ratio (AR) bandwidth are 12.1% (794.5–896.5 MHz) and 2.5% (833.5–854.5 MHz), respectively.

1. Introduction

RFID reader system in the UHF band has developed rapidly due to its numerous application [1]. The reader antenna is one of the important components in the RFID reader system, which has been designed with CP radiation for reducing the loss caused by the multipath effects between a reader and the tag. Circularly polarized reader antenna is useful in practice when tag antenna system is in rotating motion or when its respective orientation cannot be ensured

Total frequency span of 840–960 MHz is used for UHF RFID systems worldwide. In China, RFID and other low-power radio applications are regulated, allowing RFID operation with somewhat complex band restrictions from 840 to 845 MHz or 920 to 925 MHz. In this paper, a compact circularly polarized antenna for handheld UHF RFID reader operates at the bands of 840–845 MHz. Design of a compact CP microstrip antenna is attractive for handheld/portable device application. Some minimized CP antennas have been reported in literatures [2–4]. Reference [2] demonstrated that the size requirement of square-ring microstrip antenna is smaller than that of conventional design using a square patch. However, the antenna gains and 3 dB axial ratio bandwidth for such compact circular polarization (CP)

designs are also decreased, which is a severe limitation for practical applications of such antennas. Reference [3] proposed antenna with vertical ground surrounding four sides of the antenna to further reduce size and obtain high gain; however the antenna's cost is high for using a high-permittivity substrate. Reference [4] used foam instead of high-permittivity substrate to obtain the merit of low cost. The antenna in [5] is composed of two corner truncated patches and a suspended microstrip line with open-circuited termination. The main patch is fed by four probes which are sequentially connected to the suspended microstrip feed line. Reference [6] designed a circularly polarized RFID antenna using a horizontal bend line feed technology. The antenna in [7] is composed of two patches. The main patch which is truncated to generate CP radiation is fed directly by the SMA connector oriented in the same plane via the vertical part of the L-shaped ground. The parasitic patch used to enhance the bandwidth is placed at the same layer of the main patch. References [5–7] realized the UHF RFID antenna but its size is too large to facilitate the practical application. References [8, 9] developed a small circularly polarized RFID antenna, but the antenna bandwidth is so narrow that it cannot meet the requirements of UHF RFID applications.

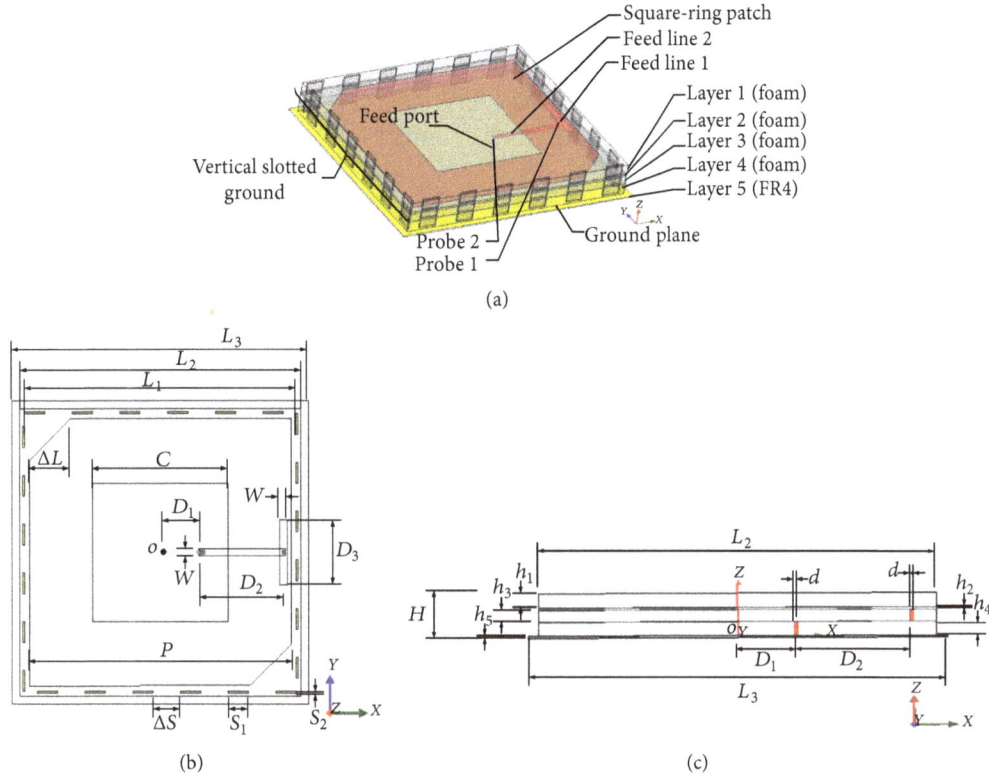

FIGURE 1: Geometry of proposed antenna. (a) 3D structure; (b) top view; (c) side view (hide vertical slotted ground). H is the total height of the five layers, $H = h_1 + h_2 + h_3 + h_4 + h_5$, and dielectric constant $\varepsilon_e = 4.4$.

TABLE 1: Values of proposed antenna parameters.

$L_1 = 106.5$ mm	$\Delta L = 15.5$ mm	$S_1 = 8$ mm	$h_3 = 5$ mm
$L_2 = 110.5$ mm	$D_1 = 15.4$ mm	$S_2 = 0.8$ mm	$h_4 = 6$ mm
$L_3 = 116.5$ mm	$D_2 = 47$ mm	$\Delta S = 18.7$ mm	$h_5 = 0.8$ mm
$P = 102.5$ mm	$D_3 = 25$ mm	$h_1 = 6$ mm	$d = 0.6$ mm
$C = 53$ mm	$W = 3$ mm	$h_2 = 1$ mm	$H = 12$ mm

In this letter, a new compact CP and low cost design of a square-ring microstrip antenna with a vertical slotted ground plane are proposed. It is found that, by properly embedding a group of slots in the finite vertical ground plane surrounding four sides of the square-ring microstrip antenna, the lowering of the resonant frequency of the antenna can be obtained. Moreover, the 3 dB axial ratio bandwidth of the proposed scheme is greater than that of the above-mentioned design [2–4].

2. Antenna Structure

Figure 1 shows the proposed CP antenna which is formed from five layers. The parameters of the optimized antenna are shown in Table 1. Layer 1 is a foam plate ($L_1 \times L_1 = 106.5 \times 106.5$ mm^2); the radiating corner truncated square-ring patch ($P \times P = 102.5 \times 102.5$ mm^2) is mounted on layer 2. The corners truncated patch is to achieve circular polarization. In the center of patch, one square slot is taken in order to achieve miniaturization of the antenna. Feed line 1 is printed

on layer 3; by the same token, feed line 2 is printed on the layer 4 ($L_2 \times L_2 = 110.5 \times 110.5$ mm^2). Feed line 1 connects the radiating patch by capacitive coupled technique. Feed line 1 is linked to feed 2 with probe 1 embedded in layer 3; similarly, feed port which is in layer 4 is attached to feed line 2 with probe 2. Layer 5 is a ground plane which is FR4 plate covered with copper sheet. Moreover, feed lines 1 and 2 are designed to Z shape in order to achieve 50 Ω impedance matching. The Z-shaped and capacitive coupled structure can obtain flexible impedance matching. To further reduce size, vertical slotted ground, which is employed by rectangular parallelepiped metallic via holes for easily fabricated, surrounding four sides of the antenna is proposed. Using a dielectric substrate with a high dielectric constant, the size of the antenna can be reduced. The dielectric constant is $\varepsilon_e = 4.4$ in this paper.

3. Simulation and Discussion

The simulation CP axial ratio and return loss parameters of the proposed antenna are displayed in Figure 2. The

TABLE 2: Performance of proposed and reference antennas.

Kinds of antenna	f_c (MHz)	Peak directivity (dB)	Peak gain (dBi)	Radiation efficiency (%)	F/B (dB)	BW1 (%)	BW2 (%)
Proposed antenna	842.5	4.1	3.8	92.8	15.0	12.1	2.5
Antenna 1	928.8	5.7	4.6	77.8	10.5	2.9	0.65
Antenna 2	1034.3	5.3	4.9	89.3	3.8	6.3	1.4
Antenna 3	1215.2	6.5	6.4	97.0	5.7	13	2.2

f_c: centre frequency of BW2; F/B: front to back ratio; BW1: 10 dB return loss impedance bandwidth; BW2: 3 dB axial ratio CP bandwidth.

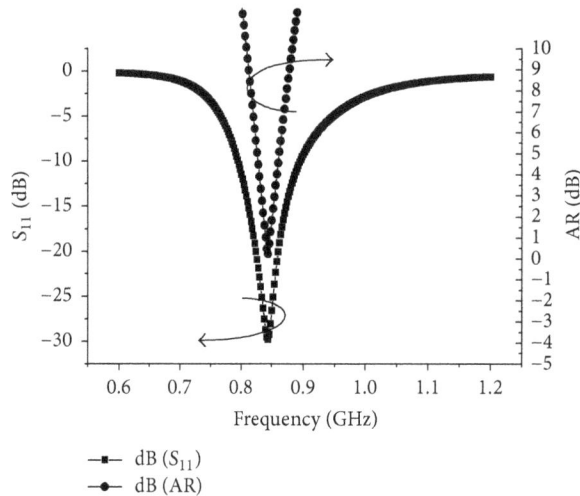

FIGURE 2: Axial ratio and return loss of the proposed antenna.

simulation results show that the bandwidth of CP and return loss operation, which are defined by 3 dB axial ration and −10 dB return loss, are about 21 MHz (2.5% at 842.5 MHz) and 102 MHz (12.1% at 842.5 MHz), respectively, at the centered 842.5 MHz, covering the bandwidth of 840–845 MHz. The radiation patterns of two principal planes at 842.5 MHz are plotted in Figure 3. The gain of 3.8 dBi is obtained at 842.5 MHz, good right-hand CP radiation is observed, and the CP isolation is more than 26 dB. The half-power beam width (HPBW) is 99° and 98°, respectively, in xoz plane and yoz plane. Due to the vertical slotted ground the front-to-back ratio (F/B) reaches 15 dB.

At the same sizes of radiating patch (parameters P) and height of substrate (parameter H), another three antenna prototypes are implemented for comparing with the proposed antenna: antenna with vertical ground and square-ring patch (antenna 1), antenna without vertical ground and with square-ring patch (antenna 2), and conventional corner truncated square patch antenna [5] (antenna 3). The simulated results of the axial ratio and return loss are shown in Figure 4. From the results obtained, it is clearly seen that the center frequency of the proposed antenna is about 9% lower than antenna 1, about 19% lower than antenna 2, and about 31% lower than antenna 3. That corresponds to an antenna size reduction of about 10%, 23%, and 44% for the proposed antenna at a fixed frequency, respectively. This characteristic suggests that the constructed vertical slotted

ground surrounding square-ring patch achieves good effect in miniaturization.

The corresponding simulated data of the proposed antenna and the other reference antennas are given in Table 2 for comparison. It is first noted that the impedance bandwidth and the CP bandwidth of the proposed antenna are larger than antenna 1 and antenna 2 and approximate to antenna 3. As known, the quality factor of the microstrip antenna dropped with operating frequency risen. Therefore, comparing proposed antenna with the reference antennas, it is found that embedded slot in vertical ground can effectively widen antenna bandwidth. Moreover, the F/B reaches 15 dB of proposed antenna, larger than any other references. It is also noted that, owing to the lowering of the resonant frequency which results in a decrease in peak directivity of the proposed antenna, the peak gain is decreased.

4. Conclusions

A novel compact circularly polarized square-ring microstrip antenna with vertical slotted ground for UHF RFID has been successfully implemented, which is designed as Z-shaped and capacitive coupled structure. The proposed design can have a size reduction of about 10% (contrasting with [2]), 23% (contrasting with [3]), and 44% (contrasting with [4]); moreover, the substrate material is foam and FR4, in order to be low cost. Furthermore, the impedance bandwidths for $S_{11} < −10$ dB

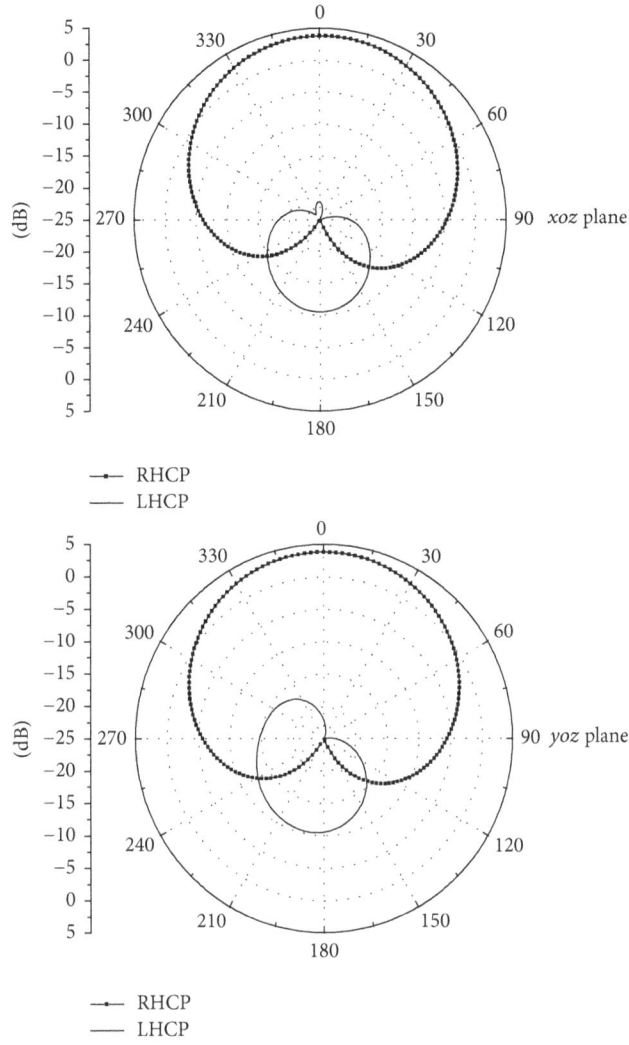

FIGURE 3: Radiation pattern of the proposed antenna.

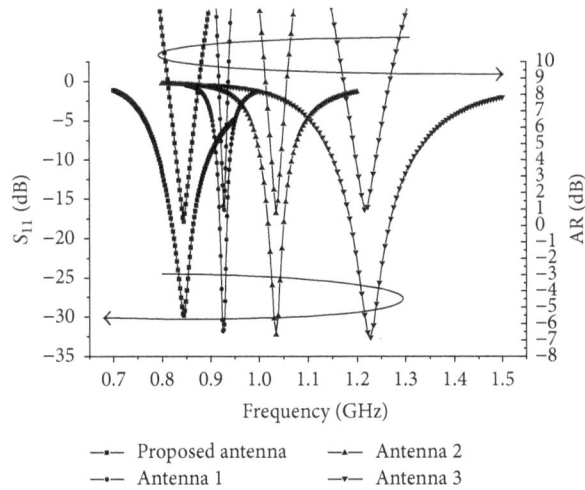

FIGURE 4: Simulated return losses and axial ratio of the proposed antenna and reference antennas.

and 3 dB AR bandwidth are 12.1% (794.5–896.5 MHz) and 2.5% (833.5–854.5 MHz), respectively. Due to the use of a vertical trench structure, the antenna achieves up to 15 dB of front-to-back ratio at 845.2 MHz.

Conflicts of Interest

The authors declare that they have no conflicts of interest.

Acknowledgments

This research work was supported by Educational Commission of Anhui Province of China (KJ2016A585, 1708085QF157, gxyq2017050, and 2015jyxm271).

References

[1] K. Finkenzeller, *RFID Handbook*, John Wiley & Sons, New York, NY, USA, 2nd edition, 2003.

[2] W.-S. Chen, C.-K. Wu, and K.-L. Wong, "Single-feed squaring-ring microstrip antenna with truncated corners for compact circular polarisation operation," *IEEE Electronics Letters*, vol. 34, no. 11, pp. 1045–1047, 1998.

[3] S. Kim, H. Park, D. Lee, and J. Choi, "A novel design of an UHF RFID reader antenna for PDA," in *Proceedings of the 2006 Asia-Pacific Microwave Conference, APMC*, pp. 1471–1473, December 2006.

[4] Z. Wang, S. Fang, and S. Fu, "A low cost miniaturized circularly polarized antenna for UHF radio frequency identification reader applications," *Microwave and Optical Technology Letters*, vol. 51, no. 10, pp. 2382–2384, 2009.

[5] N. Z. Chen, X. Qing, and L. C. Hang, "A universal UHF RFID reader Antenna," *IEEE Transactions on Microwave Theory & Techniques*, vol. 57, no. 5, pp. 1275–1282, 2009.

[6] Z. Wang, S. Fang, S. Fu, and S. Jia, "Single-fed broadband circularly polarized stacked patch antenna with horizontally meandered strip for universal UHF RFID applications," *IEEE Transactions on Microwave Theory and Techniques*, vol. 59, no. 4, pp. 1066–1073, 2011.

[7] Y. Pan, L. Zheng, H. J. Liu, J. Y. Wang, and R. L. Li, "Directly-fed single-layer wideband RFID reader antenna," *IEEE Electronics Letters*, vol. 48, no. 11, pp. 607-608, 2012.

[8] C.-Y. Sim and C.-J. Chi, "A slot loaded circularly polarized patch antenna for UHF RFID reader," *IEEE Transactions on Antennas and Propagation*, vol. 60, no. 10, pp. 4516–4521, 2012.

[9] M. K. Chang, J. R. Lin, and C. I. Deng, "A novel design of a microstrip-fed shorted square-ring slot antenna for circular polarization," *Microwave & Optical Technology Letters*, vol. 49, no. 7, pp. 1684–1687, 2010.

14

New Current-Mode Integrated Ternary Min/Max Circuits without Constant Independent Current Sources

Mona Moradi,[1] **Reza Faghih Mirzaee,**[2] **and Keivan Navi**[3]

[1]Department of Computer Engineering, Islamic Azad University, Science and Research Branch, Tehran 1477893855, Iran
[2]Department of Computer Engineering, Islamic Azad University, Shahr-e-Qods Branch, Tehran 37541-374, Iran
[3]Faculty of Electrical and Computer Engineering, Shahid Beheshti University, G.C., Tehran 1983963113, Iran

Correspondence should be addressed to Mona Moradi; mo.moradi@srbiau.ac.ir

Academic Editor: Ahmed El Wakil

Novel designs of current-mode Ternary minimum (AND) and maximum (OR) are proposed in this paper based on Carbon NanoTube Field Effect Transistors (CNTFET). First, these Ternary operators are designed separately. Then, they are combined together in order to generate both outputs concurrently in an integrated design. This integration results in the elimination of common parts when both functions are required at the same time. The third proposed current-mode integrated circuit generates both ternary operators with the usage of only 30 transistors. The new designs are composed of three main parts: (1) the part which converts current to voltage; (2) threshold detectors; and (3) the parallel paths through which the output current flows. Unlike the previously presented structure, there is no need for any constant current source within the new designs. This elimination leads to less static power dissipation. The second proposed current-mode segregated Ternary minimum operates 43% faster and consumes 40% less power in comparison with a previously presented structure.

1. Introduction

In current-mode logic (CML), logic levels are represented by current levels. It has several advantages over voltage-mode logic (VML). The merits include (1) simple wiring of the currents, making linear sum operation easier and reducing the number of active devices [1]; (2) the ease of circuit expansion [2]; (3) the usage of the direction of current as the explanation of sign, which eliminates the requirement of sign bit [1]; (4) lower noise sensitivity [3]; (5) scaling and copying the currents with a simple current mirror circuit [4]; and, last but not least, (6) high-speed operation [4]. However, despite its several advantages, high static power consumption is the main drawback of CML.

Current-mode circuits are traditionally implemented by either bipolar (bipolar CML) or MOS (MCML) devices. MCML is preferred for mixed analog-digital signal environments due to high power consumption of bipolar transistors and higher supply noise immunity of MOS devices [3]. Although the latter has been the superior technology for implementing energy-efficient circuits for many years, their suitability in today's nanometer VLSI industry is gradually fading away. This is mainly because of several critical challenges of MOS devices in nanoranges such as very high leakage currents, high power density, large parametric variations, and decreased gate control [5]. To overcome these difficulties, Carbon NanoTube Field Effect Transistor (CNTFET) is considered as the most promising successor to the MOS technology in the near future. This is mainly because of its intrinsic similarities to current technology. In addition, it benefits from the same motility of both n-type and p-type CNTFET devices, high-speed operation, low power consumption, and the realization of desired threshold voltage, which is a great feature in multiple-valued logic (MVL) circuitry [6].

MVL is an approach that utilizes more than two logic levels, while Binary logic computations are based on two values ("0" and "1"). MVL circuits benefit from fewer interconnections and pinouts, less area dissipation, and higher parallel and serial communication rates [1]. MVL is also known as an alternative design technique to Binary circuits, where the amount of interconnections is becoming a serious challenge. The huge amount of wires inside today's nanoscale chips adds undesirable parasitic effects, reduces noise tolerance, dissipates more power, and restricts the routing and placement processes of logic elements [7–9]. The increment of logic levels can be considered as a solution to these difficulties and limitations. MVL circuits will play an important role in the next generation of electronic systems [1]. The most efficient MVL system, which leads to less product cost and complexity than Binary, is Ternary logic [7].

In this paper, new Ternary minimum (Logical AND) and maximum (Logical OR) circuits are proposed. They are based on CNTFET technology and CML design technique. They are fundamental operators in mathematical and logical components and computational units such as Arithmetic Logic Unit (ALU). Since both operators are required at the same time in most processors, their circuits are combined together to share the common parts and reduce transistor-count. The new circuits are based on mixed current and voltage logics. The usage of constant current source is avoided in the new designs. Therefore, power dissipation is reduced dramatically in comparison with the previously presented structure.

The rest of the paper is organized as follows: Section 2 reviews the previously presented current-mode MVL-based works in the literature. The transistor-level implementation of the previous design is also presented in this section. New current-mode Ternary Min/Max circuits are proposed in Sections 3 and 4. Simulation results and comparisons are brought in Section 5. Finally, Section 6 concludes the paper.

2. Literature Review

New minimum and maximum circuits have been presented in [10–12] for fuzzy logic. They are based on linear addition and subtraction of input currents, and they benefit from design simplicity and low transistor-count. Although it is apparently possible to use them in any MVL system, they have not been particularly designed for Ternary logic or any other digital systems. As a result, they fail to refresh and fortify incomplete signals. One of the major features of circuits in discrete systems is the ability of generating full-swing outputs in spite of non-full-swing input signals. Therefore, the employment of these fuzzy circuits [10–12] in digital systems is neither suitable nor practical.

Some current-mode structures have been previously presented in [13] for implementing signed Ternary functions. A comparable method has been used in [14] to design an unsigned Ternary circuit, which is the main objective in this paper. The previous Ternary maximum and minimum circuits, presented in [14], are shown in Figures 1(a) and 1(b), respectively. These structures utilize constant independent current sources as threshold detectors. In case the input current (a or b) is less than the constant one (the threshold), the rest of the current charges the gate capacitor of a p-type transistor, turning it off. If the amount of input current is more than the constant one, the transistor, whose gate capacitor is discharged, switches on. The transistor-level

Table 1: The truth table of Ternary minimum and maximum.

a	b	$\sum$ in	Minimum	Maximum
0	0	0	0	0
0	1	1	0	1
0	2	2	0	2
1	0	1	0	1
1	1	2	1	1
1	2	3	1	2
2	0	2	0	2
2	1	3	1	2
2	2	4	2	2

implementation of constant independent current sources, presented in [15], is used in this paper for simulation purpose (Figure 1(c)).

In the unsigned Ternary logic, there are three positive logic values, $\{0, 1, 2\}_3$. They are represented in current-mode logic by different current levels. The unit current of $8\,\mu\text{A}$ is considered in this paper to represent logic value "1". Therefore, the logic value "2" is implemented by $16\,\mu\text{A}$, while no current flows when logic value is "0".

The final representation of the previously presented Ternary Min and Max are shown in Figure 2, where each transistor is marked with three values (numbers). They indicate the diameter of CNTs (D_{CNT}), the number of nanotubes under the gate terminal (Tube), and the pitch parameter. These designs require constant currents of $4\,\mu\text{A}$, $8\,\mu\text{A}$, $12\,\mu\text{A}$, and $28\,\mu\text{A}$ as threshold detectors, whose transistor-level implementations are included in the figures. In addition, variable input currents (a and b) are duplicated by current mirrors since two copies of them are needed in each circuit.

3. The Proposed Current-Mode Segregated Ternary Operators (CSTO)

Logical AND and OR are among the most essential and fundamental operators in digital electronics. They are, respectively, equal to the minimum and maximum mathematical functions regardless of what MVL system is used. Their function in Ternary logic is shown in Table 1. In this section, new Ternary Min and Max circuits are proposed separately. They are based on a technique where input currents are converted to voltage. Then, threshold detectors control the switching activity of the transistors, which are situated on the output paths. A unit of current ($8\,\mu\text{A}$) flows through each output path in case all of the related transistors are switched on. If the output value equals "2", the currents of two different paths are joined in order to increase the amount of current to $16\,\mu\text{A}$.

In this section, the segregated strategy is followed. Then, the separated designs are combined together to achieve an integrated circuit in the next section.

3.1. The First Approach with a Current Mirror in the Last Part. The first proposed CML Ternary Min circuit is shown in Figure 3, where each transistor is marked with three values

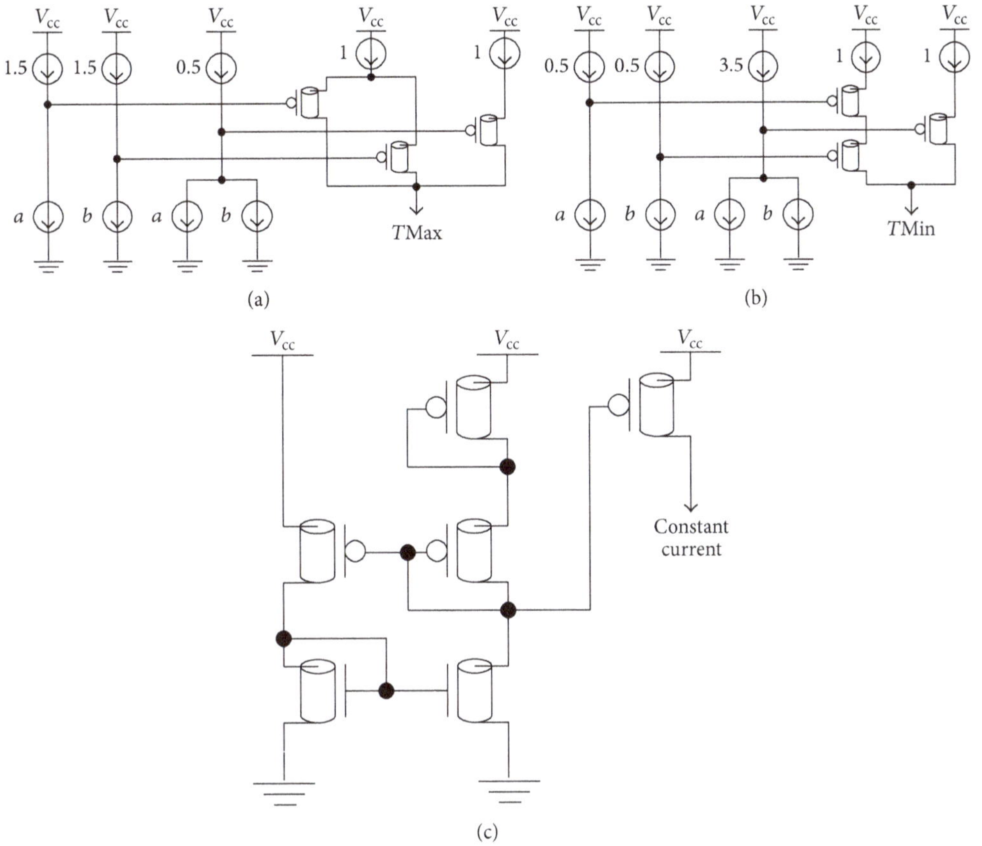

FIGURE 1: The previously presented circuits, (a) Ternary maximum [14], (b) Ternary minimum [14], (c) constant independent current source (transistor-level implementation) [15].

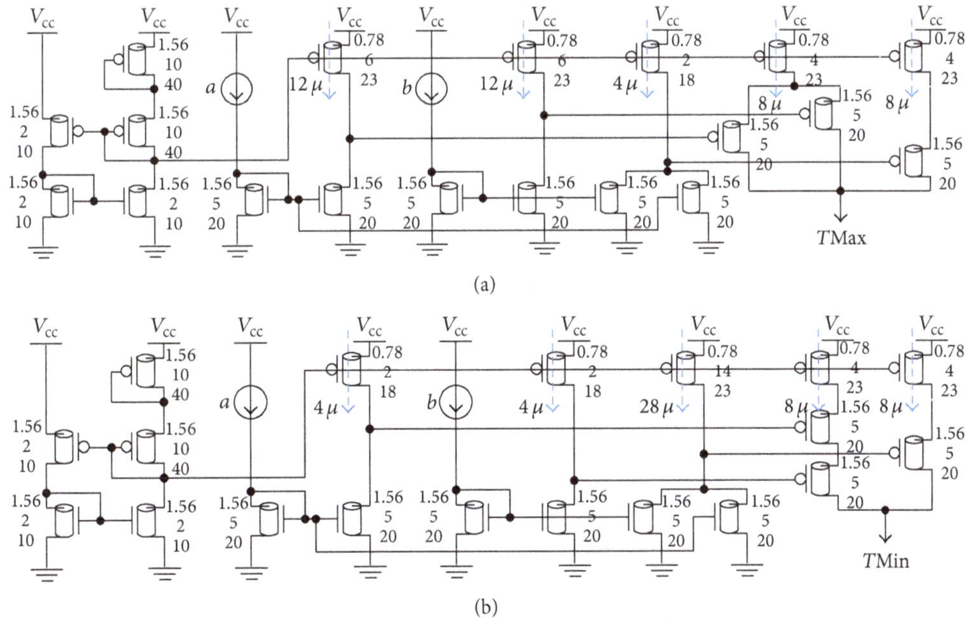

FIGURE 2: The previous current-mode circuits [14], (a) previous Ternary maximum (Previous TMax), (b) previous Ternary minimum (Previous TMin).

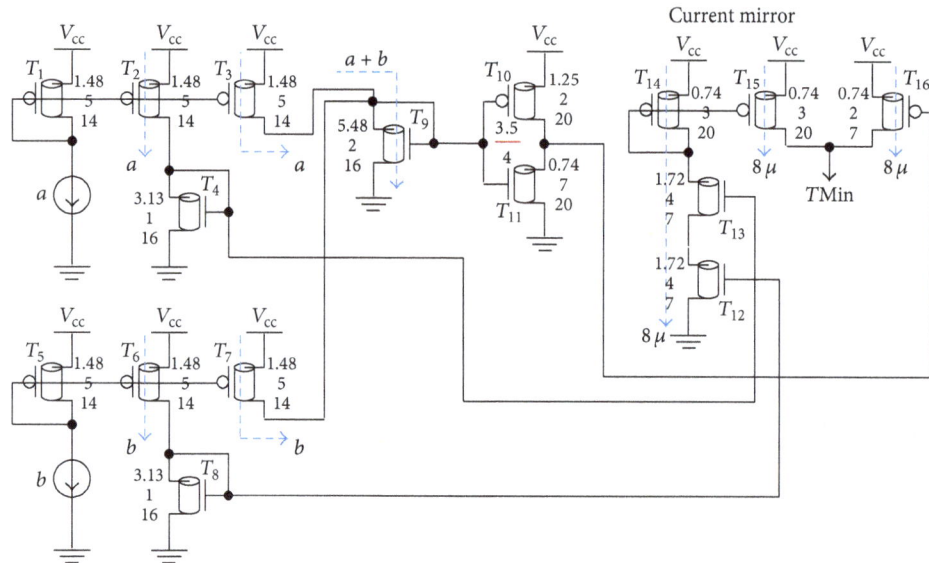

FIGURE 3: The first proposed Current-mode Segregated Ternary Min (CSTMin1) with the current mirror in the last part.

(numbers). As mentioned earlier, the numbers indicate the diameter of CNTs (in nanometer), the number of nanotubes under the gate terminal, and the pitch parameter (in nanometer). For example, there are eight nanotubes (Tubes = 8) with the diameter of 1.48 nm (D_{CNT} = 1.48 nm) under the gate of T_1. The distance between the centers of two adjacent CNTs is also 14 nm (Pitch = 14 nm).

There are two input currents, a and b. The proposed approach is on the basis of the sum of input currents. Unlike VML, it is as simple as connecting the wires (a and b) to achieve their linear addition in CML ($a + b$). Then, the summation is converted to voltage by means of a resistor. A diode-connected transistor (T_9) is used as a resistor to make this conversion possible. To obtain higher resistivity, channel (L_g) and doped CNT source- (L_{ss}) and drain-side (L_{dd}) extension regions for this transistor have been lengthened by 90 nm. The longer the channel is, the more it resists. This increment of channel length does not slow the operation because these converting transistors are not situated along the critical path of the cell.

The input currents (a and b), themselves, have to be converted to voltage as well. The transistors T_4 and T_8 are responsible for this purpose (for these two transistors: $L_g = L_{ss} = L_{dd} = 100$ nm). However, it is not feasible to take branches from the input currents due to the fact that the amount of current is divided. In CML, currents must be mirrored in case another copy it is required. In Figure 3, T_2 and T_3 (T_6 and T_7) duplicate the input current a (b) twice. In order to have the exact copy, they must have the same dimensions as T_1 (T_5). This part of the circuit (T_1 to T_9) is repeated in all of the proposed designs in this paper.

After the conversion, threshold detectors (TDs) control the switching activity of the rest of the transistors. Threshold detectors are in fact inverters with shifted voltage transfer characteristic (VTC) curves. There is only one TD in the first proposed Ternary Min circuit. The turning point of this TD

is set 3.5/4 (Figure 3). It means that the output of this inverter is "0" only if the sum of input variables becomes the largest amount, four ($\sum$ in = 4).

The last part of the circuit, which determines the output value, is the most important one. It contains different paths, through which the output currents flow. The dimensions of the transistor(s) on a specific path have to be set properly so that exactly the unit current of 8 μA flows in each path in case the related transistor(s) is/are switched on. The threshold voltage of T_{12} and T_{13} is set properly so that they switch on if $a \geq 1$ and $b \geq 1$, respectively. T_{15} copies the current to the output node (Min). In case both input variables equal logic value "2" ($\sum$ in = 4), the threshold detector switches T_{16} on, and another unit of current is added to the previous one. Hence, the total current reaches 16 μA (Min = "2").

The same approach is used to design the first proposed CML Ternary Max circuit (Figure 4). The threshold voltage of T_{16} is set "low." Therefore, it switches on as soon as $\sum$ in becomes greater than zero ($\sum$ in ≥ 1). T_{17} lets the current flow until $\sum$ in reaches four ($\sum$ in ≤ 3). Consequently, the path is active in case $1 \leq \sum$ in ≤ 3. T_{19} copies the same current to the output node (Max). There are two other parallel paths, constructed by the transistors T_{20} and T_{21}. In case only one of the inputs is "2" ($2 \leq \sum$ in ≤ 3), a unit of current is added to the previous one (the copied one). If both are "2" ($\sum$ in = 4), the former path is inactive and the added transistors (T_{20} and T_{21}) generate 16 μA current through two parallel paths.

3.2. The Second Approach without the Usage of Current Mirror in the Last Part. The second approach is almost the same as the first one except that the current mirror in the third part of the circuit is eliminated and the final currents are only generated by the p-type transistors. The second proposed CML Ternary Min and Max circuits are shown in Figures 5 and 6, respectively. In Figure 5, T_{16} and T_{17} function exactly the same way as T_{12} and T_{13} in Figure 3. However, the path

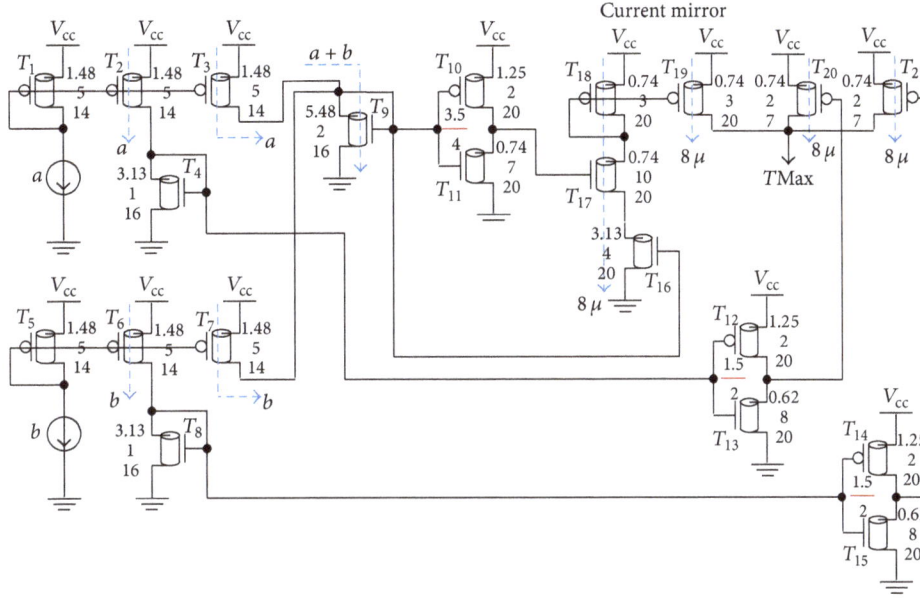

FIGURE 4: The first proposed Current-mode Segregated Ternary Max (CSTMax1) with the current mirror in the last part.

consists of only two transistors, one transistor fewer than the first approach, where current flows through two n-type transistors and a p-type one. The fewer transistors a path has, the more precise the amount of current is. In addition, the critical path shortens and the entire circuit becomes more robust.

In Figure 6, T_{20} and T_{21} are considered as the replacements of T_{16} and T_{17} in Figure 4. They let the current flow in case $1 \leq \sum \text{in} \leq 3$. Although it is possible to connect the sum of input currents directly to the gate of T_{21}, it increases robustness if the activity of T_{21} is controlled by a TD circuit. The full-swing output signal of a TD makes the following transistor switch on (or off) completely. In addition, there is no need to set the threshold voltage of this transistor precisely. It could have the same dimensions as T_{20}. It results in a simpler transistor sizing procedure as well. It is worth mentioning here that the threshold voltage of the CNTFET is an inverse function of D_{CNT} [16].

4. The Proposed Current-Mode Integrated Ternary Operators (CITO)

In some applications such as within the Arithmetic Logic Unit (ALU), both operators (AND and OR) are required at the same time. In this section, the previous segregated circuits are combined together in order to eliminate the common parts. The first proposed integrated circuit is shown in Figure 7. It is in fact the combination of CSTMin1 (Figure 3) and CSTMax1 (Figure 4), unless TDs are employed extensively to have a robust and accurate design. The common parts are the transistors which make the current to voltage conversion happen (T_1 to T_9) and the inverter with the turning point of 3.5/4.

It is also possible to share the n-type transistors (T_{28}/T_{29} and T_{33}/T_{34}). It leads us to the second integrated design

(Figure 8). While the Min circuit remains unchanged, the functionality of the Max circuit is summarized as follows (though a set of conditions):

(1) If $a \geq 1$ and $b \geq 1$: T_{26} and T_{27} switch on. A unit of current is copied to the both outputs (Min and Max).

(2) If $\sum \text{in} \geq 3$: T_{40} switches on. Another unit of current is added.

(3) If $a = $ "0" and $b \geq 1 \mid b = $ "0" and $a \geq 1$: either (T_{35} and T_{36}) or (T_{37} and T_{38}) are ON and a unit of current flows.

(4) If $a = $ "0" and $b = $ "2" $\mid a = $ "2" and $b = $ "0": either (T_{31} and T_{32}) or (T_{33} and T_{34}) are ON and another unit of current is added to (3). The output becomes "2".

The last design (Figure 9) is the combination of the second proposed segregated designs, CSTMin2 (Figure 5) and CSTMax2 (Figure 6). It has the fewest number of transistors among the integrated designs. It generates both outputs with the usage of only 30 transistors. Figure 10 shows the signal waveforms of this last design. The input pattern includes 72 transitions. The output signals are generated at the same time, and their values are exactly $0 \, \mu A$, $8 \, \mu A$, and $16 \, \mu A$, representing logic values "0", "1", and "2".

5. Simulation Results

All of the proposed designs as well as the previous ones (Figure 2) are simulated with Synopsys HSPICE and 32 nm CNTFET technology file [17, 18]. The simulations are carried out in 1 V power supply with 1 GHz operating frequency at room temperature. The complete input pattern including 72 transitions (Figure 10) is fed to the circuits to reveal the worst-case delay. During all transitions, the average power consumption is measured. The energy consumption

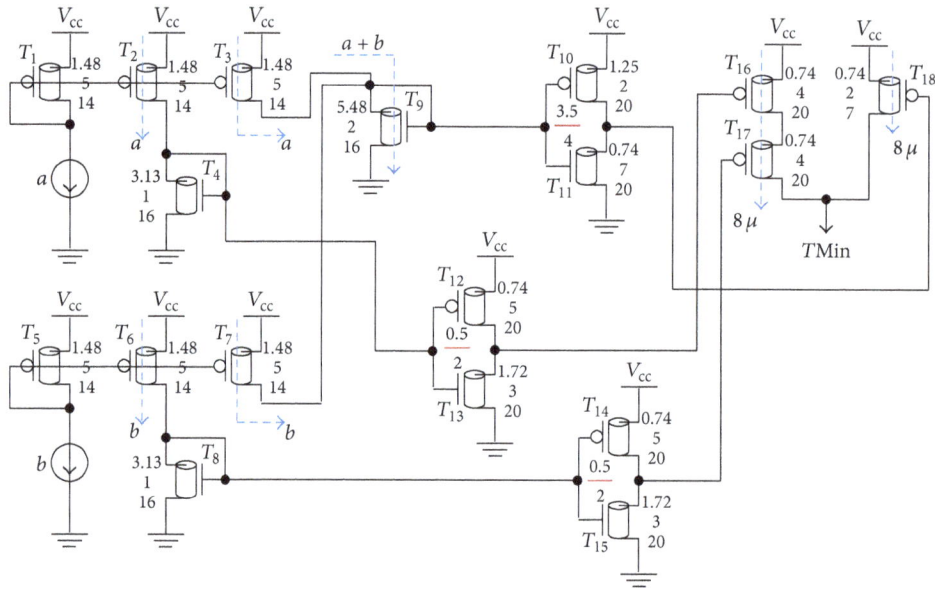

FIGURE 5: The second proposed Current-mode Segregated Ternary Min (CSTMin2) without the usage of current mirror in the last part.

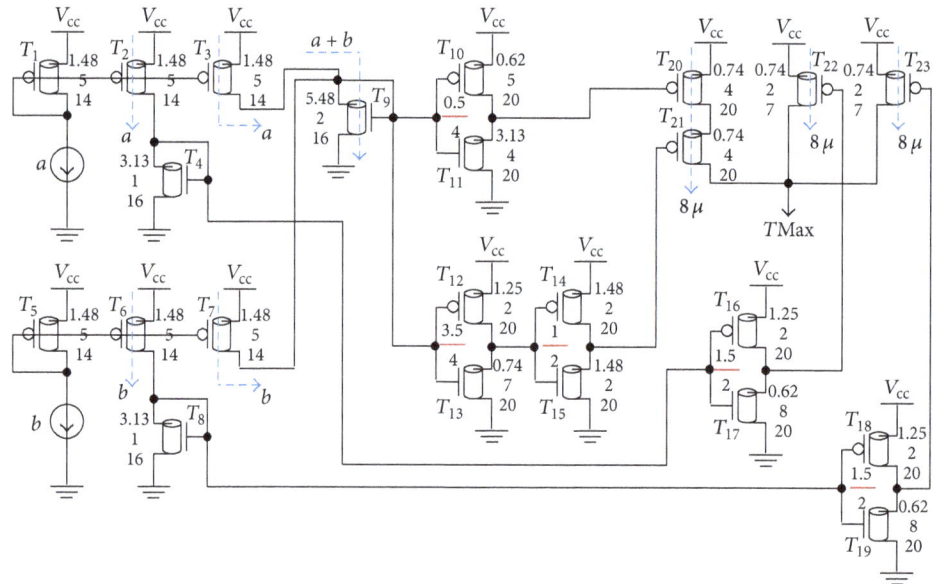

FIGURE 6: The second proposed Current-mode Segregated Ternary Max (CSTMax2) without the usage of current mirror in the last part.

(also known as PDP) is the multiplication of the maximum delay time and the average power consumption [16]. The average static power is also measured while the inputs are kept unchanged. The entire nine possible different input patterns (Table 1) are fed to the circuits to measure stand-by power dissipation. The average amount is reported as the static power.

The simulation results are exhibited in Table 2. In general, Max circuits consume more power than Min ones do. This is mainly because the output function is more likely to be "0" in Ternary AND than Ternary OR (Table 1). To talk literally, when applying a random input pattern, the probability of Ternary maximum being "0" is only 1/9, while Ternary

minimum has the same value in most cases (5/9). Therefore, less current flows in a Ternary Min circuit.

The proposed designs operate far more efficient than the previous ones. For example, the first and the second proposed segregated Ternary Min cells (CSTMin1 and CSTMin2) consume $25.64\,\mu$W (35%) and $29.37\,\mu$W (40%) less power than the previous one, respectively. Higher power consumption is mainly because of the constant current sources in the previous circuits. New designs are even faster, resulting in a great reduction in terms of PDP. Moreover, the second segregated design technique leads to less power consumption and higher efficiency due to the elimination of a few transistors and a parallel path. For instance, CSTMin2 operates

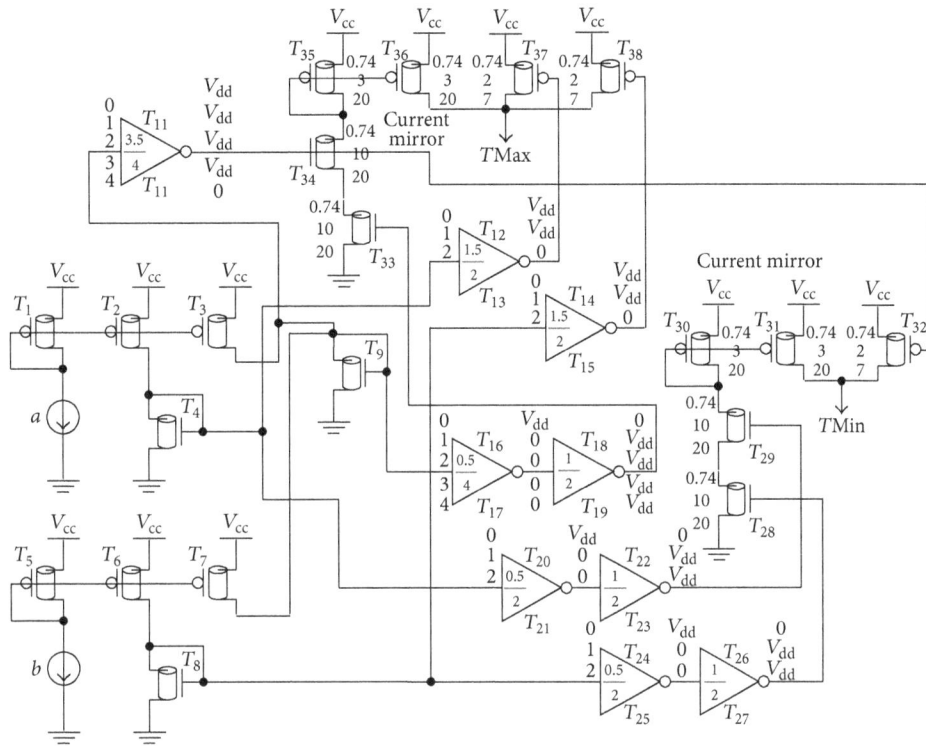

FIGURE 7: The first proposed Current-mode Integrated Ternary Min and Max (CITMin/Max1) with the current mirrors in the last parts.

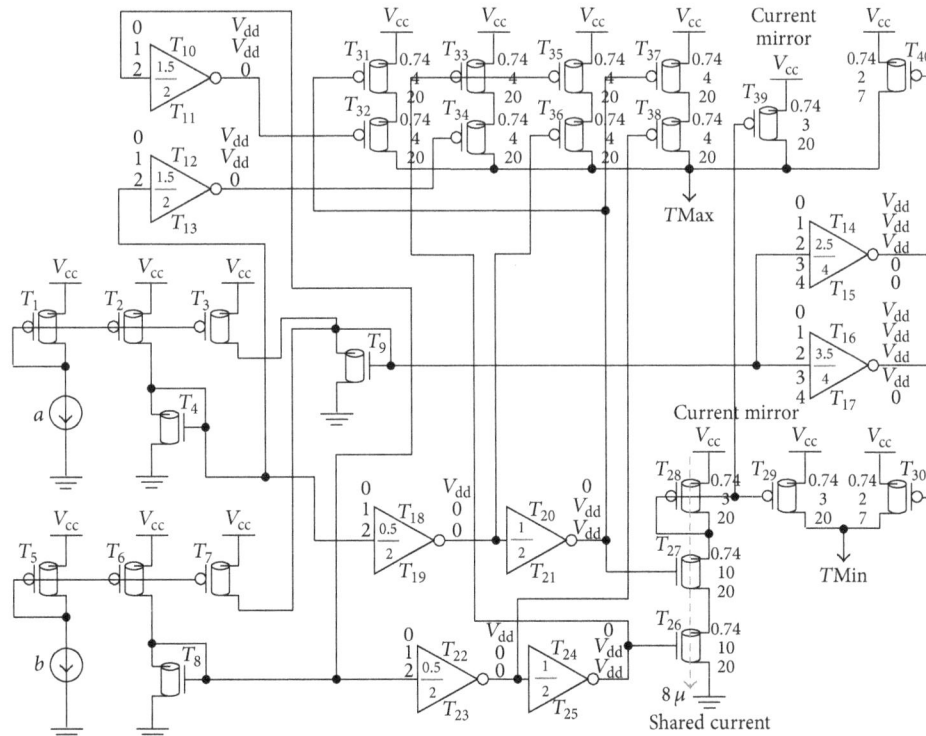

FIGURE 8: The second proposed Current-mode Integrated Ternary Min and Max (CITMin/Max2) with a shared current and the current mirrors in the last parts.

TABLE 2: The initial simulation results.

Designs	Delay (psec) TMin	Delay (psec) TMax	Average power (μW)	PDP (fJ)	Static power (μW)	Number of transistors
Previous TMin	19.64	—	72.56	1.42	72.94	19
CSTMin1	14.23	—	46.92	0.66	47.52	16
CSTMin2	11.17	—	43.19	0.48	43.71	18
Previous TMax	—	14.93	75.62	1.12	76.06	19
CSTMax1	—	13.60	56.73	0.77	57.51	21
CSTMax2	—	13.71	50.40	0.69	51.09	23
CITMin/Max1	18.48	20.68	65.06	1.34	65.69	38
CITMin/Max2	18.11	21.38	58.53	1.25	59.06	40
CITMin/Max3	17.57	18.82	54.84	1.03	55.55	30

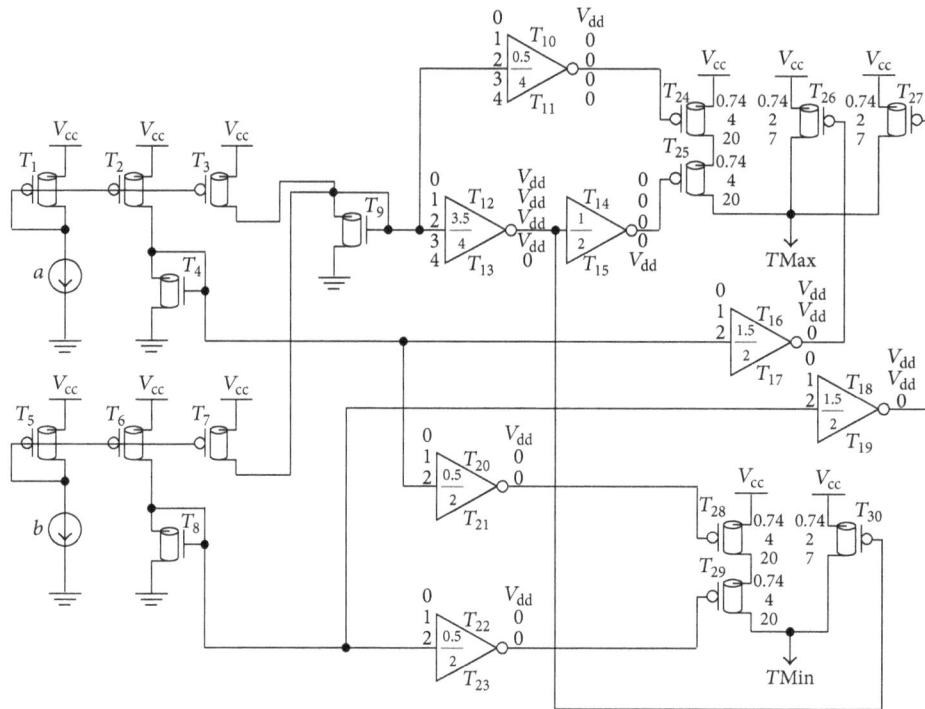

FIGURE 9: The third proposed Current-mode Integrated Ternary Min and Max (CITMin/Max3) without the usage of current mirrors in the last parts.

21% faster, consumes 8% less power, and has 35% higher performance (considering PDP) than CSTMin1. Eventually, the third integrated circuit (CITMin/Max3) has the highest performance in terms of energy. It generates both outputs with the fewest transistors needed.

Sensitivity to the variation of temperature is put under examination for the previous and some of the proposed designs. The amount of energy consumption (PDP) versus a wide range of ambient temperatures, from 0°C to 70°C, is plotted in Figure 11. The proposed designs show insignificant sensitivity to temperature variations.

One major advantage of CML over VML is that fan-out circuits do not cause speed degradation and performance loss for the current-mode circuits. This is due to the way that fan-out circuits are connected to a current-mode circuit. Instead of a direct connection, output currents are mirrored to the new branches (Figure 12). To test this capability, simulations are redone twice more. First, only the output load transistor is added to the proposed circuits. Then, four copies of the output current are also included in the examination. As it is demonstrated in Table 3, the existence of the output load transistor and the connection of the fan-out circuits do not increase cell delay. They do not add extra load to the output node. This is exactly in contrast with VML in which as the output load increases, voltage-mode circuits operate slower [6, 16].

TABLE 3: Delay parameter of the proposed designs versus the output load(s).

Designs	Without any output loads		With the output load transistor		With the output load transistor and 4 copies of the output current	
	Delay (psec) TMin	Delay (psec) TMax	Delay (psec) TMin	Delay (psec) TMax	Delay (psec) TMin	Delay (psec) TMax
CSTMin1	14.23	—	14.44	—	14.40	—
CSTMin2	11.17	—	13.45	—	13.34	—
CSTMax1	—	13.60	—	13.79	—	13.80
CSTMax2	—	13.71	—	14.03	—	13.99
CITMin/Max1	18.48	20.68	18.81	21.31	18.76	21.20
CITMin/Max2	18.11	21.38	18.46	21.72	18.53	22.08
CITMin/Max3	17.57	18.82	17.76	19.27	17.71	19.33

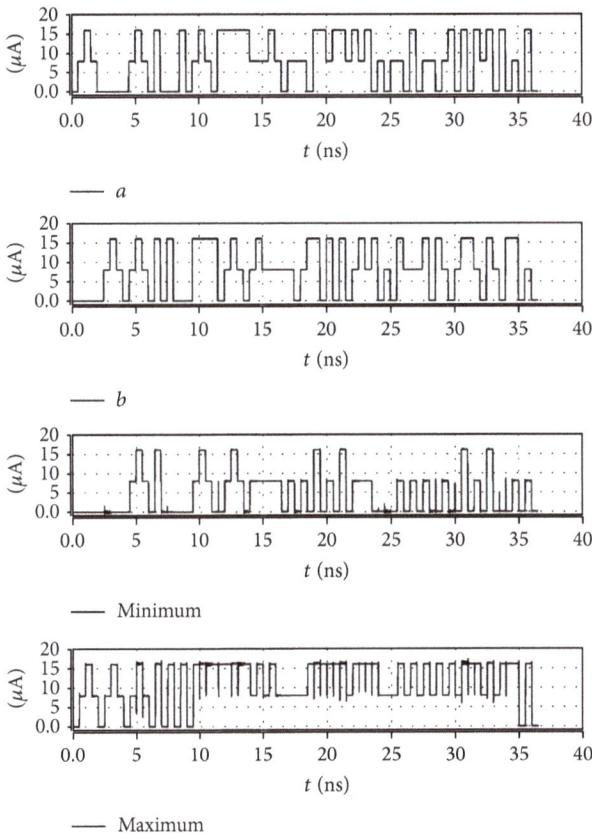

FIGURE 10: Input and output waveforms of CITMin/Max3.

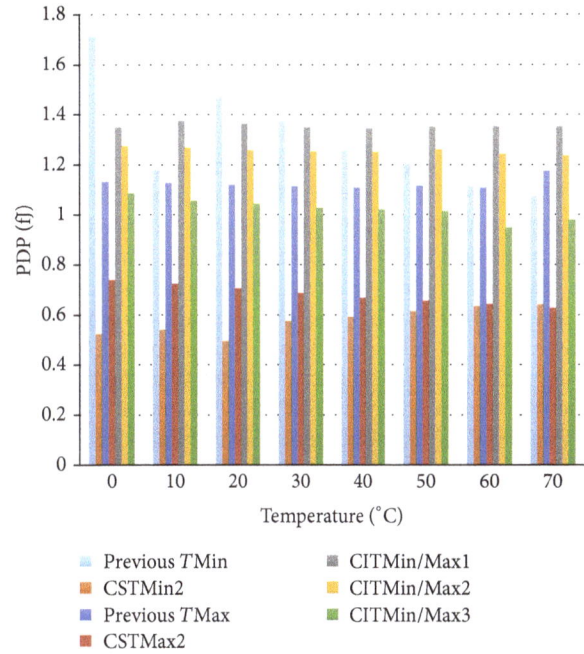

FIGURE 11: PDP versus temperature.

Finally, as it was mentioned in Section 2, digital circuits must have the ability of refreshing incomplete signals. This is an absolute fact in all discrete MVL systems. Otherwise, the signal offset intensifies and propagates within the whole system and it causes information loss eventually. Figure 13 shows how the circuit CITMin/Max3 restores an incomplete input signal. This is the ability the fuzzy circuits like the ones presented in [10–12] do not have. As a result, they are not appropriate for any digital systems such as Ternary.

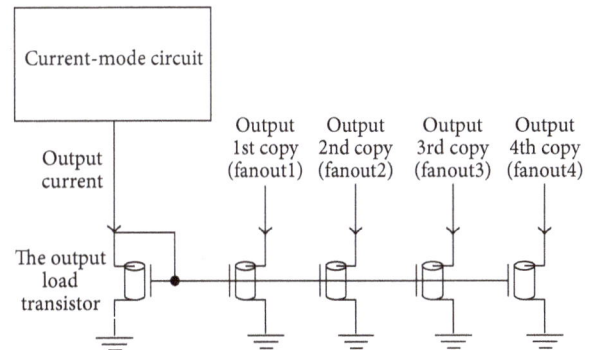

FIGURE 12: The output current is mirrored in CML to be connected to the fan-out circuits.

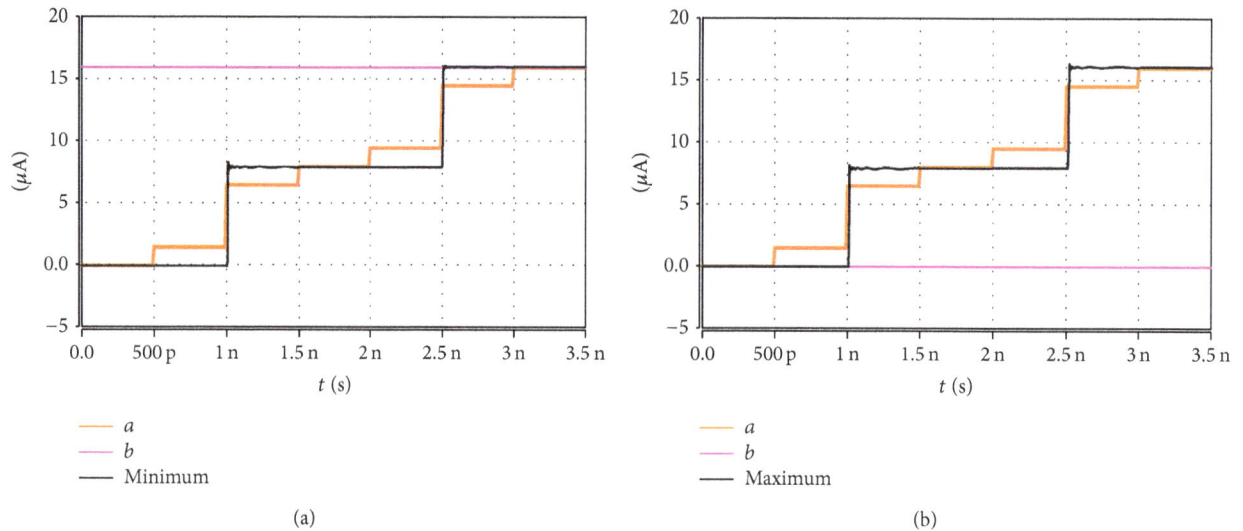

FIGURE 13: Signal restoration in CITMin/Max3, (a) Ternary minimum, (b) Ternary maximum.

6. Conclusion

In this paper, novel designs of current-mode Ternary Min and Max have been proposed. The new designs are based on mixed current and voltage logics, resulting in the elimination of constant independent current sources. Their elimination leads to higher performance for the proposed cells in comparison with the ones use constant current sources extensively as threshold detectors. The proposed CSTMin2 has approximately 66% higher efficiency than the previously presented design in terms of PDP. This paper also shows that, unlike VML, fan-out circuits have almost no effect on the delay parameter. Furthermore, Min circuits generally consume less power than the Max ones. This is due to the fact that the Ternary Min function is "0" in more input patterns than the Ternary Max one. For example, CSTMin2 consumes 7.21 μW less power than CSTMax2. Finally, the common parts could be easily integrated in order to combine two different circuits in CML. The integrated designs produce both Ternary minimum and maximum functions with almost the same speed.

Conflict of Interests

The authors declare that there is no conflict of interests regarding the publication of this paper.

References

[1] E. Dubrova, "Multiple-valued logic in VLSI: challenges and opportunities," in *Proceedings of the 17th NORCHIP Seminar (NORCHIP '99)*, pp. 340–350, Oslo, Norway, November 1999.

[2] A. Kazeminejad, K. Navi, and D. Etiemble, "CML current mode full adders for 2.5-V power supply," in *Proceedings of the 24th International Symposium on Multiple-Valued Logic*, pp. 10–14, May 1994.

[3] Y. Delican and T. Yildirim, "High performance 8-bit mux based multiplier design using MOS current mode logic," in

Proceedings of the 7th International Conference on Electrical and Electronics Engineering (ELECO '11), pp. 89–93, December 2011.

[4] T. Temel and A. Morgul, "Multi-valued logic function implementation with novel current-mode logic gates," in *Proceedings of the IEEE International Symposium on Circuits and Systems (ISCAS '02)*, vol. 1, pp. 881–884, 2002.

[5] S. Lin, Y.-B. Kim, F. Lombardi, and Y. J. Lee, "A new SRAM cell design using CNTFETs," in *Proceedings of the International SoC Design Conference*, pp. 168–171, November 2008.

[6] R. F. Mirzaee, M. H. Moaiyeri, M. Maleknejad, K. Navi, and O. Hashemipour, "Dramatically low-transistor-count high-speed ternary adders," in *Proceedings of the IEEE 43rd International Symposium on Multiple-Valued Logic (ISMVL '13)*, pp. 170–175, May 2013.

[7] S. L. Hurst, "Multiple-valued logic, its status and its future," *IEEE Transactions on Computers*, vol. 33, no. 12, pp. 1160–1179, 1984.

[8] T. Sakurai, "Perspectives on power-aware electronics," in *Proceedings of the IEEE International Solid-State Circuits Conference, Digest of Technical Papers (ISSCC '03)*, vol. 1, pp. 26–29, San Francisco, Calif, USA, February 2003.

[9] E. Özer, R. Sendag, and D. Gregg, "Multiple-valued logic buses for reducing bus energy in low-power systems," *IEE Proceedings—Computers and Digital Techniques*, vol. 153, no. 4, pp. 270–282, 2006.

[10] B. Mesgarzadeh, "A CMOS implementation of current-mode min-max circuits and a sample fuzzy application," in *Proceedings of the IEEE International Conference on Fuzzy Systems*, pp. 941–946, July 2004.

[11] G. Yosefi, S. Mirzakouchaki, and S. Neda, "Design of new CMOS current mode min and max circuits for FLC chip applications," in *Proceedings of the European Conference on Circuit Theory and Design (ECCTD '09)*, pp. 89–92, IEEE, Antalya, Turkey, August 2009.

[12] T. Temel and A. Morgul, "Implementation of multi-valued logic gates using full current-mode CMOS circuits," *Analog Integrated Circuits and Signal Processing*, vol. 39, no. 2, pp. 191–204, 2004.

[13] J. Shen, X. Chen, and m. Yao, "Design of symmetric ternary current-mode CMOS circuits," *Journal of Electronics*, vol. 14, no. 4, pp. 336–344, 1997.

[14] W. Xunwei, D. Xiaowei, and Y. Shiyan, "Design of ternary current-mode CMOS circuits based on switch-signal theory," *Journal of Electronics*, vol. 10, no. 3, pp. 193–202, 1993.

[15] Q. N. Zhou, M. Y. Yu, and Y. Z. Ye, "On-chip voltage down converter with precision CMOS current source for VLSI chip," in *Proceedings of the IEEE Conference on Electron Devices and Solid-State Circuits (EDSSC '05)*, pp. 375–378, December 2005.

[16] R. F. Mirzaee, K. Navi, and N. Bagherzadeh, "High-efficient circuits for ternary addition," *VLSI Design*, vol. 2014, Article ID 534587, 15 pages, 2014.

[17] J. Deng and H.-S. P. Wong, "A compact SPICE model for carbon-nanotube field-effect transistors including nonidealities and its application—part I: model of the intrinsic channel region," *IEEE Transactions on Electron Devices*, vol. 54, no. 12, pp. 3186–3194, 2007.

[18] J. Deng and H.-S. P. Wong, "A compact SPICE model for carbon-nanotube field-effect transistors including nonidealities and its application. Part II. Full device model and circuit performance benchmarking," *IEEE Transactions on Electron Devices*, vol. 54, no. 12, pp. 3195–3205, 2007.

15

Critical Gates Identification for Fault-Tolerant Design in Math Circuits

Tian Ban[1,2] **and Gutemberg G. S. Junior**[2]

[1]*School of Electronic and Optical Engineering, Nanjing University of Science and Technology, Nanjing 210094, China*
[2]*Department of Communications and Electronics, Institut Mines-Télécom, Télécom ParisTech, 75013 Paris, France*

Correspondence should be addressed to Tian Ban; tian.ban@njust.edu.cn

Academic Editor: Wen B. Jone

Hardware redundancy at different levels of design is a common fault mitigation technique, which is well known for its efficiency to the detriment of area overhead. In order to reduce this drawback, several fault-tolerant techniques have been proposed in literature to find a good trade-off. In this paper, critical constituent gates in math circuits are detected and graded based on the impact of an error in the output of a circuit. These critical gates should be hardened first under the area constraint of design criteria. Indeed, output bits considered crucial to a system receive higher priorities to be protected, reducing the occurrence of critical errors. The 74283 fast adder is used as an example to illustrate the feasibility and efficiency of the proposed approach.

1. Introduction

With the technology scaling, electronic circuits are becoming more and more prone to faults and defects. Reliability analysis of logic circuits is emerging as an important parameter in deep submicron electronic technologies [1, 2]. It is especially critical for systems designed to be applied in space, avionics, and biomedical applications. In order to design reliable nanoelectronic devices, different fault-tolerant strategies have been extensively researched over the past years [3, 4].

Modular redundancy is a representative method which can provide reliability enhancement to the detriment of area overhead. Motivated by the need of economical fault-tolerant designs, researchers have been committed to searching for better trade-offs between reliability and overhead [5]. A hybrid redundancy method is proposed in [6], which combines information and hardware redundancy to achieve better fault tolerance. Sensitive transistors are protected in [7] based on duplicating and sizing a subset of transistors necessary for soft error tolerance in combinational circuits. In [8], Ruano et al. presented a method to automatically apply Triple Modular Redundancy (TMR) on digital circuits. The idea is to meet the reliability constraint while reducing the area overhead of typical TMR implementation.

Although all the aforementioned works provide reductions in the area overhead when compared to classical hardware redundancy systems, they do not take account of the usage profile of the results. In fact, a designer may use this additional information to make better decisions about which are the critical blocks of a circuit and then assign the desired priorities to protect them.

This work first proposes a different approach to identify critical logic blocks in math circuits. It relies on the fact of many digital systems and applications to tolerate some loss of quality or optimality in the primary outputs. In most cases, the trade-off in area is also associated with improvement of performance like faster operations, less power consumption, and so forth. The main idea is that different errors may have different consequences for different digital applications. For instance, in a binary output word, errors located in the most significant bits tend to be more critical than errors located in the least significant bits.

This paper is organized as follows. Section 2 introduces the practical reliability concept and explains the advantages of using such metric for reliability analysis. In Section 3, a fast

TABLE 1: Reliability for the output bits of three architectures for a 4-bit adder.

Architecture	b_3	b_2	b_1	b_0	R_{nominal}	$R_{\text{practical}}$
1	99%	99%	99%	95%	92.18%	97.63%
2	95%	99%	99%	99%	92.18%	94.17%
3	98%	99%	99%	95%	91.25%	96.64%

adder circuit 74283 is applied as a case study to illustrate and validate the proposed method. As an example, the estimate of peak signal-to-noise ratio (PSNR) with different fault-prone critical gates in image processing is considered, together with both analysis and comparison of results. Finally, Section 4 outlines some conclusions and suggestions for future works.

2. Reliability Evaluation

2.1. Nominal Reliability. Let $\mathbf{y} = b_{M-1}b_{M-2}\cdots b_1 b_0$ be a vector of M bits representing the output of a circuit. The reliability of a circuit is usually defined as the probability that it produces correct outputs, that is, the probability that all $b_i \in \mathbf{y}$ are correct 0(s) and 1(s). Given that the output bits are independent, this value, also known as *nominal reliability* [2], is conventionally expressed as in (1), where R_i stands for the reliability of b_i:

$$R_{\text{nominal}} = \prod_{i=0}^{M-1} R_i. \tag{1}$$

Let us now suppose that the circuit's output "$\mathbf{y}$" is coded by the use of a binary scheme, where b_{M-1} and b_0 stand for the most significant bit (MSB) and the least significant bit (LSB), respectively. Actually, MSB is the bit position in a binary number having the greatest numerical value. Therefore, error(s) occurring in MSB(s) will result in more remarkable disparities than in any other bit. By contrast, errors in LSB(s) may even be masked by the target application.

In spite of that, nominal reliability assigns equal reliability costs to the bits of "$\mathbf{y}$" as shown in (1). In fact, two different architectures for a logic function may have the same reliability value and one may still have a higher probability to provide more acceptable results than the other does. For instance, let us suppose that a designer obtains three different architectures for a 4-bit adder in which the output is coded using a binary scheme. Besides, he has to select one among them based on the reliability of the output. The reliabilities for the output bits of such architectures are presented in Table 1.

Analyzing the nominal reliability values for the obtained architectures, *Architecture 1* and *Architecture 2* are selected as the best solutions. Indeed, no distinction can be made between these two architectures regarding the nominal reliability value. However, as the output of this circuit is coded using a binary scheme, the first architecture would provide better results (smaller disparities) than the second one. Ideally, a more desirable analysis should take account of the amount of information that each bit of an output carries (or its importance) in order to assign progressively great costs to them. To tackle this problem, a new metric to analyze the

reliability of a circuit with a multiple-bit output is presented in Section 2.2.

2.2. Practical Reliability. Practical reliability is a metric that can assess the importance of each output bit when analyzing the reliability of a circuit. It can be evaluated as shown in (2). The weight factor k_i allows a designer to adjust the importance of a specific output bit b_i to the output of the circuit. Notice that if $k_i = 1$, for all $0 \leq i \leq M - 1$, the practical reliability expression (2) becomes the nominal reliability expression (1). In this work, a standard binary representation is considered so that k_i is calculated as shown in (3). Note that (2) can also be related to the probability that an error will cause a significant disparity on the output of a circuit (a critical error).

$$R_{\text{practical}} = \prod_{i=0}^{M-1} R_i^{k_i}, \tag{2}$$

$$k_i = \frac{1}{2^{(M-1)-i}}. \tag{3}$$

Although the proposed metric does not evaluate the true reliability of a circuit, this value takes account of both the reliability and the importance of an output bit for the target application. This is of great value for practical applications. For instance, let us analyze the architectures shown in Table 1. It can be noted that the practical reliability values are different from the values obtained with nominal reliability. Actually, even the order of the best architectures changes with the proposed metric. *Architecture 2*, which was previously deemed the best architecture together with *Architecture 1*, now is viewed as the worst choice due to the low reliability value of its MSB. In fact, practical reliability punishes architectures which present low reliability in critical bits, thus providing a designer with a more realistic result based on the target application.

3. Selectively Hardening Critical Gates

We know that critical gates should be hardened first in order to increase hardware usage efficiency and, at the same time, to minimize area overhead. The main idea here is to grade gates in math circuit to be protected based on critical factors. In this work, a critical factor explores not only the probability that an error will be introduced by a gate but also how critical this error will be in the target application as shown in Section 3.1.

3.1. Identifying Critical Gates. In order to explain and validate the proposed method, the 4-bit fast adder 74283 is employed (see Figure 1). The first module (M_1) produces the generate,

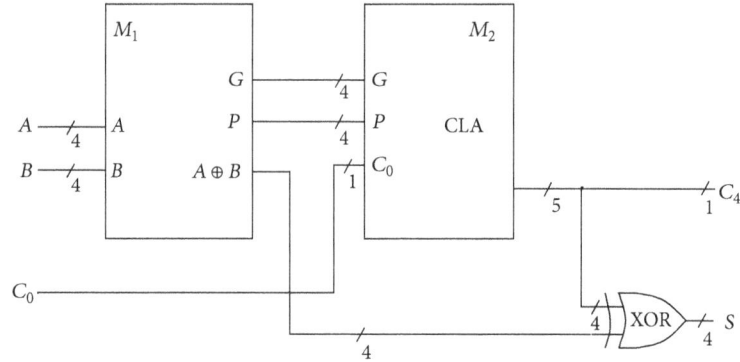

FIGURE 1: An illustration example: 4-bit adder 74283.

propagate, and XOR functions. The second module (M_2) is the carry-lookahead (CLA) realization for the carry function. Finally, the 8-bit XOR gate produces the sum function.

The fast adder 74283 has 9 inputs and 5 outputs and is composed of 36 logic gates and 4 buffers. All 40 blocks (gates and buffers) are considered as fault-prone. Further, it is supposed that these blocks (g_i ($i \in [0, 39]$)) are independent and labeled as shown in Figure 2.

The procedure of detecting which are the critical gates of this circuit takes two steps: first, a fault emulation platform, named FIFA [9], is used to inject faults due to Single Event Upsets (SEUs); next, *critical gates* are detected by analysis of errors that appear in the output vector.

The FIFA platform can generate one fault configuration per clock cycle. Further, it can inject a large number of simultaneous faults into the circuit [9]. However, in this work, it considers only the occurrence of single faults so that the platform injects just one fault each time. If the occurrence of multiple simultaneous faults is likely, the platform can be configured to deal with that.

Finally, the results, which are produced by the original and the faulty circuits, are compared bit by bit. If these results are different, it is concluded that the effects of the injected fault have been propagated to the output bits. Otherwise, it is concluded that the fault has been masked.

The fault injection emulation is performed to detect the critical factors. The idea is to inject a single fault in a gate g_i and analyze the output for all the possible input vectors. Then, for each output bit b_z, the number of errors S_z related to a single fault in g_i is evaluated (see Table 2). The columns S_{z_w} correspond to weighted versions of S_z. In our case study, as a standard binary representation is considered, S_{z_w} is obtained as shown in (4). Note that there are 2^9 possible input logic values for each faulty gate. All the simulation results are shown in Table 2.

$$S_{z_w} = 2^z \cdot S_z. \tag{4}$$

The *critical gates* are detected according to the results presented in Table 2. The more critical the gates are, the higher priorities they receive to be protected (in this case using TMR). Configuration of TMR based on this principle is more efficient in practical applications as shown in Section 3.2.

In fact, critical factors are assigned to the gates according to the number of weighted errors in Table 2. If the numbers of weighted errors are equal, gates that are closer to the primary outputs receive higher priorities. If the numbers of weighted errors and the distance to the primary outputs are both identical, gates presenting more reconvergent fan-outs are considered more critical. Gates whose three parameters are equal receive the same critical factor. Note that the rightmost column in Table 2 gives the critical factor for a gate g_i. The higher the factor number is, the more critical the gate will be. In this work, critical factors are assigned as integers $\in [0, 39]$.

3.2. Reliability Analysis and Comparison. Subsequent to obtaining the critical gates, the reliability of the redundant 74283 adder circuit is evaluated by using the SPR tool [1]. Further, the *signal reliability* of a given signal is considered as the probability that this signal carries a correct value. In fact, to assume that a binary signal x can carry incorrect information is equivalent to assuming that it can take four different values: correct zero (0_c), correct one (1_c), incorrect zero (0_i), and incorrect one (1_i).

The probabilities for occurrence of each one of these four values are represented as probability matrices shown as follows:

$$\begin{bmatrix} P\left(x = 0_c\right) & P\left(x = 1_i\right) \\ P\left(x = 0_i\right) & P\left(x = 1_c\right) \end{bmatrix} = \begin{bmatrix} x_0 & x_1 \\ x_2 & x_3 \end{bmatrix}. \tag{5}$$

The *signal reliability* of x, denoted as R_x, comes directly from (6), where $P(\cdot)$ stands for the probability function.

$$R_x = P\left(x = 0_c\right) + P\left(x = 1_c\right) = x_0 + x_3. \tag{6}$$

The SPR technique generates a matrix representing the output signal of a logical block, which explores the following information: the probability matrices representing the input signals for a given logical block, the logical function of such a block, and the probability that this block will not fail. In order to understand this procedure, let us consider a digital block b performing a logical function on a signal x to produce a signal y (see Figure 3). Now, assume that the probability that this operator will fail is represented by p, and $q = (1 - p)$

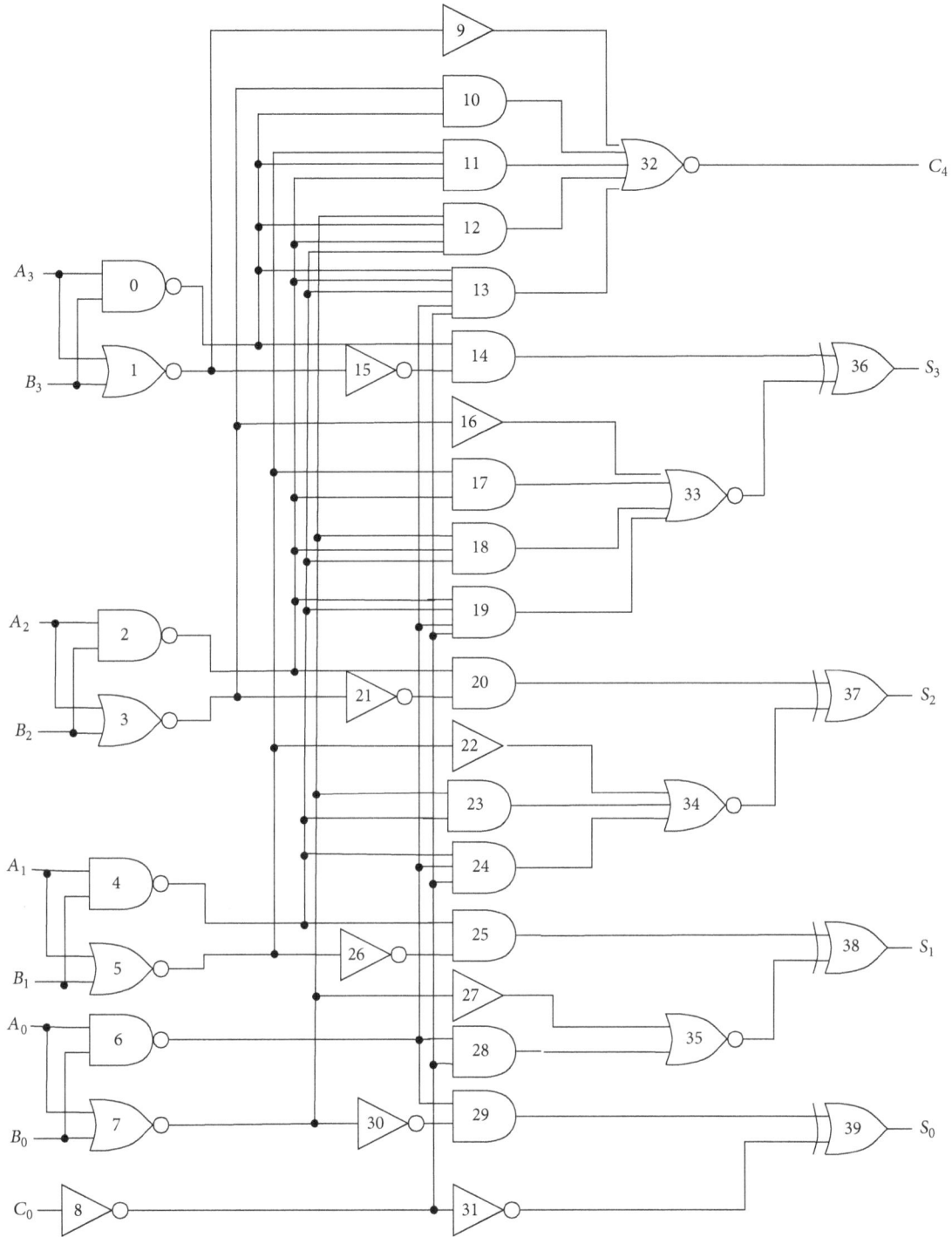

FIGURE 2: 74283 gate-level schematic.

represents the probability that it will not fail. Then, the reliability of y can be obtained by the following equation:

$$R_y = (x_0 + x_3) \cdot q + (x_1 + x_2) \cdot p. \tag{7}$$

As can be seen in (7), when the input signal is reliable, that is, $x_1 + x_2 = 0$, the reliability of the output signal is given by q, which stands for the probability of success of the logical block itself. This implies that, for fault-free inputs, the reliability of the output signal is given by the inherent reliability of the block that produces this signal.

Let us now consider hardware redundancy as the chosen redundancy technique to protect a logic block. Suppose that the area overhead constraint allows a designer to protect up to 5 gates. According to the critical factors presented in Table 2, gates g_{32}, g_1, g_3, g_0, and g_9 are selected by the proposed

TABLE 2: Error analysis for the gates of 74283.

g_i	S_0	S_{0_w}	S_1	S_{1_w}	S_2	S_{2_w}	S_3	S_{3_w}	C_4	C_{4_w}	$\sum errors_w$	Critical factor
0	0	0	0	0	0	0	384	3072	192	3072	6144	36
1	0	0	0	0	0	0	384	3072	320	5120	8192	38
2	0	0	0	0	384	1536	192	1536	96	1536	4608	33
3	0	0	0	0	384	1536	320	2560	160	2560	6656	37
4	0	0	384	768	192	768	96	768	48	768	3072	25
5	0	0	384	768	320	1280	160	1280	80	1280	4608	32
6	384	384	192	384	96	384	48	384	24	384	1920	14
7	384	384	320	640	160	640	80	640	40	640	2944	23
8	512	512	256	512	128	512	64	512	32	512	2560	22
9	0	0	0	0	0	0	0	0	320	5120	5120	35
10	0	0	0	0	0	0	0	0	288	4608	4608	34
11	0	0	0	0	0	0	0	0	272	4352	4352	31
12	0	0	0	0	0	0	0	0	264	4224	4224	29
13	0	0	0	0	0	0	0	0	272	4352	4352	31
14	0	0	0	0	0	0	512	4096	0	0	4096	27
15	0	0	0	0	0	0	384	3072	0	0	3072	24
16	0	0	0	0	0	0	320	2560	0	0	2560	21
17	0	0	0	0	0	0	288	2304	0	0	2304	20
18	0	0	0	0	0	0	272	2176	0	0	2176	18
19	0	0	0	0	0	0	288	2304	0	0	2304	20
20	0	0	0	0	512	2048	0	0	0	0	2048	17
21	0	0	0	0	384	1536	0	0	0	0	1536	13
22	0	0	0	0	320	1280	0	0	0	0	1280	12
23	0	0	0	0	288	1152	0	0	0	0	1152	10
24	0	0	0	0	320	1280	0	0	0	0	1280	12
25	0	0	512	1024	0	0	0	0	0	0	1024	7
26	0	0	384	768	0	0	0	0	0	0	768	6
27	0	0	320	640	0	0	0	0	0	0	640	4
28	0	0	384	768	0	0	0	0	0	0	768	6
29	512	512	0	0	0	0	0	0	0	0	512	2
30	384	384	0	0	0	0	0	0	0	0	384	0
31	512	512	0	0	0	0	0	0	0	0	512	1
32	0	0	0	0	0	0	0	0	512	8192	8192	39
33	0	0	0	0	0	0	512	4096	0	0	4096	27
34	0	0	0	0	512	2048	0	0	0	0	2048	15
35	0	0	512	1024	0	0	0	0	0	0	1024	8
36	0	0	0	0	0	0	512	4096	0	0	4096	28
37	0	0	0	0	512	2048	0	0	0	0	2048	16
38	0	0	512	1024	0	0	0	0	0	0	1024	9
39	512	512	0	0	0	0	0	0	0	0	512	3

method as the five candidates to be protected. The method presented in [8], under the same area overhead constraint, applies redundancy in gates g_{32}, g_{36}, g_{37}, g_{38}, and g_{39}. As the occurrence of single errors is assumed, the protected blocks are considered reliable; that is, $q = 1$.

The reliability of the output bits for the original circuit and for the redundant configurations can be obtained by the SPR technique. Table 3 shows the reliability results for the respective configurations considering $q = 0.99$ for the

gates not protected. It can be noted that both the nominal reliability and the practical reliability values are available. It is considered that the output of the 74283 adder comprises a 5-bit binary word so that the practical reliability can be evaluated from (2) and (3).

Analyzing the results presented in Table 3, it shows the effectiveness of the proposed approach. As commented above, the main idea is to take account of the impact of an error to the output of a circuit in order to prioritize the

TABLE 3: Reliability analysis of 74283 fast adder.

Reliability	No hardening	Method in [8]	Proposed method
S_0	94.07%	94.97%	94.07%
S_1	92.39%	93.26%	92.39%
S_2	91.80%	92.65%	92.43%
S_3	91.33%	92.17%	93.07%
S_4	94.60%	95.51%	97.15%
$R_{nominal}$	68.93%	72.24%	72.63%
$R_{practical}$	87.29%	88.89%	90.65%

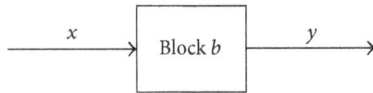

FIGURE 3: Generation of the output signal y from the input signal x processed by the digital block b.

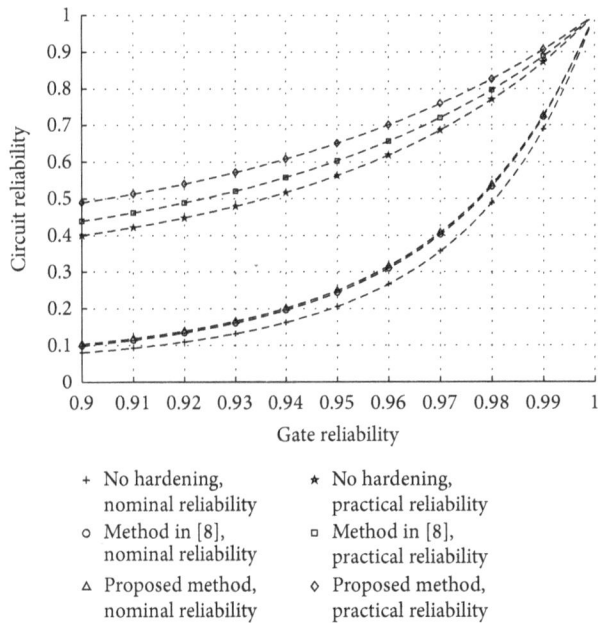

FIGURE 4: Simulation results for the 74283 fast adder.

reliability enhancement of the most important bits for the application. Indeed, the proposed hardening method shows a notable increase in the reliability of the most significant bits of the circuit (see Table 3). For instance, the reliabilities of S_0 and S_1 (LSBs) do not present any increase compared to the original circuit. Besides, the reliability of S_4 (MSB) presents the highest improvement as expected, once it is considered the most critical bit for this application.

Furthermore, it can be noted that, under the same area overhead, the nominal reliability increases by almost the same amount with both methods (see Figure 4). In fact, nominal reliability assigns equal reliability costs to the output bits of the 74283. This means that the output bits are considered as having the same importance to the system, so that the nominal reliability value does not distinguish in which bit the reliability was actually increased. In spite of that, practical reliability results can handle this problem and can indeed provide a sharper distinction between these two hardened architectures as shown in Figure 4.

4. Conclusion

In this paper, we presented a method to selectively apply hardening method to arithmetic circuits. Critical constituent gates are detected by taking account of not only the probability of error occurrence but also the impact of such error to the system. Indeed, bits considered critical to the target application receive higher priorities to be protected when the proposed method is employed.

Simulation results show the effectiveness of the proposed approach. This indicates that such critical gates should be hardened with priorities in order to increase hardware usage efficiency and to minimize area overhead simultaneously. The results could also be combined to approximate computing algorithms dedicated to fault-tolerant design [10]. Future works include approximating logic design based on gate grading results.

Conflicts of Interest

The authors declare that there are no conflicts of interest regarding the publication of this paper.

Acknowledgments

This work was partially supported by National Natural Science Foundation of China (Grant no. 61401205).

References

[1] D. T. Franco, M. C. Vasconcelos, L. Naviner, and J.-F. Naviner, "Signal probability for reliability evaluation of logic circuits," *Microelectronics Reliability*, vol. 48, no. 8-9, pp. 1586–1591, 2008.

[2] P. Zhu, J. Han, L. Liu, and F. Lombardi, "A stochastic approach for the analysis of dynamic fault trees with spare gates under probabilistic common cause failures," *IEEE Transactions on Reliability*, vol. 64, no. 3, pp. 878–892, 2015.

[3] T. Ban and L. Naviner, "Progressive module redundancy for fault-tolerant designs in nanoelectronics," *Microelectronics Reliability*, vol. 51, no. 9–11, pp. 1489–1492, 2011.

[4] P. K. Samudrala, J. Ramos, and S. Katkoori, "Selective triple modular redundancy (STMR) based single-event upset (SEU) tolerant synthesis for FPGAs," *IEEE Transactions on Nuclear Science*, vol. 51, no. 5, pp. 2957–2969, 2004.

[5] V. Hamiyati Vaghef and A. Peiravi, "Node-to-node error sensitivity analysis using a graph based approach for VLSI logic circuits," *Microelectronics Reliability*, vol. 55, no. 1, pp. 264–271, 2015.

[6] D. A. Tran, A. Virazel, A. Bosio et al., "A new hybrid fault-tolerant architecture for digital CMOS circuits and systems," *Journal of Electronic Testing*, vol. 30, no. 4, pp. 401–413, 2014.

[7] A. T. Sheikh, A. H. El-Maleh, M. E. S. Elrabaa, and S. M. Sait, "A fault tolerance technique for combinational circuits based on selective-transistor redundancy," *IEEE Transactions on Very Large Scale Integration (VLSI) Systems*, vol. 25, no. 1, pp. 224–237, 2016.

[8] O. Ruano, J. A. Maestro, and P. Reviriego, "A methodology for automatic insertion of selective TMR in digital circuits affected by SEUs," *IEEE Transactions on Nuclear Science*, vol. 56, no. 4, pp. 2091–2102, 2009.

[9] L. A. B. Naviner, J.-F. Naviner, G. G. Dos Santos Jr., E. C. Marques, and N. M. Paiva, "FIFA: a fault-injection-fault-analysis-based tool for reliability assessment at RTL level," *Microelectronics Reliability*, vol. 51, no. 9–11, pp. 1459–1463, 2011.

[10] H.-J. Wunderlich, C. Braun, and A. Schll, "Fault tolerance of approximate compute algorithms," in *Proceedings of the 34th VLSI Test Symposium*, p. 1, Las Vegas, Nev, USA, 2016.

16

A Mixed-Signal Programmable Time-Division Power-On-Reset and Volume Control Circuit for High-Resolution Hearing-Aid SoC Application

Chengying Chen [ID],[1] **Liming Chen,**[1] **and Jun Yang**[2]

[1]*School of Opto-Electronic and Communication Engineering, Xiamen University of Technology, Fujian, Xiamen 361024, China*
[2]*School of Automation, Foshan University, Foshan 528000, China*

Correspondence should be addressed to Chengying Chen; chenchengying363@163.com

Academic Editor: M. Jamal Deen

A mixed-signal programmable Time-Division Power-On-Reset (TD-POR) circuit based on 8-bit Successive Approximation Analog-to-Digital Converter (SAR ADC) for accurate control in low-power hearing-aid System on Chip (SoC) is presented in this paper. The end-of-converter (EOC) signal of SAR ADC is used as the mode-change signal so that the circuit can detect the battery voltage and volume voltage alternately. And the TD-POR circuit also has brown-out reset (BOR) detection capability. Through digital logic circuit, the POR, BOR threshold, and delay time can be adjusted according to the system requirement. The circuit is implemented in SMIC 0.13 μm 1P8M CMOS process. The measurement results show that, in 1 V power supply, the POR, BOR, and volume control function are accomplished. The detection resolution is the best among previous work. With 120 Hz input signal and 15 kHz clock, the ADC shows that Signal to Noise plus Distortion Ratio (SNDR) is 46.5 dB and Effective Number Of Bits (ENOB) is 7.43 bits. Total circuit power consumption is only 86 μw for low-power application.

1. Introduction

With the development of VLSI integrated circuits, SoC becomes more and more complex, which not only consists of transitional analog circuits, but also comprises registers, memory, and Digital Signal Processing (DSP) circuits. When SoC is powered up, some of their internal nodes are with intermediate metastable states. Therefore at the initial stage of operation, a Power-On-Reset (POR) circuit is needed. The POR signal should hold circuits in the reset state until the power supply reaches a steady-state level when all the circuits can operate correctly. And, in many applications, when supply voltage suddenly drops, the SoC may malfunction. The POR circuit also needs to generate a brown-out reset (BOR) signal for warning or reset. The parameter definition of POR and BOR is shown in Figure 1.

V_{POR} and V_{BOR} are POR and BOR thresholds, respectively, which determine the system to enter or leave the stable state. DT_{POR} and DT_{BOR} are the delay time to generate the POR and BOR pulse when the supply voltage reaches V_{POR} or drops

under V_{BOR}, respectively. DT_{POR} and DT_{BOR} must be set long enough to confirm all the modules have entered steady or reset state. T_{BOR} and T_{POR} are the POR and BOR pulse width. As long as the DT_{POR} and DT_{BOR} meet the system specification, T_{POR} and T_{BOR} have no specific requirements. Transitional POR circuits are pure analog circuit that are based on delay cell and Schmitt Trigger [1–4]. Delay time is set by resistor and capacitor of delay cell occupying large chip area. The resistance and capacitance devices are affected by temperature, voltage, and other parameters in the process of integrated circuit, and the uncertainty of the delay time can be easily caused. At the same time, the fixed POR and BOR threshold are generated by the bandgap-based reference, which has a large temperature drift. So it can greatly affect the detection accuracy of the POR and BOR voltage that can hardly meet the high-resolution requirement.

The zinc-air battery that supplies hearing-aid SoC has a large voltage variation from 1.4 V to 0.7 V. When supply voltage reaches or drops to the right threshold, to ensure the accurate power-on control and low-voltage warning,

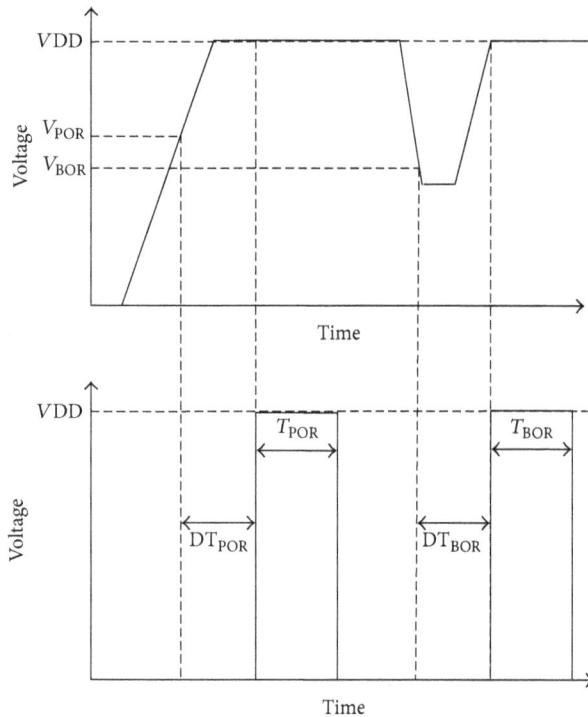

FIGURE 1: Parameter definition of POR and BOR.

channel for high frequency power supply ripple to ground and relieve the noise effect.

A complete SAR ADC conversion cycle requires 12 clock cycles, where three clock cycles are for sampling, eight cycles are for a binary searching, and half cycle is for EOC signal. Common mode voltage 0.5 V and reference voltage 0.6 V are provided by the LDO circuit. Clock generator provides 15 kHz frequency signal for ADC. EOC is divided into two outputs: one is input to the clock port of a D flip-flop. And the output of the D flip-flop controls input stage of MUX which completes the input switching. At the same time the output of the D flip-flop goes through an inverter and then generates Time-Division multiplex flag signal Vc_bat_mux. And Vc_bat_mux is output to the digital logic circuit (when Vc_bat_mux is high, the input voltage signal is the volume signal; when Vc_bat_mux is low, the input voltage signal is the battery signal). The second output of EOC goes through a delay buffer to clock ports of eight D flip-flops, which maintains the ADC 8-bit output data for a conversion cycle. Meanwhile, the output of the delay buffer is delayed by seven clock cycles and generates a digital logic read flag signal Data_en, to ensure the output can be read by digital logic circuit correctly. Finally the 8-bit ADC output codes Out$\langle 7{:}0 \rangle$ are maintained by D flip-flops for a SAR ADC conversion cycle until the next EOC signal arrives.

The inputs of digital logic circuit are 8-bit ADC output code, Data_en, Vc_bat_mux, 4 bit V_{POR}, and 2 bit V_{BOR}. The final outputs of digital logic circuits are POR counting signal cnt_en and POR (BOR) signal Soft_rst. Firstly V_{POR} is set after initialization. When the battery voltage starts to rise and exceeds V_{POR}, cnt_en starts counting. If the voltage is greater than V_{POR} and lasts for 30 ms, Soft_rst turns to high value; when the battery voltage drops to V_{BOR} and also lasts for 30 ms, Soft_rst turns to low value and makes system enter the reset state. According to signal Vc_bat_mux, TD-POR circuit determines whether the battery voltage or volume voltage is read at this time. When Vc_bat_mux is high, the 8-bit ADC output codes will be output to the DSP for volume control. And when Vc_bat_mux is low, the circuit runs into the POR mode. Timing relationships of Data_en, Out$\langle 7{:}0 \rangle$, and Vc_bat_mux are shown in Figure 3, vc0, vc1, ... represented the volume voltage, and bat0, bat1, ... indicated the battery voltage.

the POR with ADC must be needed. Moreover V_{POR} and V_{BOR} as well as DT$_{POR}$ and DT$_{BOR}$ should be programmable due to battery characteristic. Meanwhile the customers may adjust the volume of hearing-aid device according to the environment and their hearing condition. And the volume adjustment usually has 32 stages. For precision, an 8-bit ADC with 256 quantized steps is used to divide the adjustment stages. Since the power determines the operation duration of hearing-aid SoC and both supply voltage and volume voltage are 1 V, to minimize the power consumption and realize accurate control, a mixed-signal programmable TD-POR and volume control circuit based on SAR ADC for hearing-aid SoC is presented in this paper. The EOC signal of SAR ADC is used as the mode-change signal so that the circuit can detect the battery voltage and volume voltage alternately, which reduces both circuit size and total power consumption.

2. System Design Considerations

Figure 2 shows the architecture of the proposed TD-POR and volume control circuit, which consists of Time-Division input stage, 8-bit SAR ADC, clock generator, LDO, and digital logic circuit. The input stage comprises a MUX, two resistors ($R1$, $R2$), and a MOS capacitor. MUX controlled by EOC signal inputs battery voltage (bat_level) and volume voltage (vc_level) alternately. The ratio of $R1$ and $R2$ is $2{:}3$. Since in 1 V power supply the voltage range of battery and volume is 0-1 V, the two resistors' mechanism reduces the voltage detection range to 0-0.6 V to minimize the detection error. And the MOS capacitor in input stage is used to provide a

3. Circuit Design

The core of POR circuit is an 8-bit SAR ADC as shown in Figure 4, including sample/hold circuit, analog-to-digital converter (DAC), comparator, successive approximation logic circuit, and timing generator [5–9].

Figure 5 shows the circuit of 8-bit charge redistribution DAC. To meet the requirement of full-swing input, only one sampling capacitor (Cs) of 32 times unit capacitor is adopted [10].

The DAC has two phases as sampling phase and charge redistribution phase. When DAC is in sampling phase, the switches $S0{\sim}S7$ are connected with GND. The switch Ssample is connected with VIN. And Svcm is also on, so the

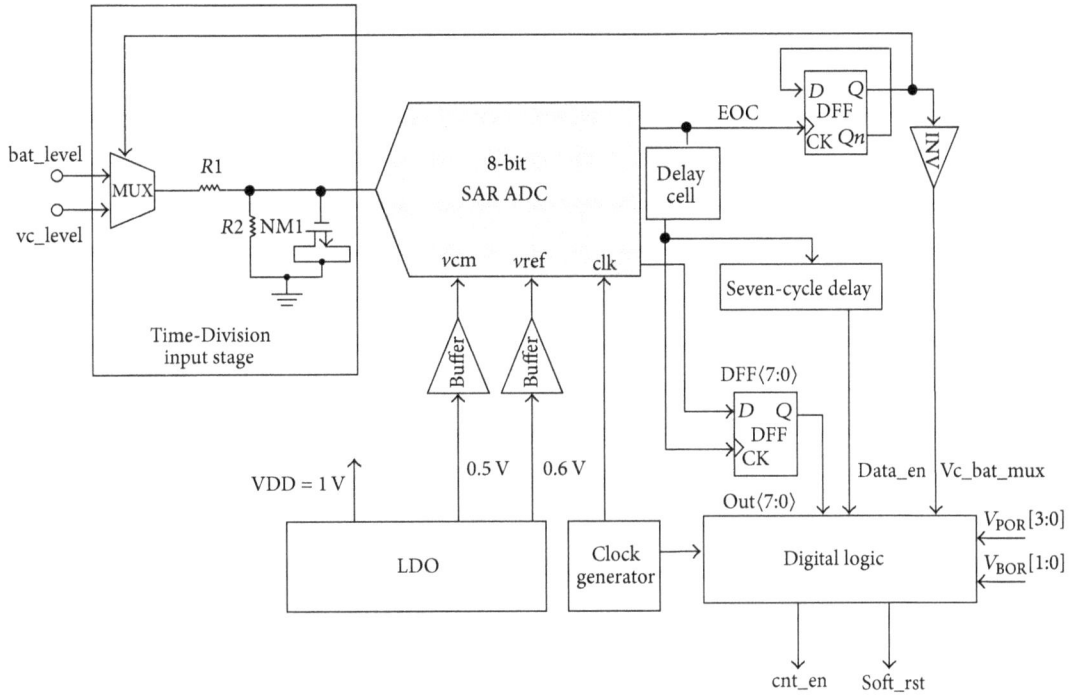

FIGURE 2: Block diagram of Time-Division Power-On-Reset and volume control circuit.

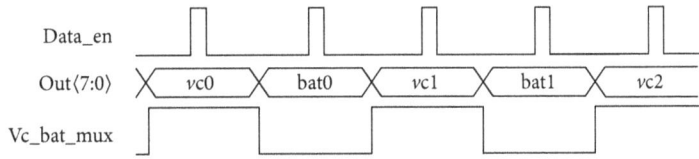

FIGURE 3: Analog circuit outputs timing.

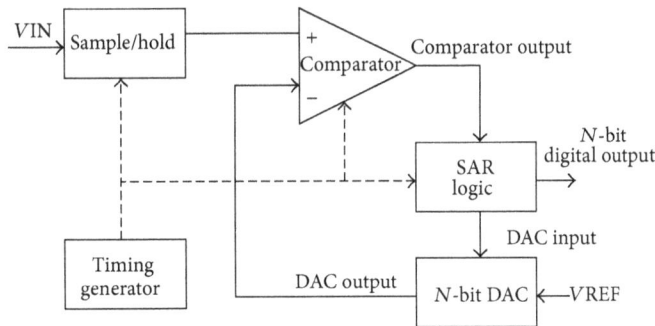

FIGURE 4: Block diagram of SAR ADC.

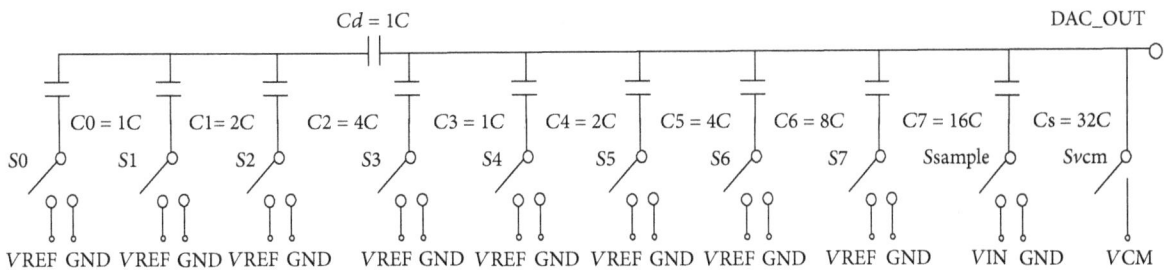

FIGURE 5: DAC structure.

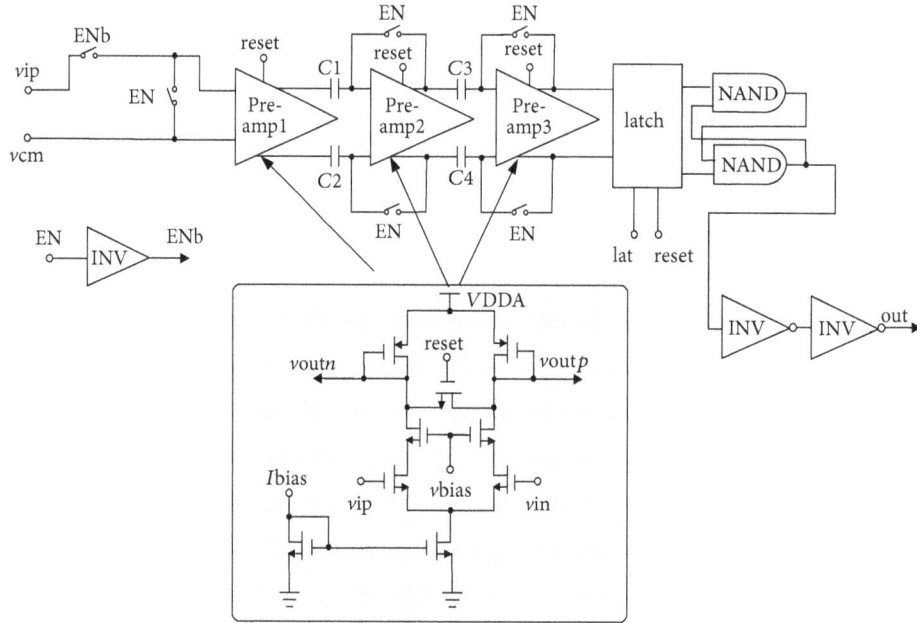

FIGURE 6: Comparator structure and circuit.

lower plate and the upper plate of Cs are connected with VIN and VCM, respectively. Therefore the output of DAC is

$$V_x = \frac{Q_x}{C_t} = \frac{32C\,(vcm - vin) + vcm\,(7C//1C + 31C)}{7C//1C + 63C}$$
$$= -\frac{256}{511}vin + vcm. \tag{1}$$

When DAC is in charge redistribution phase, the Most Significant Bit (MSB/7th bit) is initialized as "1" and the switch $S7$ is connected to reference voltage $VREF$. At this time if the input voltage (VIN) is greater than half of $VREF$, the comparator output is "0" and the MSB stays as "1"; otherwise the MSB changes to "0." After that the 6th bit is initialized as "1" and $S6$ is connected to $VREF$. Then the 6th bit can be "1" or "0" which depends on the output of comparator. The DAC needs eight cycles until the Least Significant Bit (LSB) is determined. And the DAC output is

$$V_x = \frac{256}{511}\left(-vin + \sum_{i=1}^{8}\frac{b_i}{2^{9-i}}vref\right) + vcm, \tag{2}$$

where b_i is the i bit output of DAC.

The comparator is shown in Figure 6, which comprises three preamplifiers and an output latch. The three preamplifiers with input offset reduction and output offset reduction techniques can relieve the effect of offset voltage efficiently. The output latch is adopted to change the differential output into single output and hold the output [11, 12].

SAR logic circuit is the most important part of SAR ADC. It realizes the feedback control of DAC. A complete ADC conversion requires twelve clock cycles, including two sampling

cycles, nine successive approximation cycles, and one end-signal generation cycle. The SAR logic circuit comprises D flip-flops, inverters, AND gates, and JK flip-flops, as shown in Figure 7. $F0{\sim}F7$ are 8-bit SAR logic registers made up of JK flip-flops. FS, GA, and GB constitute the start-up circuit, and timing generator consists of $FA{\sim}FJ$. EN is start-up signal lasting for two clock cycles. Vc is the comparator output. EOC is the end-of-converter signal. $D7{\sim}D0$ are DAC input signal and $b7{\sim}b0$ are the digital output of ADC [13].

The timing of POR signal is shown in Figure 8. The signal sample is a three-cycle sampling signal. EN is the start-up signal of SAR logic circuit. And reset and lat are the reset signal of preamplifier and latch, respectively.

4. Experiment Result

The proposed TD-POR circuit has been implemented in $0.13\,\mu m$ CMOS technology with 1 V supply. The chip is shown in Figure 9. The power consumption and the core chip area are $86\,\mu W$ ($28\,\mu W$ @ LDO, $25\,\mu W$ @ SAR ADC, $20\,\mu W$ @ oscillator, and $13\,\mu W$ @ others) and $0.175\,mm^2$ ($0.675\,mm^2$ including IO cells), respectively. Figures 10 and 11 show the measurement results. When the input signal Vp-p is 800 mV, the input frequency is 120 Hz and the clock frequency is 15 kHz, the FFT spectrum of ADC output signal is analyzed, the SNDR is 46.5 dB, and ENOB reaches 7.43 bits. As shown in Figure 11(a), when the voltage exceeds V_{POR} (800 mV), cnt_en turns to high value and starts counting. When its counting exceeds 30 ms, which meets the POR condition, Soft_rst turns to high value and completes the power-on process. In Figure 11(b) when the battery voltage drops below V_{BOR} (910 mV), and cnt_en counts for 30 ms; Soft_rst turns to low value to reset the system.

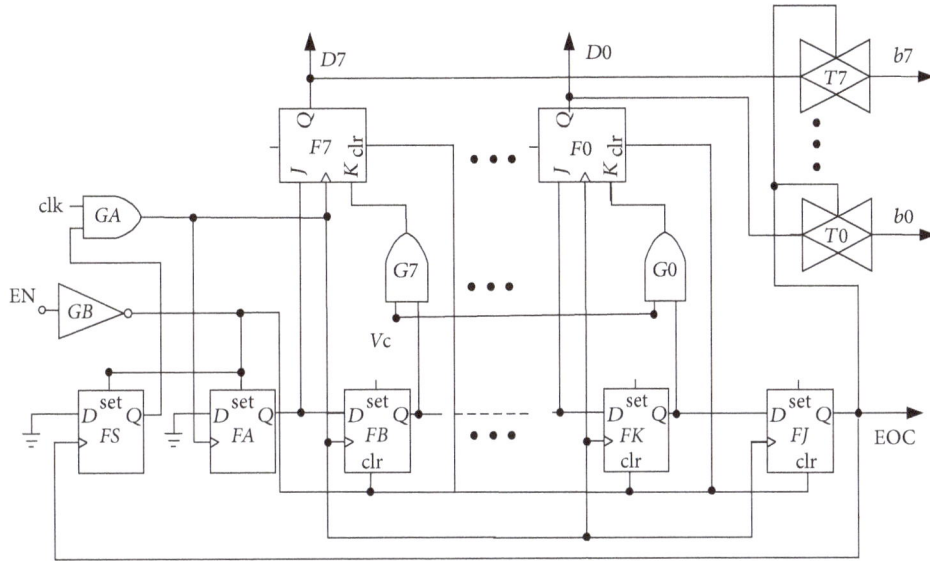

FIGURE 7: SAR logic circuit.

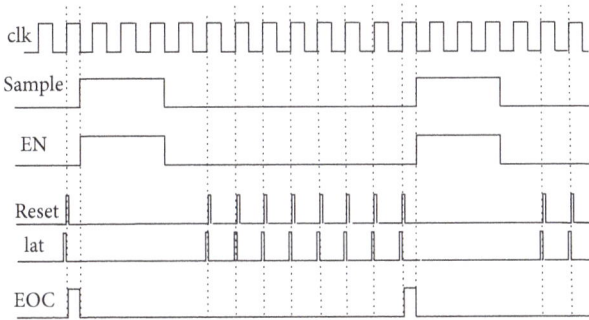

FIGURE 8: Timing of POR circuit.

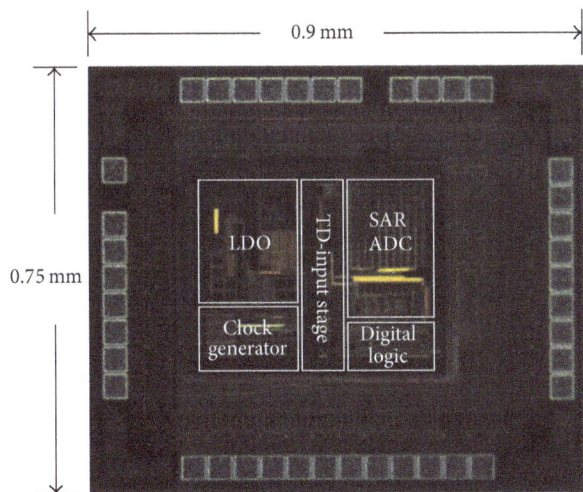

FIGURE 9: Chip of the proposed circuit.

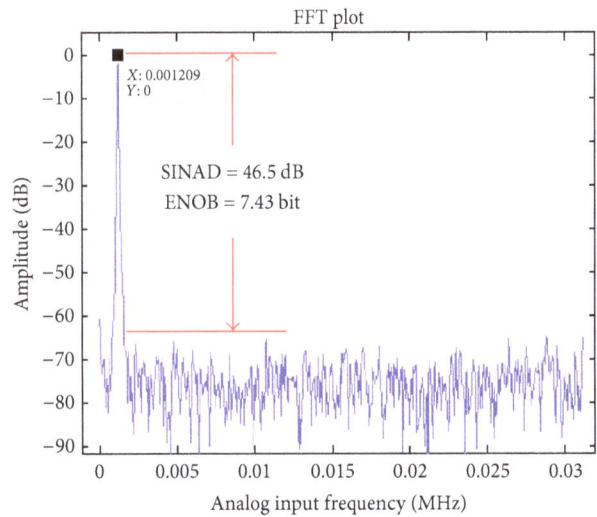

FIGURE 10: FFT measurement result of SAR ADC.

POR circuits are also listed. The circuit designed in this paper is the only one POR circuit, which can be configured with V_{POR}/V_{BOR} and delay time, and has high flexibility. It also has the highest detection accuracy. But due to the ADC-based structure the chip area is larger than other circuits.

In order to optimize power consumption, because the frequency of both POR and volume control signal is low, a suitable clock signal frequency is used to reduce DAC activity, thus decreasing the dynamic switching power. On the other hand, near-threshold transistors are adopted to design preamplifiers and comparators that save much power than their saturation opponents. Since the power dissipation of traditional analog volume control circuit is in hundreds of microwatts, our work puts forward a Time-Division multiplexing way, which completes the two functions of POR/BOR

The performance summary of the proposed Time-Division Power-On-Reset and volume control circuit is given in Table 1, and the performance comparisons with published

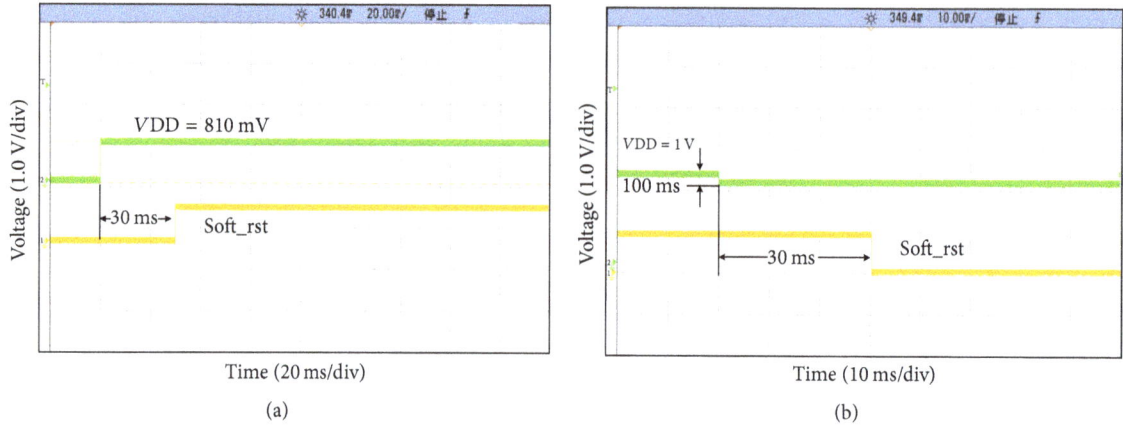

FIGURE 11: Function of measurement: (a) Soft_rst for a POR event; (b) Soft_rst for a BOR event.

TABLE 1: Comparison with previous work.

Parameters	This work	[1]	[2]	[4]
Process	CMOS 0.13 μm	CMOS 0.18 μm	CMOS 0.5 μm	CMOS 0.25 μm
Supply	1 V	1.8 V	1.8–5 V	2.5 V
BOR detection	Yes	Yes	Yes	No
V_{POR}/V_{BOR}	Programmable	Fixed	Fixed	Fixed
Delay time (30 ms in this paper)	Programmable	Fixed	Fixed	Fixed
ADC function	Yes	No	No	No
Other functions	Volume control	No	No	No
Detection resolution	<4 mV	NA	NA	NA
Power consumption	25 μW (core)	1.8 μW	115.5 μW	125 μW
Active size	0.175 mm^2	0.012 mm^2	0.001925 mm^2	0.0396 mm^2

and volume control through a single circuit, and it also has excellent power consumption and detection performance.

5. Conclusion

A mixed-signal programmable TD-POR and volume control circuit for hearing-aid SoC is proposed in this paper. It uses the EOC signal of SAR ADC as the mode-change signal that makes circuit detect the battery voltage and volume voltage alternately. And the POR and BOR function are both included. The mixed-signal structure ensures flexible configuration of V_{POR}/V_{BOR} and delay time, which greatly improve the stability of the application system. The circuit is implemented in SMIC 0.13 μm 1P8M CMOS process. The measurement results show that, in 1 V power supply, the POR, BOR, and volume control function are accomplished. With 120 Hz input frequency and 15 kHz clock frequency, the ADC shows that SNDR is 46.5 dB and ENOB is 7.43 bits. Total circuit power consumption is 86 μw. And the circuit shows the best detection resolution among all references.

Conflicts of Interest

There are no conflicts of interest related to this paper.

Acknowledgments

This work is supported by the National Natural Science Foundation of China (Grant no. 61704143), Young and Middle-Aged Teacher Education Research Project of Fujian Province (Grant no. JAT170428), and High-Level Talent Project of Xiamen University of Technology (Grant no. YKJ17019R).

References

[1] H.-B. Le, X.-D. Do, S.-G. Lee, and S.-T. Ryu, "A long reset-time power-on reset circuit with brown-out detection capability," *IEEE Transactions on Circuits and Systems II: Express Briefs*, vol. 58, no. 11, pp. 778–782, 2011.

[2] K. Shu, E. Sánchez-Sinencio, J. Silva-Martínez, and S. H. K. Embabi, "A 2.4-GHz monolithic fractional-N frequency synthesizer with robust phase-switching prescaler and loop capacitance multiplier," *IEEE Journal of Solid-State Circuits*, vol. 38, no. 6, pp. 866–874, 2003.

[3] ADM709 Data Sheet, Analog device, Power Supply Monitor With Reset.

[4] T. Yasuda, M. Yamamoto, and T. Nishi, "A power-on reset pulse generator for low voltage applications," in *Proceedings of the 2001 IEEE International Symposium on Circuits and Systems, ISCAS 2001*, pp. 598–601, Australia, May 2001.

[5] X. Han, Q. Wei, H. Yang, and H. Wang, "A single channel, 6-bit 410-MS/s 3bits/stage asynchronous SAR ADC based on resistive DAC," *Journal of Semiconductors*, vol. 36, no. 5, Article ID 055010, 2015.

[6] R. Ma, W. Bai, and Z. Zhu, "An energy-efficient and highly linear switching capacitor procedure for SAR ADCs," *Journal of Semiconductors*, vol. 36, no. 5, Article ID 055014, 2015.

[7] Z. Zhu, Y. Xiao, and X. Song, "V_{CM}-based monotonic capacitor switching scheme for SAR ADC," *IEEE Electronics Letters*, vol. 49, no. 5, pp. 327–329, 2013.

[8] Y. Zhu, C.-H. Chan, and U.-F. Chio, "A 10-bit 100-MS/s reference-free SAR ADC in 90 nm CMOS," *IEEE Journal of Solid-State Circuits*, vol. 45, no. 6, pp. 1111–1121, 2010.

[9] L. Qin, *Research and design of 11-bit SAR ADC based on reused terminating capacitor switching procedure. [Master thesis]*, Zhejiang University, 2012.

[10] G.-Y. Huang, S.-J. Chang, C.-C. Liu, and Y.-Z. Lin, "A 1-μW 10-bit 200-kS/s SAR ADC with a bypass window for biomedical applications," *IEEE Journal of Solid-State Circuits*, vol. 47, no. 11, pp. 2783–2795, 2012.

[11] P. Harpe, E. Cantatore, and A. van Roermund, "A 10b/12b 40 kS/s SAR ADC with data-driven noise reduction achieving up to 10.1b ENOB at 2.2 fJ/conversion-step," *IEEE Journal of Solid-State Circuits*, vol. 48, no. 12, pp. 3011–3018, 2013.

[12] A. Agnes, E. Bonizzoni, P. Malcovati, and F. Maloberti, "A 9.4-ENOB 1V 3.8μW 100kS/s SAR ADC with time-domain comparator," in *Proceedings of the 2008 IEEE International Solid State Circuits Conference, ISSCC*, pp. 246-237, USA, February 2008.

[13] N. Verma and A. P. Chandrakasan, "An ultra low energy 12-bit rate-resolution scalable SAR ADC for wireless sensor nodes," *IEEE Journal of Solid-State Circuits*, vol. 42, no. 6, pp. 1196–1205, 2007.

17

A Novel Method for Constructing Grid Multi-Wing Butterfly Chaotic Attractors via Nonlinear Coupling Control

Yun Huang

Asset Management Department, Chongqing University of Posts and Telecommunications, Chongqing 400065, China

Correspondence should be addressed to Yun Huang; hyztt1@sohu.com

Academic Editor: George S. Tombras

A new method is presented to construct grid multi-wing butterfly chaotic attractors. Based on the three-dimensional Lorenz system, two first-order differential equations are added along with one linear coupling controller, respectively. And a piecewise linear function, which is taken into the linear coupling controller, is designed to form a nonlinear coupling controller; thus a five-dimensional chaotic system is produced, which is able to generate gird multi-wing butterfly chaotic attractors. Through the analysis of the equilibrium points, Lyapunov exponent spectrums, bifurcation diagrams, and Poincaré mapping in this system, the chaotic characteristic of the system is verified. Apart from the research above, an electronic circuit is designed to implement the system. The circuit experimental results are in accordance with the results of numerical simulation, which verify the availability and feasibility of this method.

1. Introduction

Since Lorenz found the first chaotic model in 1963 [1], people have had a great interest to construct the chaotic attractors with different shape and quantity. At present, the existing chaotic attractors could be divided into two types: the multi-scroll chaotic attractors (including double-scroll [2], multi-scroll [3, 4], gird multi-scroll [5], and multi-directional multi-scroll chaotic attractors [6]) and the multi-wing butterfly chaotic attractors (containing two-wing [7–11], four-wing [12–16], multi-wing [17–20], and grid multi-wing [21, 22] butterfly chaotic attractors). The approach had grown pretty mature constructing multi-scroll chaotic attractors. However, it is rarely researched for the method constructing multi-wing and grid multi-wing butterfly chaotic attractors.

In recent years, [17–22] reported the latest research achievement about the multi-wing and grid multi-wing butterfly chaotic attractors. Reference [17] constructed a class of hyperchaotic systems generating 2^n-wing butterfly hyperchaotic attractors by coordinate transition and absolute value transition. References [18, 21] proposed the grid multi-wing butterfly chaotic attractors by constructing heteroclinic loops. Reference [7] presented the first and second kinds of Lorenz-type systems to simplify the algebraic form of the Lorenz system while keeping the butterfly structure of Lorenz attractor. Reference [19] constructed a multi-wing butterfly chaotic system based on the first and second kinds of Lorenz-type systems. By designing some piecewise functions to take place of the state variables of the Lorenz system directly, [20, 22] proposed some multi-wing and grid multi-wing butterfly chaotic systems. Though the above literatures proposed some methods constructing multi-wing and grid multi-wing butterfly chaotic attractors, it is very difficult to construct the new multi-wing and grid multi-wing butterfly chaotic attractors via these approaches. Therefore, it is necessary to design the novel methods constructing multi-wing and grid multi-wing butterfly chaotic attractors.

In this paper, through combining a linear coupling controller with a piecewise linear function skillful, a new method is presented to construct grid multi-wing butterfly chaotic attractors. Based on the three-dimensional Lorenz system, two first-order differential equations are added along with one linear coupling controller, respectively. And a piecewise linear function, which is taken into the linear coupling controller, is designed to form a nonlinear coupling controller; thus a five-dimensional chaotic system is produced, which is able to generate gird multi-wing butterfly chaotic attractors.

The system is easy to be realized by analog circuits and its algebraic form is also simple.

2. Constructing the Grid Multi-Wing Butterfly Chaotic Attractors via the Nonlinear Coupling Control

The mathematic model of Lorenz chaotic system is shown as follows:

$$\dot{x}_1 = a\left(x_2 - x_1\right),$$
$$\dot{x}_2 = cx_1 - x_2 - x_1 x_3, \qquad (1)$$
$$\dot{x}_3 = x_1 x_2 - bx_3,$$

where $a = 10$, $b = 8/3$, and $c = 28$.

Constructing grid multi-wing butterfly chaotic attractors via the nonlinear coupling control, the method is shown as follows.

Step 1. In order to construct gird multi-wing butterfly chaotic attractors easier, the state variables need normalization processing in system (1). The method of the normalization processing is to make the peak value of the all state variables less than one via scale transformation. Let the scaling factors of the state variables x_1, x_2 and x_3 be γ_1, γ_2, and γ_3, respectively. According to the numerical simulation results of the Lorenz system [1], we can choose $\gamma_1 = 16$, $\gamma_2 = 23$, and $\gamma_3 = 40$. Therefore, system (1) is changed as follows:

$$\dot{X}_1 = \frac{a\gamma_2}{\gamma_1} X_2 - aX_1,$$
$$\dot{X}_2 = \frac{c\gamma_1}{\gamma_2} X_1 - X_2 - \frac{\gamma_1 \gamma_3}{\gamma_2} X_1 X_3, \qquad (2)$$
$$\dot{X}_3 = \frac{\gamma_1 \gamma_2}{\gamma_3} X_1 X_2 - bX_3.$$

Step 2. Two first-order differential equations about state variables X_4 and X_5 are added based on system (2). The exact approach is as follows: First, the first and second equations are copied and taken as the fourth and fifth equations in system (2), and then the state variable X_1 is replaced with X_4 in the fourth equation of system (2), and the state variable X_5 is used to take place of the state variable X_2 in the fifth equation of system (2). Therefore, a five-dimensional autonomous system is obtained:

$$\dot{X}_1 = \frac{a\gamma_2}{\gamma_1} X_2 - aX_1,$$
$$\dot{X}_2 = \frac{c\gamma_1}{\gamma_2} X_1 - X_2 - \frac{\gamma_1 \gamma_3}{\gamma_2} X_1 X_3,$$
$$\dot{X}_3 = \frac{\gamma_1 \gamma_2}{\gamma_3} X_1 X_2 - bX_3, \qquad (3)$$
$$\dot{X}_4 = \frac{a\gamma_2}{\gamma_1} X_2 - aX_4,$$
$$\dot{X}_5 = \frac{c\gamma_1}{\gamma_2} X_1 - X_5 - \frac{\gamma_1 \gamma_3}{\gamma_2} X_1 X_3.$$

Step 3. Two linear coupling controllers u_1 and u_2 are added in the fourth and fifth equations of system (3), respectively. Then system (3) is changed as follows:

$$\dot{X}_1 = \frac{a\gamma_2}{\gamma_1} X_2 - aX_1,$$
$$\dot{X}_2 = \frac{c\gamma_1}{\gamma_2} X_1 - X_2 - \frac{\gamma_1 \gamma_3}{\gamma_2} X_1 X_3,$$
$$\dot{X}_3 = \frac{\gamma_1 \gamma_2}{\gamma_3} X_1 X_2 - bX_3, \qquad (4)$$
$$\dot{X}_4 = \frac{a\gamma_2}{\gamma_1} X_2 - aX_4 + u_1,$$
$$\dot{X}_5 = \frac{c\gamma_1}{\gamma_2} X_1 - X_5 - \frac{\gamma_1 \gamma_3}{\gamma_2} X_1 X_3 + u_2,$$

where the coupling controllers $u_1 = h(X_1 - X_4)$ and $u_2 = h(X_2 - X_5)$. h is gain parameter.

Step 4. A new piecewise linear function is designed:

$$f(x) = x - \sum_{n=-N}^{M} \text{sgn}(x + 2n + 1) + (M - N + 1), \qquad (5)$$

where $N, M \in \{0, 1, 2, \ldots\}$. Through replacing the state variables X_4 and X_5 with the piecewise linear function $f(x)$ in system (4), a novel system is obtained as follows:

$$\dot{X}_1 = a_1 X_2 - a_2 X_1,$$
$$\dot{X}_2 = a_3 X_1 - X_2 - a_4 X_1 X_3,$$
$$\dot{X}_3 = a_5 X_1 X_2 - a_6 X_3, \qquad (6)$$
$$\dot{X}_4 = a_1 X_2 - a_2 f(X_4) + u_1',$$
$$\dot{X}_5 = a_3 X_1 - f(X_5) - a_4 X_1 X_3 + u_2',$$

where the system parameter $a_1 = a\gamma_2/\gamma_1$, $a_2 = a$, $a_3 = c\gamma_1/\gamma_2$, $a_4 = \gamma_1 \gamma_3/\gamma_2$, $a_5 = \gamma_1 \gamma_2/\gamma_3$, $a_6 = b$, and the nonlinear coupling controllers $u_1' = h[X_1 - f(X_4)]$ and $u_2' = h[X_2 - f(X_5)]$

$$f(X_4) = X_4 - \sum_{n=-N_1}^{M_1} \text{sgn}(X_4 + 2n + 1)$$
$$+ (M_1 - N_1 + 1),$$
$$\qquad (7)$$
$$f(X_5) = X_5 - \sum_{n=-N_2}^{M_2} \text{sgn}(X_5 + 2n + 1)$$
$$+ (M_2 - N_2 + 1),$$

where $N_1, M_1, N_2, M_2 \in \{0, 1, 2, \ldots\}$.

Let the system parameters $a = 10$, $b = 8/3$, $c = 28$, $\gamma_1 = 16$, $\gamma_2 = 23$, $\gamma_3 = 40$, $h = 5$; that is, $a_1 = 14.38$,

$a_2 = 10$, $a_3 = 19.48$, $a_4 = 27.83$, $a_5 = 9.20$, $a_6 = 2.67$. Under the action of the nonlinear coupling controllers u_1' and u_2', system (6) is able to generate $2(N_1 + M_1 + 2) \times (N_2 + M_2 + 2)$-wing butterfly chaotic attractors, as shown in Figure 1. When $N_1 = M_1 = N_2 = M_2 = 0$, system (6) creates 4×2-wing butterfly chaotic attractors as shown in Figure 1(a). When $N_1 = M_1 = M_2 = 0$ and $N_2 = 1$, system (6) generates 4×3-wing butterfly chaotic attractors as shown in Figure 1(b). When $N_1 = N_2 = 1$ and $M_1 = M_2 = 0$, system (6) creates 6×3-wing butterfly chaotic attractors as shown in Figure 1(c), and the time series about the state variable X_4 is shown in Figure 1(e). When $N_1 = N_2 = M_2 = 1$ and $M_1 = 0$, system (6) generates 6×4-wing butterfly chaotic attractors as shown in Figure 1(d).

3. Basic Dynamic Characteristic

3.1. Equilibrium Point. Let $\dot{X}_1 = \dot{X}_2 = \dot{X}_3 = \dot{X}_4 = \dot{X}_5 = 0$; that is,

$$a_1 X_2 - a_2 X_1 = 0,$$

$$a_3 X_1 - X_2 - a_4 X_1 X_3 = 0,$$

$$a_5 X_1 X_2 - a_6 X_3 = 0, \qquad (8)$$

$$a_1 X_2 - a_2 f\left(X_4\right) + u_1' = 0,$$

$$a_3 x_1 - f\left(x_5\right) - a_4 x_1 x_3 + u_2' = 0.$$

According to solution for the equation set (8), the equilibrium points of system (6) are obtained:

$$E_{n_1 n_2}^1 = \left(0, 0, 0, 2n_1, 2n_2\right),$$

$$E_{n_1 n_2}^2 = \left(A, \frac{Aa_2}{a_1}, \frac{A^2 a_2 a_5}{a_1 a_6}, A + 2n_1, \frac{Aa_2}{a_1} + 2n_2\right),$$

$$E_{n_1 n_2}^3$$

$$\qquad (9)$$

$$= \left(-A, -\frac{Aa_2}{a_1}, \frac{A^2 a_2 a_5}{a_1 a_6}, -A + 2n_1, -\frac{Aa_2}{a_1} + 2n_2\right),$$

where $A = \sqrt{(a_1 a_3 a_6 - a_2 a_6)/a_2 a_4 a_5}$, $n_1 = -(N_1 + 1), -N_1, \ldots, 0, \ldots, M_1$ and $n_2 = -(N_2 + 1), -N_2, \ldots, 0, \ldots, M_2$.

For the equilibrium points $E^* = (X_1^*, X_2^*, X_3^*, X_4^*, X_5^*)$, system (6) is linearized and the Jacobian matrix is defined as

$$\mathbf{J}$$

$$= \begin{pmatrix} -a_2 & a_1 & 0 & 0 & 0 \\ a_3 - a_4 X_3^* & -1 & -a_4 X_1^* & 0 & 0 \\ a_5 X_2^* & a_5 X_1^* & -a_6 & 0 & 0 \\ h & a_1 & 0 & -a_2 - h & 0 \\ a_3 - a_4 X_3^* & h & -a_4 X_1^* & 0 & -1 - h \end{pmatrix}, \qquad (10)$$

where X_1^*, X_2^*, X_3^*, X_4^*, and X_5^* are the coordinates of the equilibrium points $E_{n_1 n_2}^1$, $E_{n_1 n_2}^2$, and $E_{n_1 n_2}^3$.

Substituting the equilibrium points $E_{n_1 n_2}^1 = (0, 0, 0, 2n_1, 2n_2)$ into the characteristic equation $\det(\mathbf{J} - \lambda \mathbf{I}) = 0$, we get the following eigenvalues: $\lambda_1 = -6.00$, $\lambda_2 = -15.00$, $\lambda_3 = -22.83$, $\lambda_4 = 11.83$, and $\lambda_5 = -2.67$. $\lambda_1, \lambda_2, \lambda_3$, and λ_5 are negative real numbers, and λ_4 is a positive real number. So the equilibrium points $E_{n_1 n_2}^1$ are unstable saddle.

Substituting the equilibrium points $E_{n_1 n_2}^2 = (A, Aa_2/a_1, A^2 a_2 a_5/a_1 a_6, A + 2n_1, Aa_2/a_1 + 2n_2)$ into the characteristic equation $\det(\mathbf{J} - \lambda \mathbf{I}) = 0$, we get the following eigenvalues: $\lambda_1 = -15.00$, $\lambda_2 = -6.00$, $\lambda_{3,4} = 0.09 \pm 10.20i$, and $\lambda_5 = -13.86$. λ_1, λ_2, and λ_5 are negative real numbers, and $\lambda_{3,4}$ are a pair of conjugate complex eigenvalues with positive real parts. Through computation, we know that the equilibrium points $E_{n_1 n_2}^3$ have the same eigenvalues as the equilibrium points $E_{n_1 n_2}^2$. Therefore, the equilibrium points $E_{n_1 n_2}^2$ and $E_{n_1 n_2}^3$ are unstable saddle-foci of index 2.

3.2. Lyapunov Exponent Spectrum, Bifurcation Diagram, and Poincaré Mapping. Let $N_1 = N_2 = 1$ and $M_1 = M_2 = 0$; the Lyapunov exponent of system (6) and its bifurcation diagram which varies with the coefficient a_6 and its Poincaré map are shown in Figure 2. The range of values for the coefficient a_6 is 0 to 3. Figures 2(a) and 2(b) show that when $a_6 \in [0.75, 3]$, system (6) has one positive Lyapunov exponent, so it is in the chaotic state. From Figure 2(c), we can see that the Poincaré mapping spread out from multiple directions. This shows the complicated dynamic behavior of system (6).

4. Circuit Design of the Grid Multi-Wing Butterfly Chaotic Attractors

According to system (6), the circuit of grid multi-wing butterfly chaotic attractors is designed as shown in Figure 3.

Let $N_1 = N_2 = M_2 = 1$ and $M_1 = 0$; the piecewise linear function (7) is changed as follows

$$f\left(X_4\right) = X_4 - \sum_{n=-1}^{0} \text{sgn}\left(X_4 + 2n + 1\right),$$

$$\qquad (11)$$

$$f\left(X_5\right) = X_5 - \sum_{n=-1}^{1} \text{sgn}\left(X_5 + 2n + 1\right) + 1.$$

The circuit diagram of the piecewise linear function (11) is shown in Figure 4.

In Figures 3 and 4, all the operational amplifiers are selected as UA741CN. Their supply voltage $E = \pm 15\,\text{V}$ and saturated voltage $V_{\text{sat}} \approx \pm 13.5\,\text{V}$. All the multipliers are of type AD633JN and their gain is 0.1. E^+ is the positive pole of the supply voltage E; that is, $E^+ = 15\,\text{V}$. E^- is the negative pole of the supply voltage E; that is, $E^- = -15\,\text{V}$.

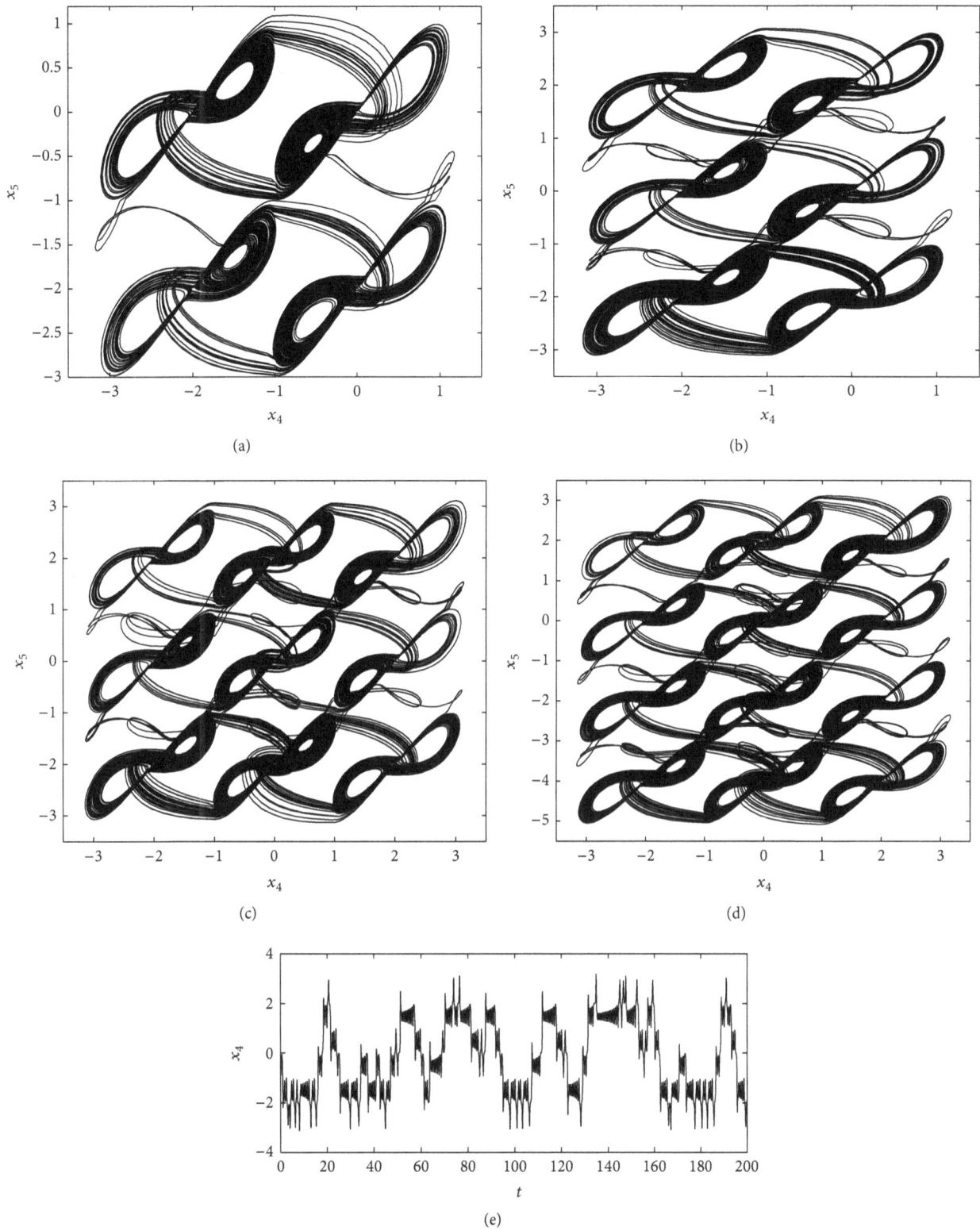

FIGURE 1: Grid multi-wing butterfly chaotic attractors and time series of system (6): (a) 4×2; (b) 4×3; (c) 6×3; (d) 6×4; (e) time series about the state variable X_4.

(a)

(b)

(c)

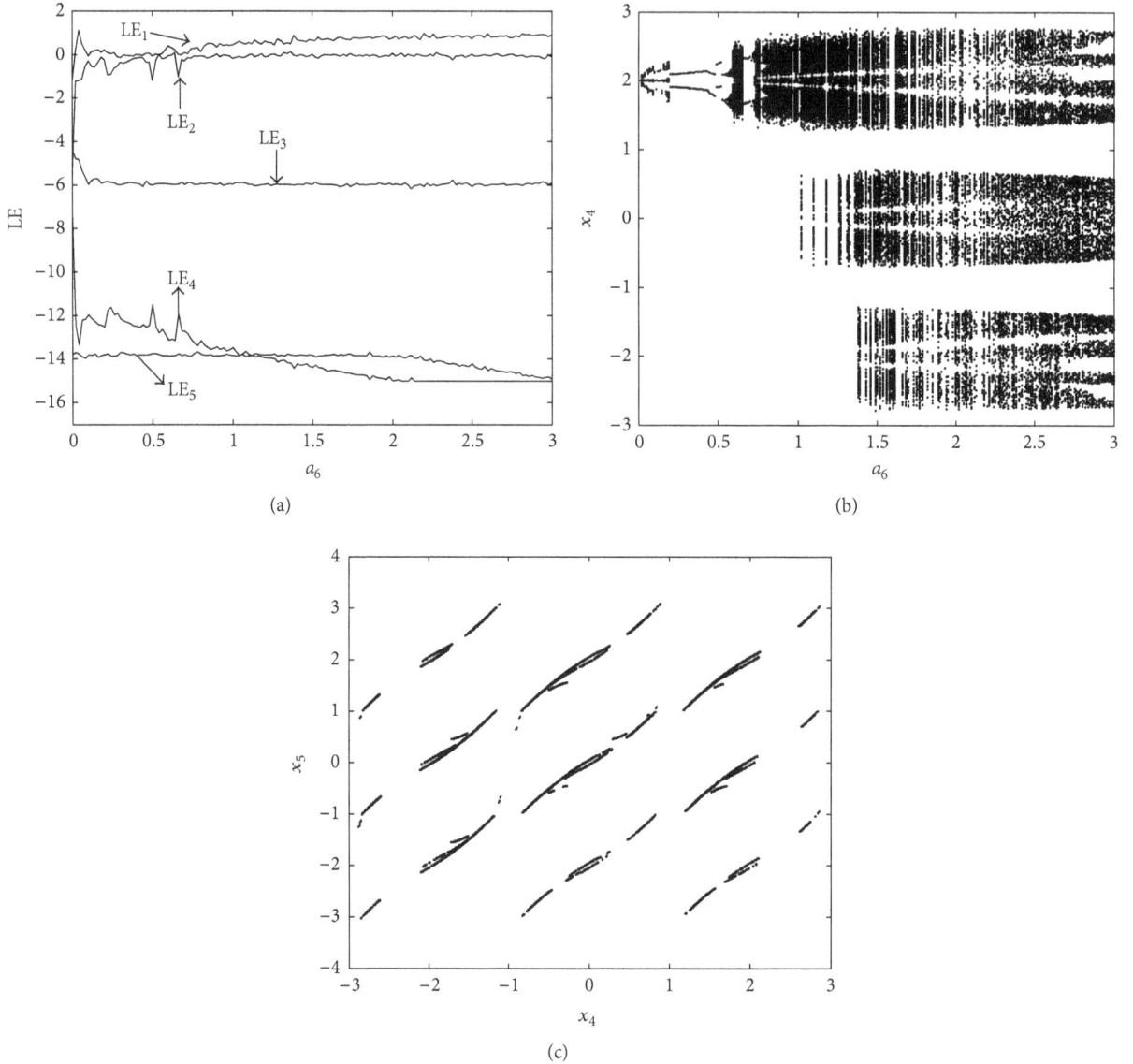

Figure 2: Lyapunov exponent spectrum, bifurcation diagram, and Poincaré mapping of system (6): (a) Lyapunov exponent spectrum; (b) bifurcation diagram; (c) Poincaré mapping.

According to Figure 3, the circuit equation can be obtained as follows:

$$\frac{dX_1}{d\tau} = \frac{1}{R_0 C_0} \left(\frac{R_{15}}{R_1} X_2 - \frac{R_{15}}{R_2} X_1 \right),$$

$$\frac{dX_2}{d\tau} = \frac{1}{R_0 C_0} \left(-\frac{R_{15}}{10 R_3} X_1 X_3 + \frac{R_{15}}{R_4} X_1 - \frac{R_{15}}{R_5} X_2 \right),$$

$$\frac{dX_3}{d\tau} = \frac{1}{R_0 C_0} \left(\frac{R_{15}}{10 R_6} X_1 X_2 - \frac{R_{15}}{R_7} X_3 \right),$$

$$\frac{dX_4}{d\tau} = \frac{1}{R_0 C_0} \left(\frac{R_{15}}{R_8} X_2 + \frac{R_{15}}{R_9} X_1 - \frac{R_{15}}{R_{10}} f(X_4) \right),$$

$$\frac{dX_5}{d\tau} = \frac{1}{R_0 C_0} \left(-\frac{R_{15}}{10 R_{11}} X_1 X_3 + \frac{R_{15}}{R_{12}} X_1 + \frac{R_{15}}{R_{13}} X_2 \right.$$

$$\left. - \frac{R_{15}}{R_{14}} f(X_5) \right).$$

(12)

To observe the output wave experimentally, the time scale transformation must be executed for τ, that is, let $\tau = \tau_0 t$ and $\tau_0 = 10^{-3}$, and (12) can be changed as follows:

$$\frac{dX_1}{d\tau} = \frac{10^{-3}}{R_0 C_0} \left(\frac{R_{15}}{R_1} X_2 - \frac{R_{15}}{R_2} X_1 \right),$$

$$\frac{dX_2}{d\tau} = \frac{10^{-3}}{R_0 C_0} \left(-\frac{R_{15}}{10 R_3} X_1 X_3 + \frac{R_{15}}{R_4} X_1 - \frac{R_{15}}{R_5} X_2 \right),$$

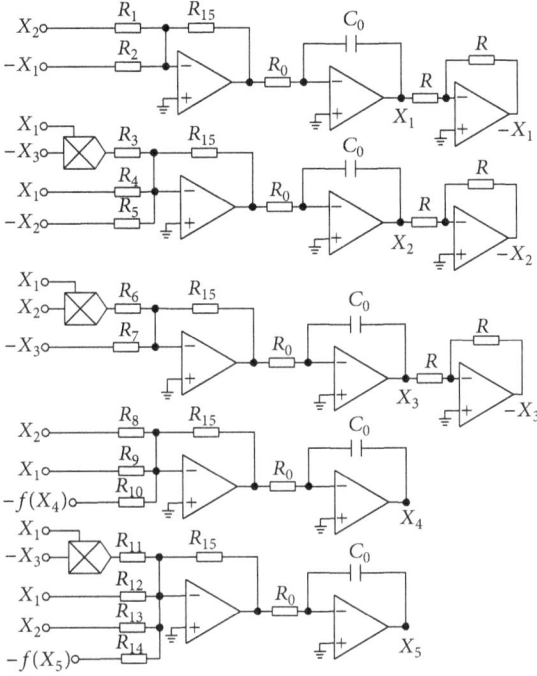

FIGURE 3: Circuit diagram of grid multi-wing butterfly chaotic attractors.

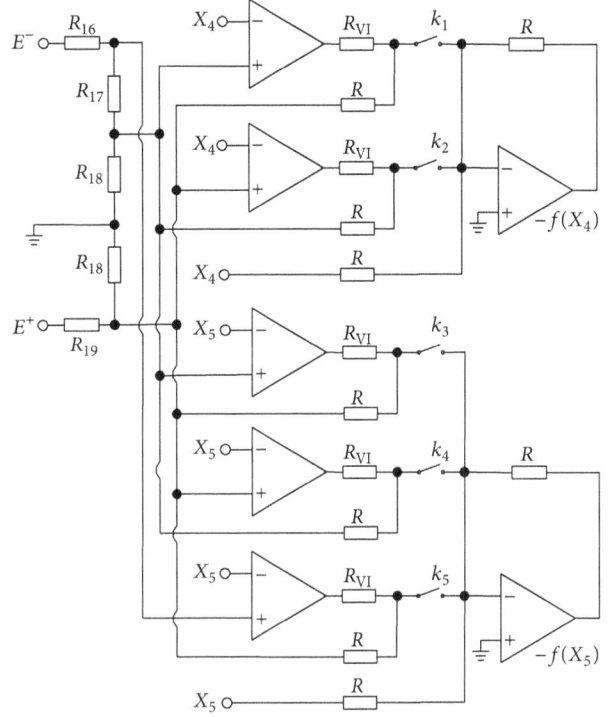

FIGURE 4: Circuit diagram of the piecewise linear function (11).

$$\frac{dX_3}{d\tau} = \frac{10^{-3}}{R_0 C_0} \left(\frac{R_{15}}{10 R_6} X_1 X_2 - \frac{R_{15}}{R_7} X_3 \right),$$

$$\frac{dX_4}{d\tau} = \frac{10^{-3}}{R_0 C_0} \left(\frac{R_{15}}{R_8} X_2 + \frac{R_{15}}{R_9} X_1 - \frac{R_{15}}{R_{10}} f(X_4) \right),$$

$$\frac{dX_5}{d\tau} = \frac{10^{-3}}{R_0 C_0} \left(-\frac{R_{15}}{10 R_{11}} X_1 X_3 + \frac{R_{15}}{R_{12}} X_1 + \frac{R_{15}}{R_{13}} X_2 \right.$$

$$\left. - \frac{R_{15}}{R_{14}} f(X_5) \right).$$

$$(13)$$

Let $R_0 = 10 \, k\Omega$, $C_0 = 10 \, nF$, $R = 10 \, k\Omega$, and $R_{15} = 100 \, k\Omega$; according to system (6) and equation (13), we can obtain $R_1 = R_8 = 69.54 \, k\Omega$, $R_2 = 100 \, k\Omega$, $R_3 = R_{11} = 3.59 \, k\Omega$, $R_4 = R_{12} = 51.33 \, k\Omega$, $R_5 = 1 \, M\Omega$, $R_6 = 10.87 \, k\Omega$, $R_7 = 362.32 \, k\Omega$, $R_9 = R_{13} = 200 \, k\Omega$, $R_{10} = 66.67 \, k\Omega$, and $R_{14} = 166.67 \, k\Omega$.

In Figure 4, let $R_{VI} = 135 \, k\Omega$. When k_1, k_2, k_3, k_4, and k_5 are switched on, the circuit equation can be obtained as follows:

$$- f(X_4)$$

$$= -\frac{R}{R} X_4$$

$$+ \frac{R}{R_{VI}} |V_{sat}| \, \text{sgn} \left(X_4 - \frac{R_{18}}{R_{16} + R_{17} + R_{18}} E^- \right)$$

$$- \frac{R}{R} \cdot \frac{R_{18}}{R_{18} + R_{19}} E^+$$

$$+ \frac{R}{R_{VI}} |V_{sat}| \, \text{sgn} \left(X_4 - \frac{R_{18}}{R_{18} + R_{19}} E^+ \right) - \frac{R}{R}$$

$$\cdot \frac{R_{18}}{R_{16} + R_{17} + R_{18}} E^-$$

$$= -X_4 + \text{sgn} \left(X_4 + \frac{15 R_{18}}{R_{16} + R_{17} + R_{18}} \right)$$

$$- \frac{15 R_{18}}{R_{18} + R_{19}} + \text{sgn} \left(X_4 - \frac{15 R_{18}}{R_{18} + R_{19}} \right)$$

$$+ \frac{15 R_{18}}{R_{16} + R_{17} + R_{18}},$$

$$- f(X_5)$$

$$= -\frac{R}{R} X_5$$

$$+ \frac{R}{R_{VI}} |V_{sat}| \, \text{sgn} \left(X_5 - \frac{R_{18}}{R_{16} + R_{17} + R_{18}} E^- \right)$$

$$- \frac{R}{R} \cdot \frac{R_{18}}{R_{18} + R_{19}} E^+$$

$$+ \frac{R}{R_{VI}} |V_{sat}| \, \text{sgn} \left(X_5 - \frac{R_{18}}{R_{18} + R_{19}} E^+ \right) - \frac{R}{R}$$

$$\cdot \frac{R_{18}}{R_{16} + R_{17} + R_{18}} E^-$$

$$+ \frac{R}{R_{VI}} |V_{sat}| \, \text{sgn} \left(X_5 - \frac{R_{17} + R_{18}}{R_{16} + R_{17} + R_{18}} E^- \right)$$

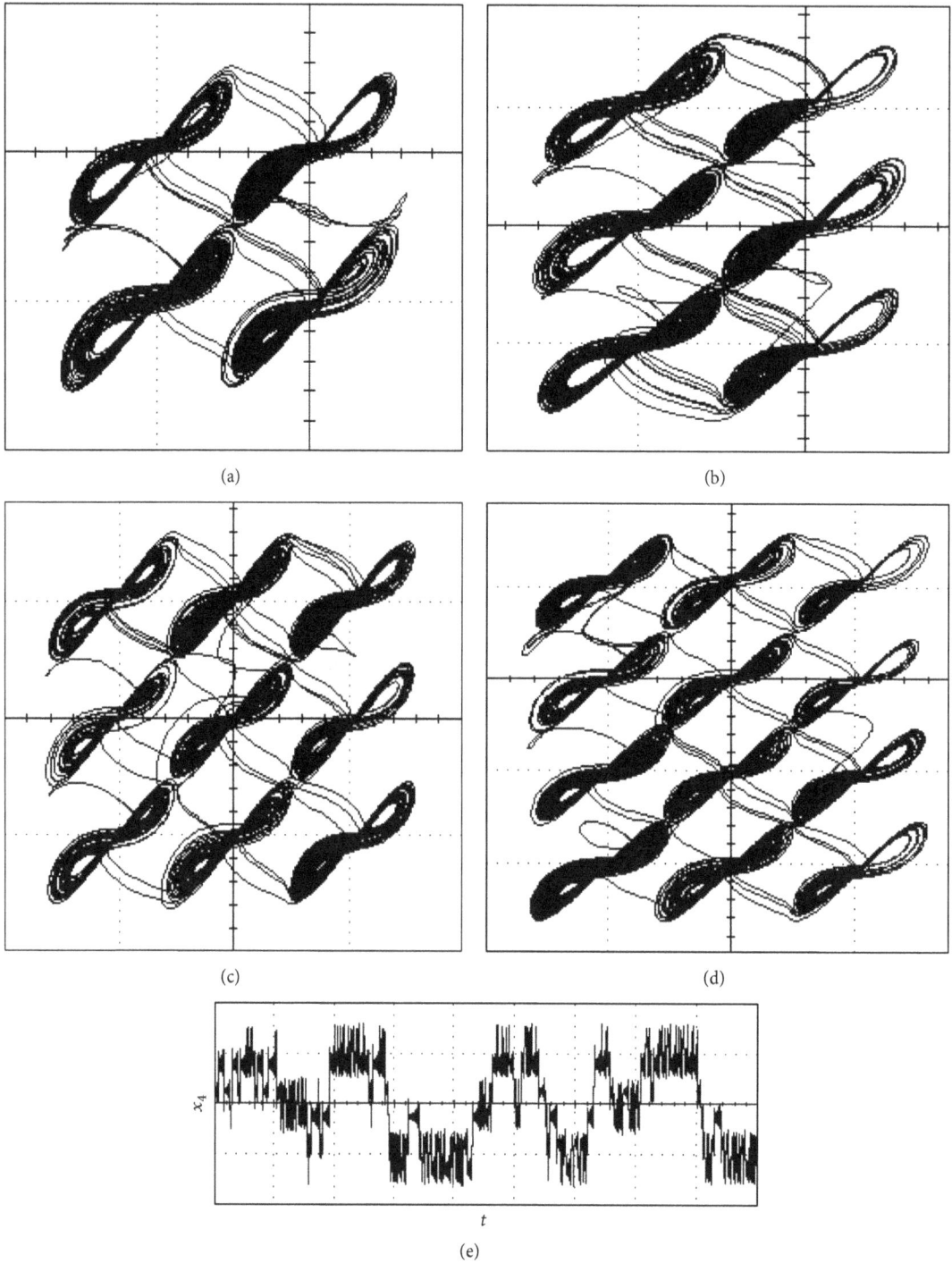

(a)

(b)

(c)

(d)

(e)

FIGURE 5: The results of circuit experiment: (a) 4×2; (b) 4×3; (c) 6×3; (d) 6×4; (e) time series about the state variable X_4.

$$-\frac{R}{R} \cdot \frac{R_{18}}{R_{18} + R_{19}} E^+$$

$$= -X_5 + \text{sgn}\left(X_5 + \frac{15R_{18}}{R_{16} + R_{17} + R_{18}}\right)$$

$$-\frac{15R_{18}}{R_{18} + R_{19}} + \text{sgn}\left(X_5 - \frac{15R_{18}}{R_{18} + R_{19}}\right)$$

$$+ \frac{15R_{18}}{R_{16} + R_{17} + R_{18}}$$

$$+ \text{sgn}\left(X_5 + \frac{15\left(R_{17} + R_{18}\right)}{R_{16} + R_{17} + R_{18}}\right) - \frac{15R_{18}}{R_{18} + R_{19}}.$$

$$(14)$$

According to the piecewise linear function (11) and equation (14), we can choose $R_{16} = 12\,\mathrm{k\Omega}$, $R_{17} = 2\,\mathrm{k\Omega}$, $R_{18} = 1\,\mathrm{k\Omega}$, and $R_{19} = 14\,\mathrm{k\Omega}$.

According to Figures 3 and 4, the gird multi-wing butterfly chaotic attractors are obtained via circuit simulation software Multisim 10.0, as shown in Figure 5. When k_1, k_3 are switched on and k_2, k_4, and k_5 are switched off, the circuit generates 4×2-wing butterfly chaotic attractors as shown in Figure 5(a). When k_1, k_3, and k_4 are switched on and k_2 and k_5 are switched off, the circuit creates 4×3-wing butterfly chaotic attractors as shown in Figure 5(b). When k_1, k_2, k_3, and k_4 are switched on and k_5 is switched off, the circuit generates 6×3-wing butterfly chaotic attractors as shown in Figure 5(c), and the time series about the state variable X_4 is shown in Figure 5(e). When k_1, k_2, k_3, k_4, and k_5 are switched on, the circuit creates 6×4-wing butterfly chaotic attractors as shown in Figure 5(d).

From Figures 5 and 1, we can see that the circuit experimental results are in agreement with the results of numerical simulation.

5. Conclusion

A new method is presented to construct grid multi-wing butterfly chaotic attractors via nonlinear coupling control in this paper. A five-dimensional grid multi-wing butterfly chaotic system is constructed via this approach. Through adjusting the nonlinear coupling controllers and the piecewise linear functions, the 4×2, 4×3, 6×3, and 6×4-wing butterfly chaotic attractors are obtained. Through the theoretical analysis and numerical simulation, the complex dynamic characteristics of the five-dimensional grid multi-wing butterfly chaotic system are shown. Also, the system has been implemented by designing an electronic circuit.

Competing Interests

The author declares that there is no conflict of interests regarding the publication of this article.

Acknowledgments

This work is supported by the Technology Research Projects of The Chongqing Education Committee (Grant nos. KJ130509, KJ1400410, and KJ130520).

References

[1] E. N. Lorenz, "Deterministic nonperiodic flow," *Journal of the Atmospheric Sciences*, vol. 20, no. 1, pp. 130–141, 1963.

[2] L. O. Chua, M. Komuro, and T. Matsumoto, "The double scroll family. I. Rigorous proof of chaos," *IEEE Transactions on Circuits and Systems*, vol. 33, no. 11, pp. 1072–1097, 1986.

[3] L. J. Ontañón-García, E. Jiménez-López, E. Campos-Cantón, and M. Basin, "A family of hyperchaotic multi-scroll attractors in $\mathbb{R}^n$," *Applied Mathematics and Computation*, vol. 233, pp. 522–533, 2014.

[4] Y. Lin, C. Wang, and H. He, "A simple multi-scroll chaotic oscillator employing CCIIs," *Optik*, vol. 126, no. 7-8, pp. 824–827, 2015.

[5] F. Yu, C. Wang, and H. He, "Grid multiscroll hyperchaotic attractors based on colpitts oscillator mode with controllable grid gradient and scroll numbers," *Journal of Applied Research and Technology*, vol. 11, no. 3, pp. 371–380, 2013.

[6] C. H. Wang, H. Xu, and F. Yu, "A novel approach for constructing high-order Chua's circuit with multi-directional multi-scroll chaotic attractors," *International Journal of Bifurcation and Chaos*, vol. 23, no. 2, Article ID 1350022, 2013.

[7] A. S. Elwakil, S. Özoğuz, and M. P. Kennedy, "Creation of a complex butterfly attractor using a novel Lorenz-type system," *IEEE Transactions on Circuits and Systems. I. Fundamental Theory and Applications*, vol. 49, no. 4, pp. 527–530, 2002.

[8] C. Han, S. Yu, and G. Wang, "A sinusoidally driven Lorenz system and circuit implementation," *Mathematical Problems in Engineering*, vol. 2015, Article ID 706902, 11 pages, 2015.

[9] J. Lü and G. Chen, "A new chaotic attracor coined," *International Journal of Bifurcation and Chaos*, vol. 12, no. 3, pp. 659–662, 2002.

[10] Q. Han, C.-X. Liu, L. Sun, and D.-R. Zhu, "A fractional order hyperchaotic system derived from a Liu system and its circuit realization," *Chinese Physics B*, vol. 22, no. 2, Article ID 020502, 2013.

[11] R. Wang, H. Sun, J.-Z. Wang, L. Wang, and Y.-C. Wang, "Applications of modularized circuit designs in a new hyperchaotic system circuit implementation," *Chinese Physics B*, vol. 24, no. 2, Article ID 020501, 2015.

[12] S. Dadras and H. R. Momeni, "A novel three-dimensional autonomous chaotic system generating two, three and four-scroll attractors," *Physics Letters. A*, vol. 373, no. 40, pp. 3637–3642, 2009.

[13] S. Dadras, H. R. Momeni, and G. Qi, "Analysis of a new 3D smooth autonomous system with different wing chaotic attractors and transient chaos," *Nonlinear Dynamics*, vol. 62, no. 1-2, pp. 391–405, 2010.

[14] J. Liu, "A four-wing and double-wing 3D chaotic system based on sign function," *Optik*, vol. 125, no. 23, pp. 7089–7095, 2014.

[15] J. Ma, Z. Chen, Z. Wang, and Q. Zhang, "A four-wing hyperchaotic attractor generated from a 4D memristive system with a line equilibrium," *Nonlinear Dynamics*, vol. 81, no. 3, pp. 1275–1288, 2015.

[16] A. Zarei, "Complex dynamics in a 5-D hyper-chaotic attractor with four-wing, one equilibrium and multiple chaotic attractors," *Nonlinear Dynamics*, vol. 81, no. 1-2, pp. 585–605, 2015.

[17] B. Yu and G. Hu, "Constructing multiwing hyperchaotic attractors," *International Journal of Bifurcation and Chaos*, vol. 20, no. 3, pp. 727–734, 2010.

[18] S. Yu, J. Lü, X. Yu, and G. Chen, "Design and implementation of grid multiwing hyperchaotic Lorenz system family via switching control and constructing super-heteroclinic loops," *IEEE Transactions on Circuits and Systems I: Regular Papers*, vol. 59, no. 5, pp. 1015–1028, 2012.

[19] S. Yu, W. K. S. Tang, J. Lü, and G. Chen, "Design and implementation of multi-wing butterfly chaotic attractors via lorenz-type systems," *International Journal of Bifurcation and Chaos*, vol. 20, no. 1, pp. 29–41, 2010.

[20] M.-W. Luo, X.-H. Luo, and H.-Q. Li, "A family of four-dimensional multi-wing chaotic system and its circuit implementation," *Acta Physica Sinica*, vol. 62, no. 2, Article ID 020512, 2013.

18

Modified Droop Method based on Master Current Control for Parallel-Connected DC-DC Boost Converters

Muamer M. Shebani ⓘD, Tariq Iqbal ⓘD, and John E. Quaicoe

Department of Electrical and Computer Engineering, Faculty of Engineering and Applied Science,
Memorial University of Newfoundland, St. John's, NL, Canada

Correspondence should be addressed to Muamer M. Shebani; mms137@mun.ca

Academic Editor: Jit S. Mandeep

Load current sharing between parallel-connected DC-DC boost converters is very important for system reliability. This paper proposes a modified droop method based on master current control for parallel-connected DC-DC boost converters. The modified droop method uses an algorithm for parallel-connected DC-DC boost converters to adaptively adjust the reference voltage for each converter according to the load regulation characteristics of the droop method. Unlike the conventional droop method, the current feedback signal (master current) for one of the parallel-connected converters is used in the inner loop controller for all converters to avoid any differences in the time delay of the control loops for the parallel-connected converters. The algorithm ensures that the load current sharing is identical to the load regulation characteristics of the droop method. The proposed algorithm is tested with a mismatch in the parameters of the parallel converters. The effectiveness of the proposed algorithm is verified using Matlab/Simulink simulation.

1. Introduction

In comparison to a single, high power, centralized power converter, parallel connection of low power converters offers several advantages. Some of these advantages are associated with the system performance such as higher efficiency, better dynamic response, and better load regulation. The other advantages are related to the system, which include expandability of output power and ease of maintenance [1].

For the operation of parallel DC-DC converters, the conventional droop method provides true redundancy because there is no interconnection between modules. However, effective load current sharing and voltage regulation are the main drawbacks of the droop method. A novel droop method proposed by Kim et al. [2] adaptively regulates a reference voltage to improve the load current sharing and properly regulate the output voltage. The output voltage variation in parallel-connected converters is caused by various conditions such as changes in load, input power, or measurement error in the voltage feedback signal [2, 3]. In general, the circulating current between modules could be initiated by a small

mismatch in the output voltage which leads to unequal load current sharing between modules. Anand and Fernandes in [4] propose a modified droop controller to overcome the mismatch in output voltage due to the error in measurement of the voltage feedback signal. The circulating current measurement between converters is used to modify the nominal voltage which reduces the error of the modules output voltage.

The operating point for parallel-connected converter is modified randomly and thus a new set of parameters for the controller must be calculated again. To overcome this issue, several control laws have been proposed in the literature [5–9]. One of these control laws is a nonlinear control, which enhances the power quality for condition of different loads. Mazunder and Kamisetty in [7] gave an experimental validation of the proposed control scheme in [5] for parallel-connected buck converters. The performance of the proposed control is demonstrated under the transient and steady state for two parallel converters. The operation of the two modules uses interleaving mode of operation which is achieved by phase shifting the ramp signals of the two modules by one half cycle.

One of the commonly employed methods for active current sharing control of parallel converters is the master-slave current control that uses an analog wireless communication or an intercommunication link between converters [10]. The master-slave current sharing control of parallel converters is demonstrated using analog wireless communication in [11]. It uses one converter to operate as master controller. The chosen converter operates in voltage controlled mode to regulate the output voltage which tracks the reference voltage. The other converters operate in current controlled mode to regulate their output currents. Those converters operate as slaves because the master controller obtains their reference current value based on the total load current. Slave converters receive the new reference current through a high-speed communication link [12, 13]. The high-speed communication between converters is used to minimize the time delay and improve the system performance. The high-speed communication link such as digital communication scheme increases the total cost, which makes it appropriate in medium and high-power applications. However, in some power applications, an analog controller with connecting wires experiences noise which makes it applicable only for low power applications where converters are located close to each other.

To achieve wireless power sharing between converters, a droop method with virtual resistance (VR) has been proposed [14]. The method implements tertiary level optimization control for parallel-connected DC-DC converter to enhance the efficiency of the droop method with VR. The decision variable, which is VR, is used to adjust the load current sharing between the converters. Stability analysis is used to examine the effect of varying VR on the dynamic performance. In [15] an improved droop method for parallel-connected converters is presented to enhance the current sharing accuracy. Although the other modified droop methods adjust the output voltage by making the slope of the droop method steeper such as virtual resistance, the improved droop method adjusts the output voltage set point for each converter based on a generated digital signal from a module experiencing the highest current. A few specific current set points in the improved droop method are selected in advance. If the module with the highest current reaches the set point, a digital signal is sent by the module to the other modules to regulate their output voltage. However, the module which sends a digital signal does not respond to a signal of the other module at the same set point. Once all the modules have generated one signal at the same set point, the output voltage regulation for that set point is terminated and a new process is started.

In general, any mismatch in the output voltage level of the parallel-connected DC-DC boost converter could initiate a circulating current. To avoid an initiation of circulating current, a synchronous switching for parallel-connected DC-DC boost converter could be used [16]. The synchronous switching forces the parallel-connected converters to operate in two operating modes only. This ensures that the switching of the parallel-connected converters is synchronized. Lab measurement of two parallel connected MPPT modules with no intercommunication link between the two modules is shown in Figure 1. In this setup, there is no synchronous

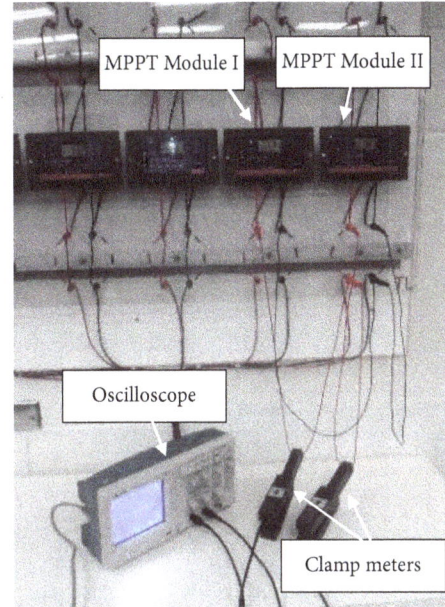

FIGURE 1: Two parallel-connected MPPT modules.

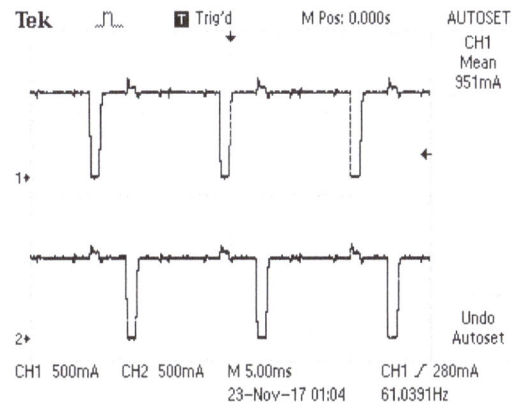

FIGURE 2: Output current for the two parallel-connected MPPT modules.

switching for the two parallel connected 260 W PV modules with individual MPPT.

The outputs of the two MPPT modules are connected in parallel to the load. Also, the two clamp meters in Figure 1 monitor the currents from both MPPT modules. During a sunny day, the measurements of the currents for each module are shown in Figure 2.

It can be observed from Figure 2 that when a switch closes in one of the parallel-connected MPPT modules, the other MPPT module experiences higher current at the same time and vice versa. This is a clear evidence of circulating currents between the two MPPT modules. The asynchronous switching causes the undesired higher current for each one of the two MPPT modules. Moreover, there is no intercommunication link between the two MPPT modules to synchronize the switching and avoid circulating current.

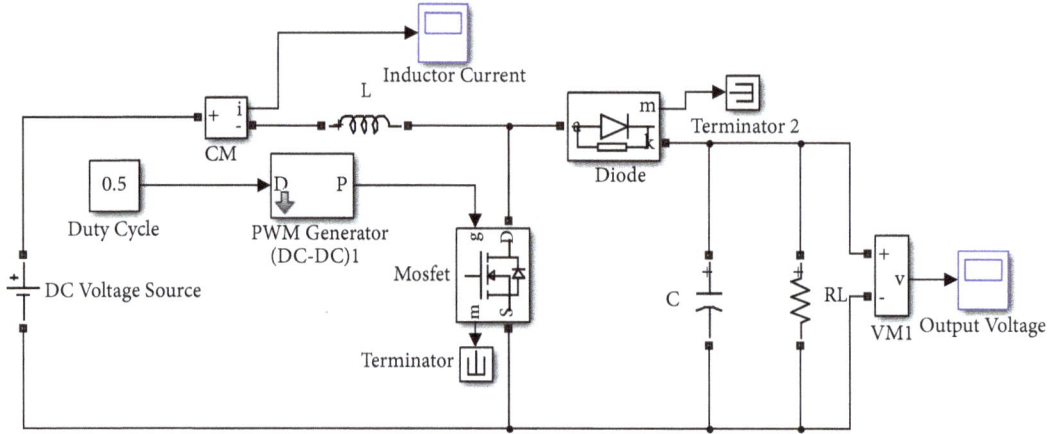

FIGURE 3: DC-DC boost converter.

In this paper, synchronous switching for the parallel-converter is achieved by intercommunication link between the modules. The proposed method based on master current control for parallel-connected DC-DC boost converters can synchronize the output voltage level during any change in the load condition. In this method, a linear control was implemented; thus the two parallel-connected converters are controlled using voltage controlled mode as the outer control loop and its reference voltage is adjusted by using the proposed algorithm. The algorithm inputs are the voltage measurement from the common dc bus of the parallel-connected converters and the current measurement of one converter (master current). However, the inner current loop control takes the chosen current measurement (master current) as feedback current signal for the two control loops. Therefore, the load current sharing is achieved based on the droop method. As the control loops for the parallel-connected converters have the same time delay, the output voltage levels for the parallel-connected converters are changed synchronously. This ensures no mismatch in the output voltage and no circulating current during changes in the load, in addition to minimizing the ripple in the output current waveforms.

2. DC-DC Boost Converter Design

The equivalent circuit of a DC-DC boost converter is shown in Figure 3. The DC-DC boost converter circuit consists of a DC power supply, inductor, MOSFET switch, diode, capacitor, and load resistance.

From the operating point of view, the DC-DC boost converter could be operated in Continuous Conduction Mode (CCM) and Discontinuous Condition Mode (DCM) [17]. The operation mode depends on the converters' parameters. For CCM, the parameters for the boost converter such capacitance, C, inductance, L, and duty cycle, D, can be obtained by the following expressions.

$$D = 1 - \frac{V_{in}}{V_{out}} \tag{1}$$

TABLE 1: Operating values for boost converter.

Parameters	DC-DC Boost Converter
Switching frequency f_s	25 KHz
Inductance L	15.986 mH
Capacitance C	128.646 μF
Power P	144 W

$$C = \frac{(I_{out} * D)}{(\delta * V_{out} * f)} \tag{2}$$

$$L = \frac{(V_{in} * D)}{(\delta * I_{in} * f)} \tag{3}$$

where the ripple limit (δ) is (1%). The boost converter parameters are obtained as shown in Table 1.

3. Design of PI Controller Using SISO Tool in Matlab

By considering a CCM, the boost converter in Figure 1 is switched between two states (ON/OFF). The inductor current (I_L) and the output voltage (V_{out}) waveforms with PWM at duty cycle of 0.5 are shown in Figure 4.

As shown in Figure 4, during the ON state the inductor current is increased, but the output voltage is decreased. Moreover, for the OFF state, the MOSFET switch is opened resulting in a decrease in the inductor current and increase in the output voltage. During one cycle (ON and OFF states), the state space averaging technique is employed over one switching cycle [18], and it is given by

$$\begin{bmatrix} \dot{i}_L \\ \dot{v}_c \end{bmatrix} = \begin{bmatrix} 0 & \dfrac{-(1-D)}{L} \\ \dfrac{(1-D)}{C} & \dfrac{-1}{RC} \end{bmatrix} \begin{bmatrix} i_L \\ v_c \end{bmatrix} + \begin{bmatrix} \dfrac{1}{L} \\ 0 \end{bmatrix} V_{in} \tag{4}$$

$$v_{out} = \begin{bmatrix} 0 & 1 \end{bmatrix} \begin{bmatrix} i_L \\ v_c \end{bmatrix} \tag{5}$$

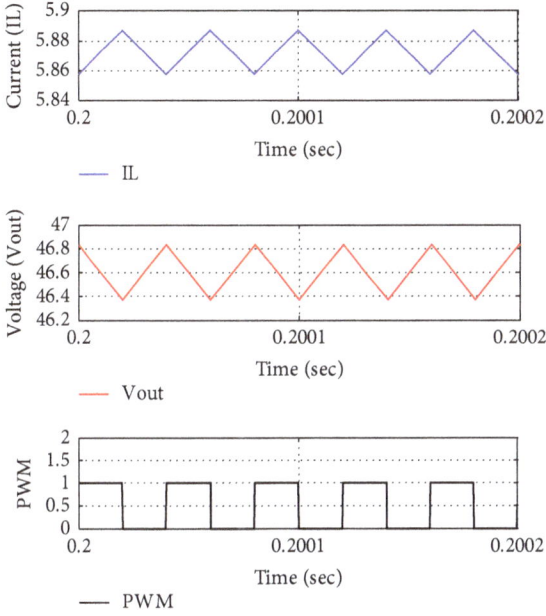

FIGURE 4: Waveforms of inductor current (I_L) and output voltage (V_{out}) in CCM.

For a small perturbation, the standard linearization technique is used as shown below:

$$i_L = i_L + \tilde{i}_L$$
$$v_c = v_c + \tilde{v}_c$$
$$d = D + \tilde{d}$$
$$v_{in} = v_{in} + \tilde{v}_{in}$$

(6)

During the small signal analysis, the input voltage is considered to be constant. The DC-DC boost converter small signal model can be given by

$$\begin{bmatrix} \dot{\tilde{i}}_L \\ \dot{\tilde{v}}_c \end{bmatrix} = \begin{bmatrix} 0 & \dfrac{-(1-D)}{L} \\ \dfrac{(1-D)}{C} & \dfrac{-1}{RC} \end{bmatrix} \begin{bmatrix} \tilde{i}_L \\ \tilde{v}_c \end{bmatrix} + \begin{bmatrix} \dfrac{V_c}{L} \\ \dfrac{-I_L}{C} \end{bmatrix} \tilde{d}$$

(7)

$$V_{out} = \begin{bmatrix} 0 & 1 \end{bmatrix} \begin{bmatrix} \tilde{i}_L \\ \tilde{v}_c \end{bmatrix}$$

(8)

The transfer function $\tilde{v}_0/\tilde{d}$ can be obtained from (7) and (8). Similarly, by replacing (8) with (9), the transfer function $\tilde{i}_{in}/\tilde{d}$ can be determined.

$$i_{in} = \begin{bmatrix} 1 & 0 \end{bmatrix} \begin{bmatrix} \tilde{i}_L \\ \tilde{v}_c \end{bmatrix}$$

(9)

TABLE 2: Parameters of PI controller for boost converter.

Gains		Type of Controller	
		PI_{id} Controller	PI_{vi} Controller
Converter	Proportional gain k_p	0.0809484	0.016874
	Integral gain k_i	20.756	12.98

The transfer functions G_{vi} for the outer loop and the inner loop G_{id} are given as

$$G_{vi} = \frac{-16\,(S - 235.8)}{(S + 971.7)}$$

(10)

$$G_{id} = \frac{3092.9\,(S + 971.7)}{(S^2 + 485.8 * S + 1.146 * 10^5)}$$

(11)

where v is the output voltage, i is the inductor current, and d is the duty cycle. However, Figure 5 shows the basic structure of PI_{vi} controller for the outer loop and PI_{id} for the inner loop.

Several methods such as Ziegler-Nicholas, loop shaping method, and frequency response can be used to design a PI controller [19, 20]. However, for simplicity, Sisotool from Matlab/Simulink is used to tune the controller and evaluate the suitability of stability. For the outer loop, the Sisotool command of the transfer functions G_{vi} is used in the control system toolbox to provide a graphical user interface for the bode plot and root locus as shown in Figure 6.

The zero in the right z plane for root locus of the boost converter does not affect the absolute stability of the system [21, 22]. By modifying the pole-zero pattern of the feedback controller, the closed-loop frequency response can be changed until the desired repose is obtained. To determine the PI_{vi} gains, the position of the root locus is modified online. The gains for the proportional and integral controller for the outer loop are given in Table 2. Similarly, the Sisotool command is used for the inner loop transfer function G_{id}. The graphical user interface is generated for the root locus and bode plot as shown in Figure 7.

For rise time of 0.002971 seconds, settling time of 0.0164 seconds, and overshoot less than 8%, the gains for PI_{id} are obtained for the DC-DC boost converter. The gains for the PI_{vi} and PI_{id} controllers are given in Table 2.

4. Proposed Method

A schematic diagram of two parallel-connected DC-DC boost converters is shown in Figure 8.

The two parallel-connected DC-DC boost converters in Figure 8 are connected to a common DC, with the load connected directly to the common DC bus. The output voltage and current for converters I and II are V_1, I_1 and V_2, I_2, respectively. Because converter I and converter II are connected to the common DC bus, the output voltages V_1 and V_2 are equal. Therefore, the output voltage of each converter is equal to the load voltage, and the load current is the sum of the output currents of both converters.

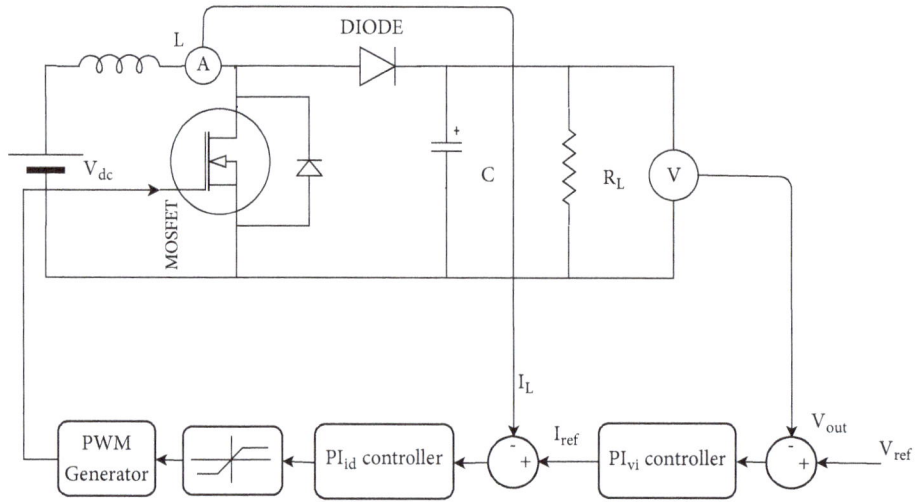

FIGURE 5: Basic structure of PI_{vi} and PI_{id} controllers for the two loops.

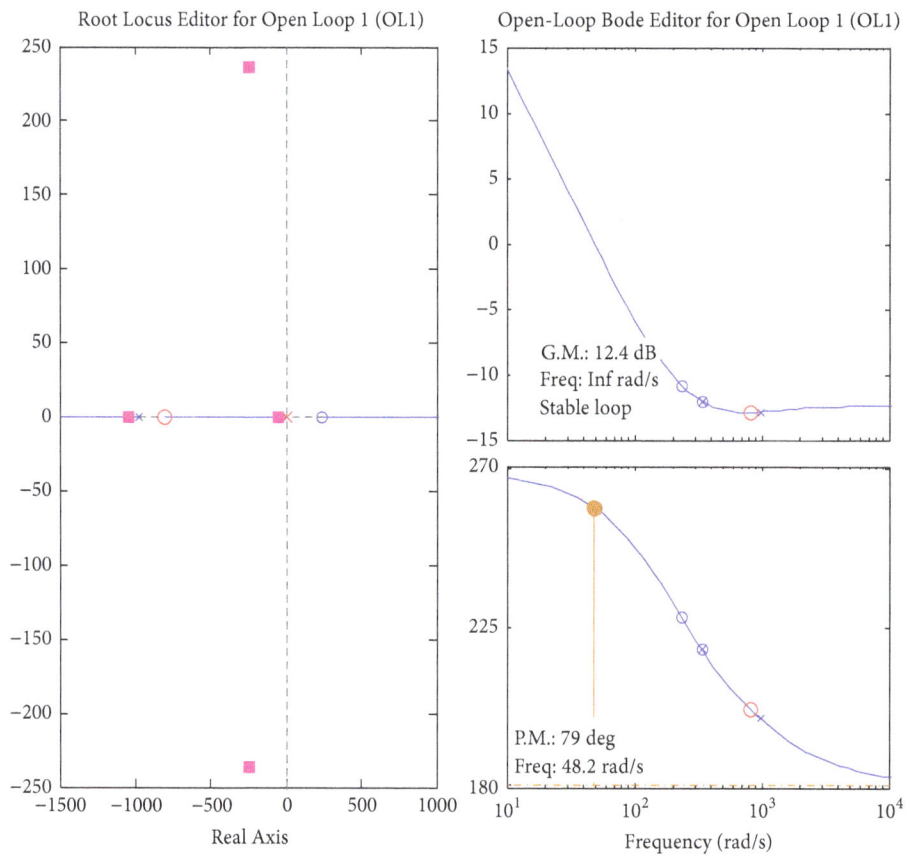

FIGURE 6: Root locus and open loop bode plots for the outer loop of boost converter.

Zero-circulating current in Figure 8 can be achieved by having zero mismatch in the output voltage during the changes in the load or the input power. The proposed method ensures synchronization in the output voltage level by adjusting the reference voltage according to the droop method. The outer voltage loop for each converter regulates the output voltage which would be in the level of the adjusted reference voltage. The inner control loop uses one of the converters' measured current signal as feedback current signal for the parallel-connected converters (master current control). The inner current loop needs interconnection current measurement between modules. However, to overcome the mismatch in the output voltage, the algorithm is used to adjust the reference voltage for the parallel-connected converters. Figure 9

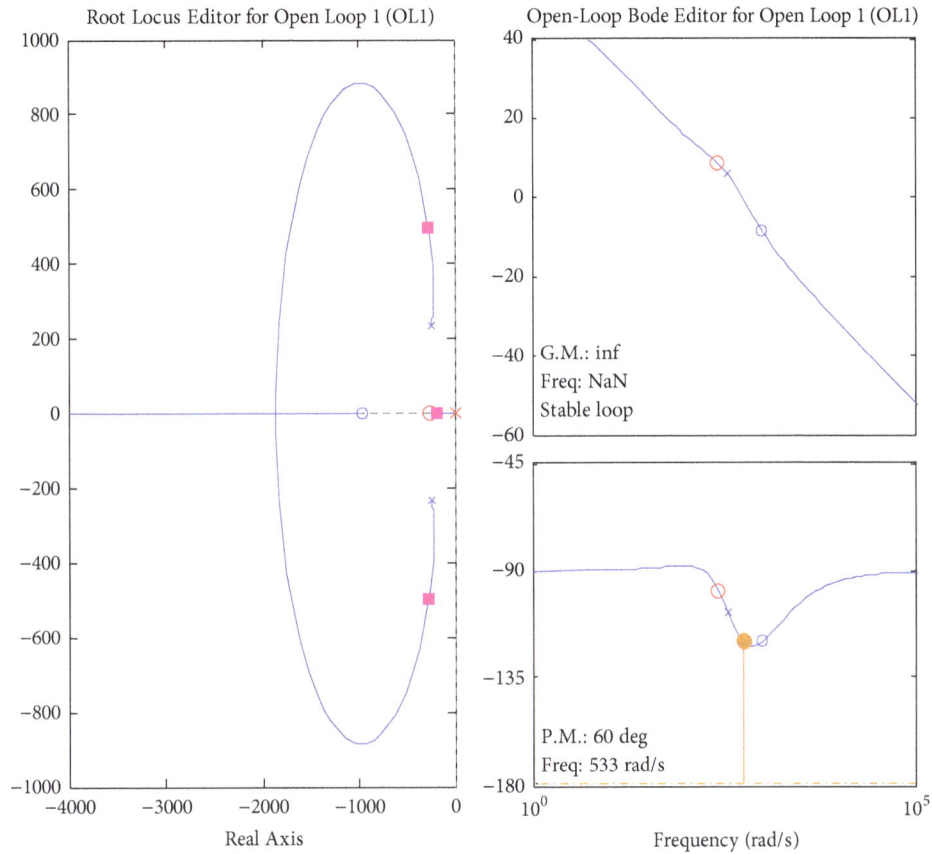

FIGURE 7: Root locus and open loop bode plots for the inner loop of boost converter.

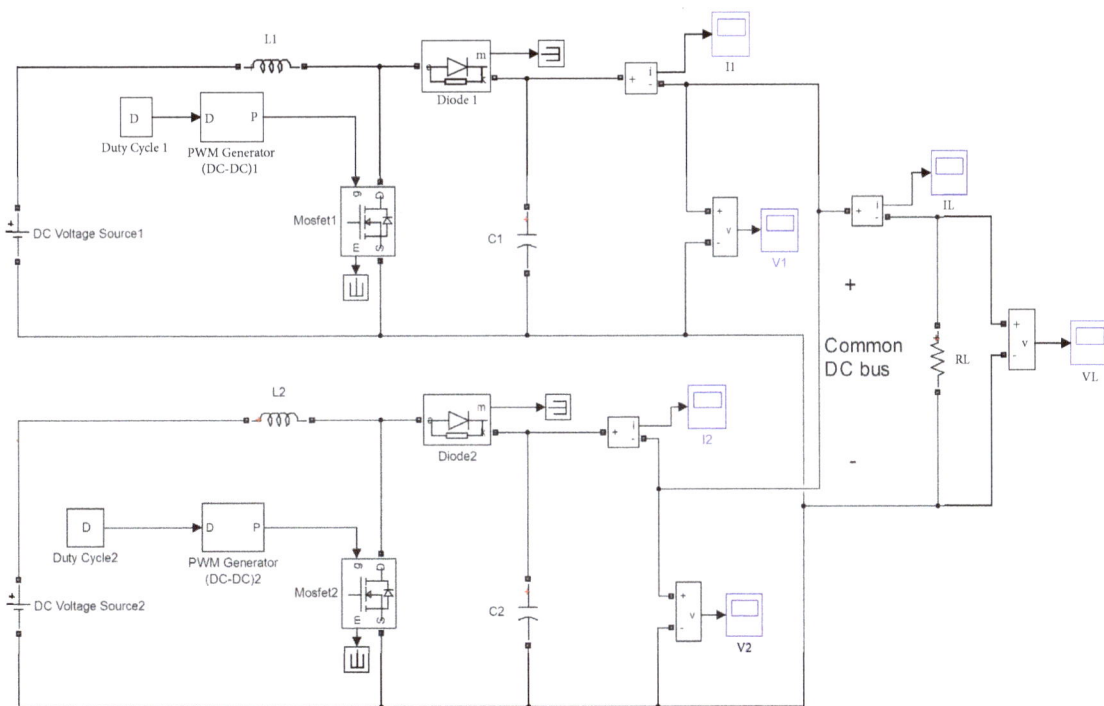

FIGURE 8: Schematic diagram of the two boost parallel-connected converters.

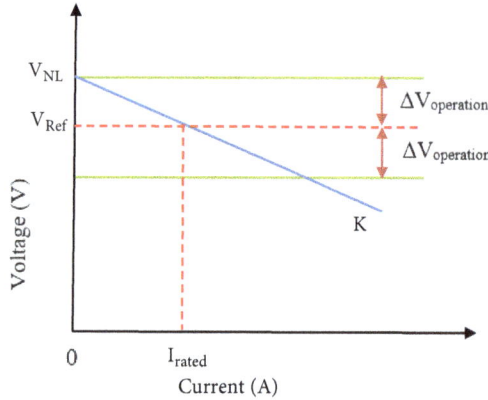

FIGURE 9: Load regulation characteristic of the droop method with gain of K.

shows the general load regulation characteristics of the droop method [23, 24].

As indicated in Figure 9, V_{NL} is the no-load voltage, V_{ref} is the reference voltage (rated operation voltage), $\Delta V_{operation}$ is the range of voltage that would allow the algorithm to adjust the reference voltage, and K is the droop gain of one of the converters. Therefore, if the output voltage is not within the operation range of the voltage $\Delta V_{operation}$, the reference voltage will be constant and equal to the rated voltage of the converter. From Figure 9, the output droops as the output current increases by

$$V_{out} = V_{NL} - K * I_{out} \tag{12}$$

where V_{out} is the output voltage at the DC bus, and I_{out} is the output current for one of the parallel converters.

Figure 10 shows the block diagram of the two parallel-connected boost converters with the proposed algorithm.

The proposed algorithm inputs are the measured output voltage at the common bus, the output current of one of the converters, and the reference voltage. The load regulation characteristics for the chosen converter are used to adjust the reference voltage during changes in load. Figure 11 shows the flowchart of the proposed algorithm.

From Figure 11, the first step in the proposed algorithm is to check if the output voltage is within the operating zone for the algorithm. If the absolute ΔV is not less than or equal to $\Delta V_{operation}$, ΔV_{ref} is set to zero, and the reference voltage remains at its rated value.

However, if the absolute ΔV is less than or equal to $\Delta V_{operation}$, the output current for the chosen converter is compared with the droop current (I_{droop}). The droop current is calculated based on the load regulation characteristics for the chosen converter as follows:

$$I_{droop} = \frac{(V_{NL} - V_{out})}{K} \tag{13}$$

$$\Delta I_{droop} = I_{droop} - I_{out} \tag{14}$$

ΔI_{droop} is the difference between the calculated droop current and the actual measurement of the chosen converter

current. If the absolute ΔI_{droop} is less than or equal to a tolerance error ($1 * e^{-15}$), the operating point will be the same as the calculated operating point based on the load regulation characteristics for the chosen converter. Therefore, the reference voltage will remain at its rated value.

However, if the absolute ΔI_{droop} is not less than or equal to tolerance error ($1 * e^{-15}$), the reference voltage will be adjusted based on the load regulation characteristics for the chosen converter as

$$V_{droop} = V_{NL} - I_{out} * K \tag{15}$$

$$\Delta V_{ref} = V_{ref} - V_{droop} \tag{16}$$

ΔV_{ref} is the difference between the reference voltage and the calculated droop voltage based on the load regulation characteristics for the chosen converter. If ΔV_{ref} is a positive value, it will be subtracted from the reference voltage as shown in the block diagram of the parallel-connected converters, and it means that the output voltage needs to be lowered according to the load regulation characteristics of the droop method. In contrast, if ΔV_{ref} is negative value, it will be added to the reference voltage to increase the output voltage level.

5. Simulation Results and Discussion

Figure 12 shows droop characteristics of two identical parallel-connected DC-DC converters.

The load current sharing is identical because the droop gains of both converters are equal. The main reason for having the same droop gains is that there is no mismatch in the power stage of both converters. At different levels of the output voltage, the output current of the two converters would be shared equally. The simulation of the first case considers the two parallel-connected boost converters with no mismatch or zero mismatch in the power stage. The algorithm for the proposed method will adjust the reference voltage when the load changes. The PI controller parameters in Table 2 are used to obtain the results. In the first examined case, the load is changed from 16 Ω to 8 Ω by adding parallel resistances at 0.2 sec. The load resistance is further reduced from 8 Ω to 5.33 Ω at 0.4 sec. Figure 13 shows the output voltage at the common dc bus.

The output voltage in Figure 13 is adjusted by the proposed algorithm according to the load regulation characteristics of the droop method. The proposed algorithm adjusts the reference voltage when the load changes and provides synchronization of the output voltage level of the parallel converters. Therefore, the load current sharing will be equal based on the identical load regulation characteristics for the two converters as shown in Figure 10. The output current for each converter and the total load current are shown in Figure 14.

As shown in Figures 12, 13, and 14, when the load resistance is 16 Ω, the proposed algorithm adjusts the reference voltage according to the droop method. Because of the load resistance value of 16 Ω, the output voltage level is regulated by the algorithm to be 48.7V, and the output current of each

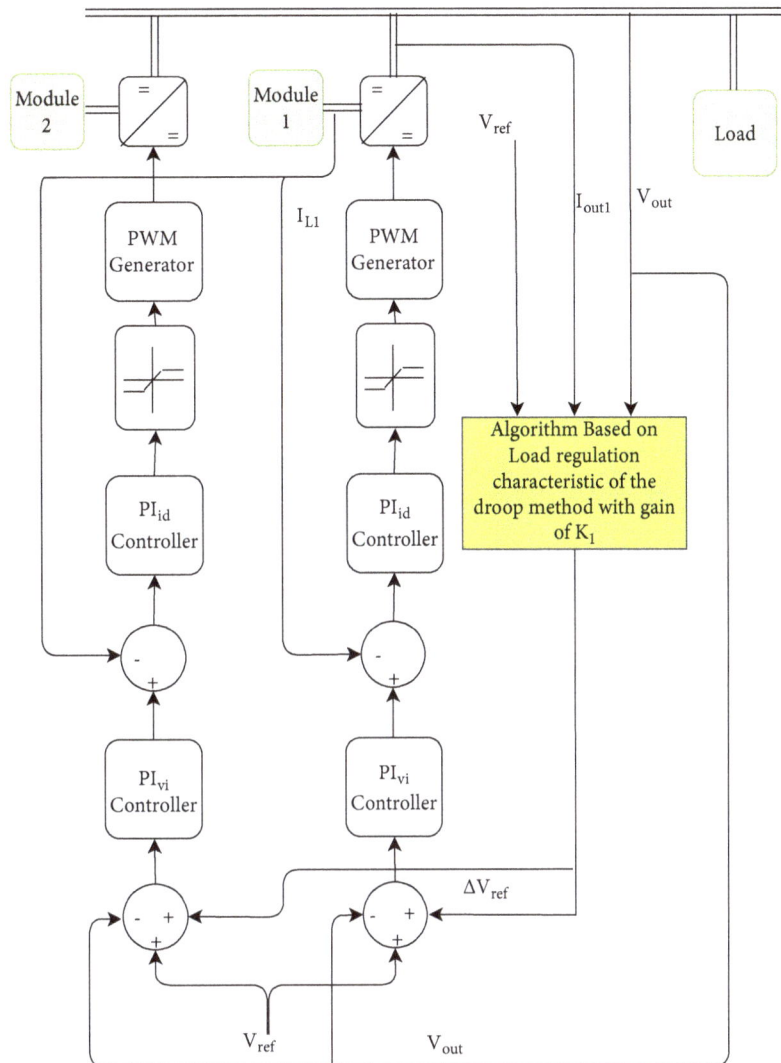

FIGURE 10: Block diagram of two parallel-connected converters with their control loops and the proposed algorithm.

converter is 1.5 A. Furthermore, when the load is changed from 16 Ω to 8 Ω at 0.2 sec., the proposed algorithm adjusts the reference voltage due to the new change in the load resistance. The output voltage level becomes 48 V, and the output current of each converter is 3 A for the period between 0.2 sec and 0.4 sec. At 0.4 sec. when the load resistance is further reduced to 5.33 Ω, the algorithm regulates the output voltage level to be 47.4 V, and the load current sharing of each converter is 4.425A. With no mismatch in the load regulation characteristics, the proposed method with its algorithm regulates the output voltage level synchronously which results in no mismatch in the output voltage and hence no circulating current.

However, the parameters for the two converters with mismatch in power stage of 10% and 20% of the DC-DC boost converter in Table 1 are given in Table 3. These parameters are implemented in cases two and three.

A second case with 10% mismatch in the load regulation characteristics of the two converters is shown in Figure 15.

TABLE 3: Operating values for boost converter.

Parameters	DC-DC Boost Converter with 10% mismatch	DC-DC Boost Converter with 20% mismatch
Switching frequency f_s	25 KHz	25 KHz
Inductance L	14.533 mH	13.322 mH
Capacitance C	141.51 μF	154.375μF
Power P	158.4 W	178.8 W

From Figure 15, the load current sharing between the two converters is not identical because of the 10% mismatch in the power stage of the two converters. The droop gains of both converters are different. Based on the mismatch in the load regulation characteristics, the two converters share the load current unequally. Similar to the first case, the load resistance is changed from 16 Ω to 8 Ω at 0.2 sec., and, at 0.4 sec., the

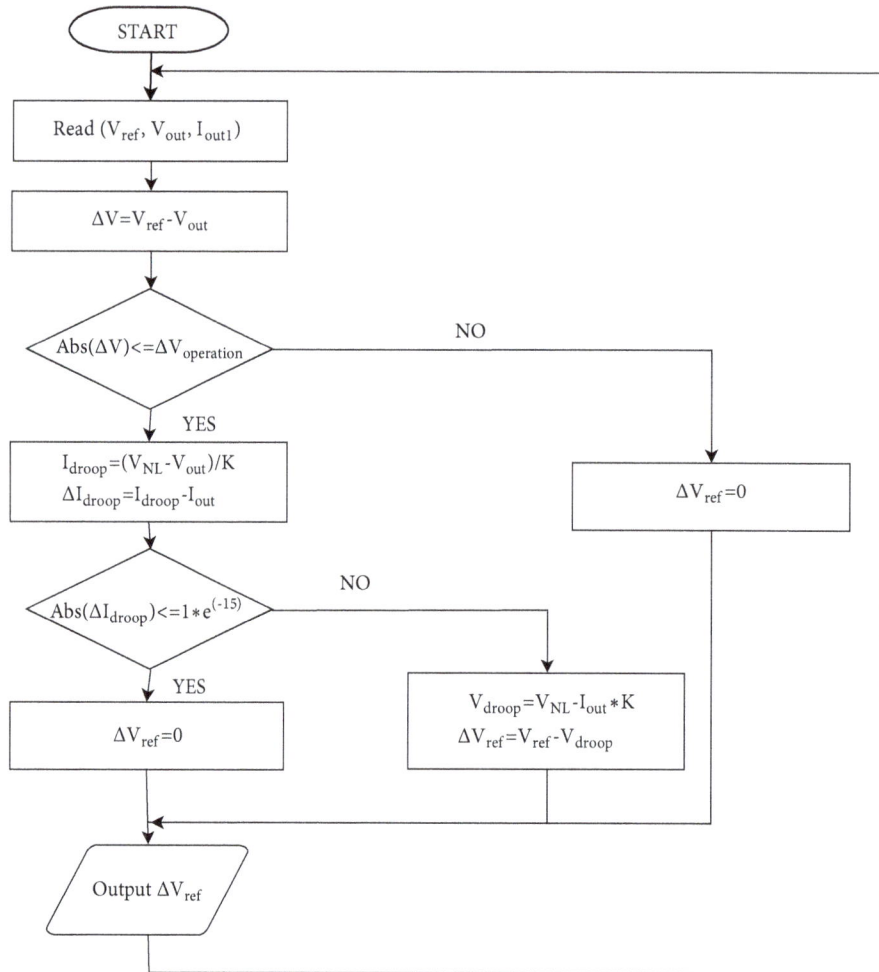

FIGURE 11: Flowchart of the proposed algorithm.

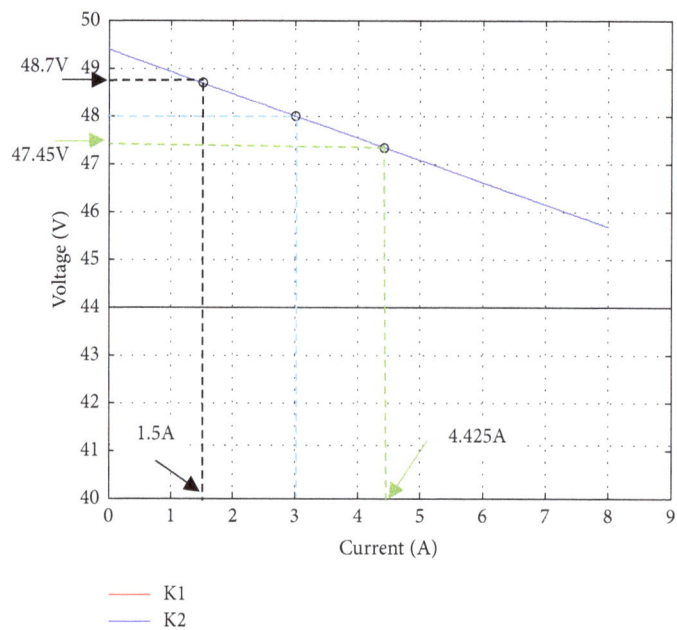

FIGURE 12: Zero mismatch in load regulation characteristic of the droop method for the two converters.

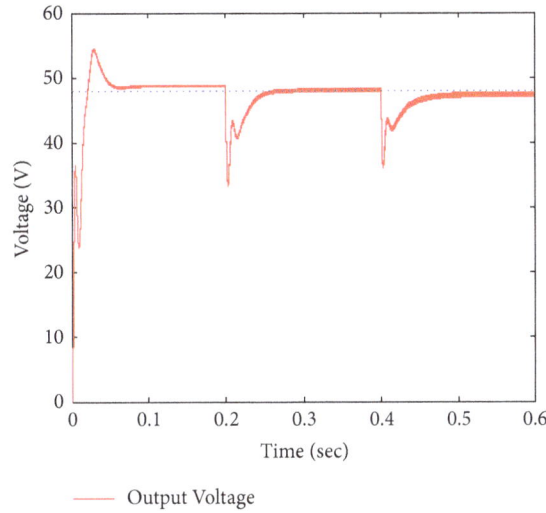

FIGURE 13: Output voltage at the common dc bus.

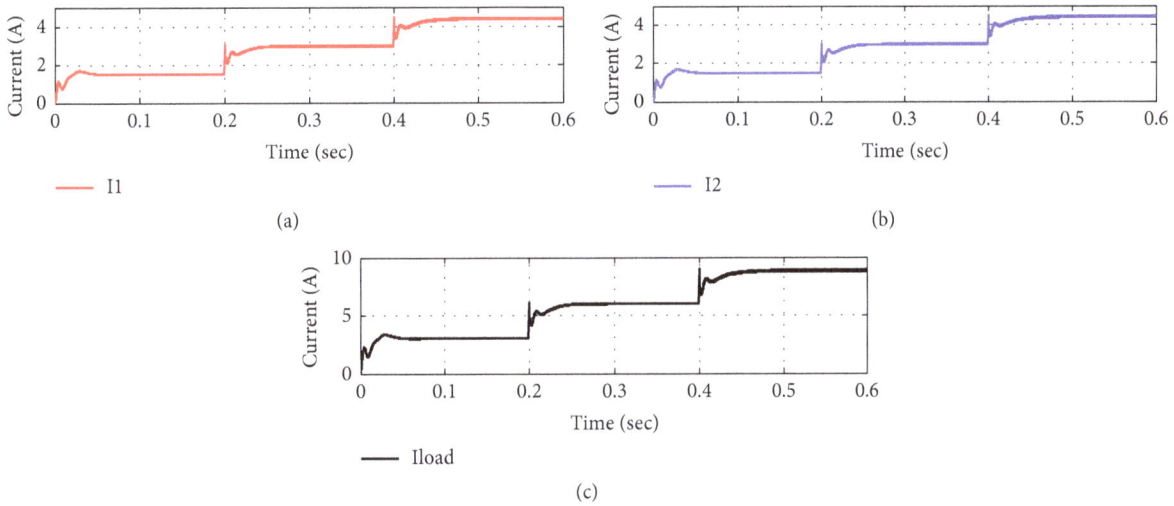

FIGURE 14: The output current from the simulation results of no mismatch case: (a) converter I output current; (b) converter II output current; (c) load current.

load resistance is decreased further from 8 Ω to 5.333 Ω. The output voltage at the common bus with 10% mismatch of the load regulation characteristic of the droop method is shown in Figure 16.

From Figure 16, the proposed algorithm adjusts the reference voltage according to load regulation characteristic of the two converters and it shows that the output voltage is regulated for changes in the load resistance. The output current for each converter and the load current are shown in Figure 17.

When the load resistance is 16 Ω, the regulated output voltage is 48.7 V and the load current sharing for converter I and converter II is 1.4 A and 1.6 A, respectively. Due to the change in the load resistance from 16 Ω to 8 Ω at 0.2 sec., the output voltage level is changed synchronously by the proposed algorithm for both converters to be 48 V. Converters I and II contribute 2.85 A and 3.15 A, respectively, to the total load current. Similarly, the change in the load resistance from

8 Ω to 5.33 Ω causes the proposed algorithm to regulate the output voltage at 47.4 V. The contributions to the load current from converters I and II are 4.2 A and 4.65 A, respectively.

A third case with 20 % mismatch of the load regulation characteristics is shown in Figure 18.

With a 20 % mismatch in the load regulation characteristics, the two converters share the load current unequally. The third case is also simulated with the changes in the load resistance from 16 Ω to 8 Ω at 0.2 sec and from 8 Ω to 5.33 Ω at 0.4 sec. As shown in Figure 19, the output voltage level is regulated by the proposed algorithm.

The output voltage is regulated to be 48.7 V when the load resistance is 16 Ω. When the load resistance is changed to 8 Ω, the algorithm adjusts the output voltage to the new level of 48 V. With transition in the load resistance from 8 Ω to 5.33 Ω at 0.4 sec., the algorithm regulates the output voltage level to be 47.4 V. The load current is shared based on the 20% mismatch

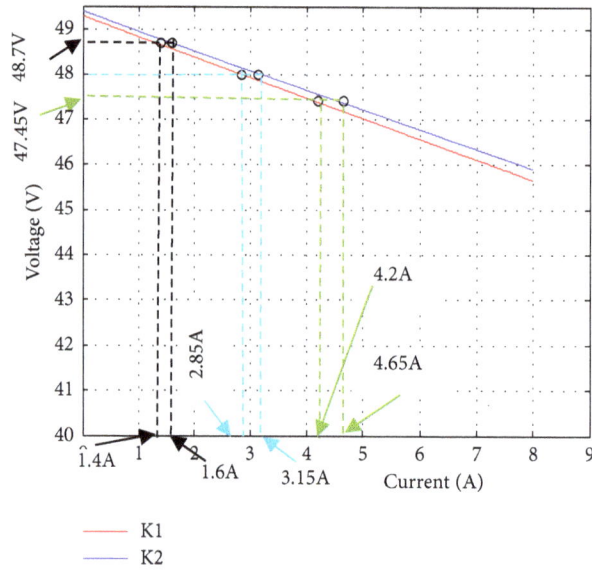

FIGURE 15: 10% mismatch in load regulation characteristic of the droop method for the two converters.

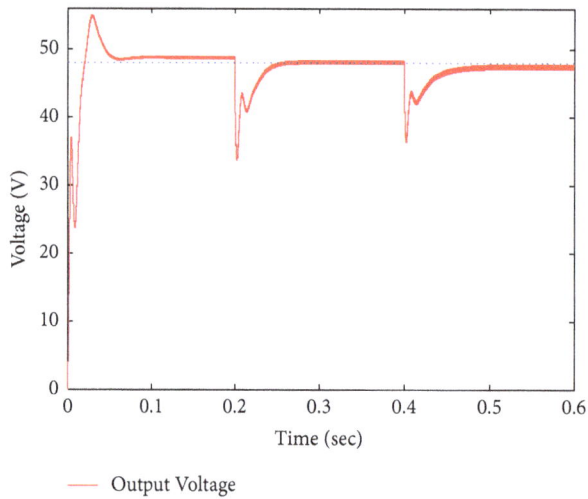

FIGURE 16: The output voltage at the common dc bus with 10% mismatch in the load regulation characteristics of the droop method.

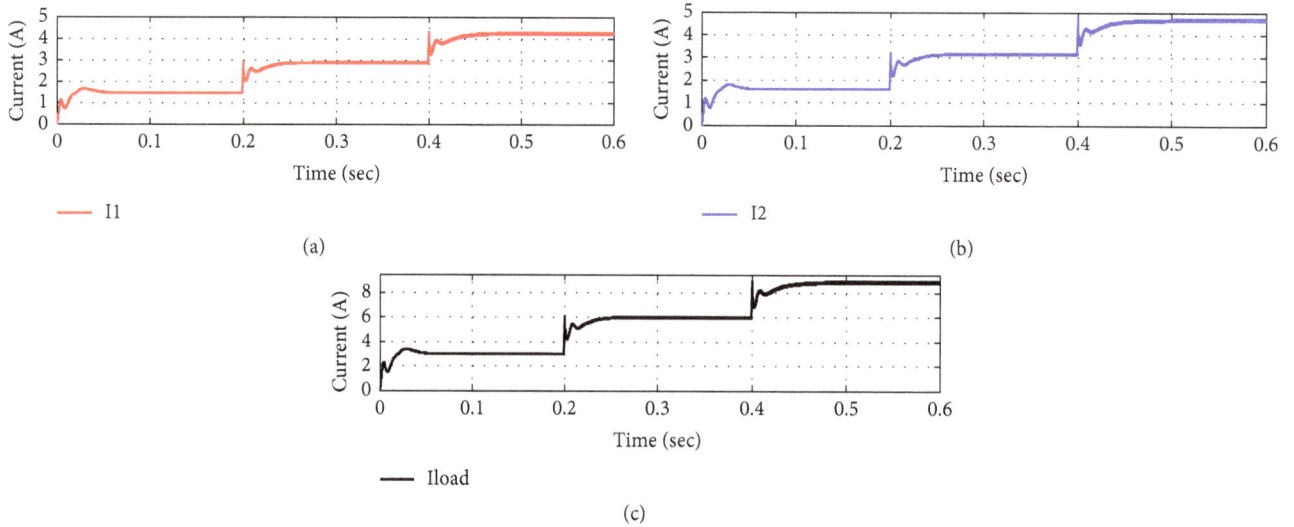

FIGURE 17: The output current from the simulation results of 10% mismatch case: (a) converter I output current; (b) converter II output current; (c) load current.

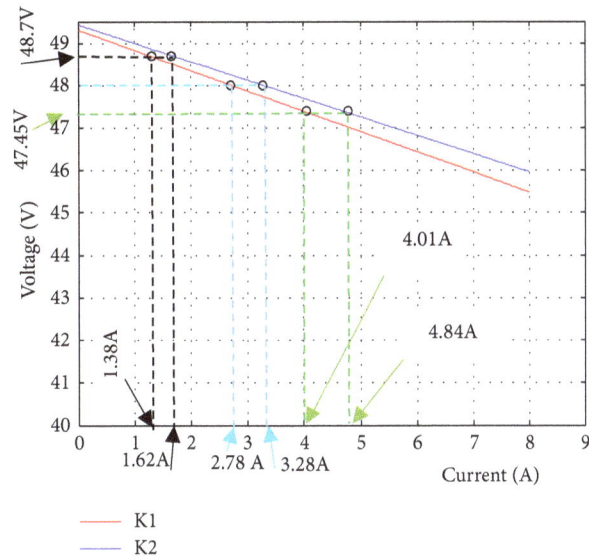

FIGURE 18: 20% mismatch in load regulation characteristic of the droop method for the two converters.

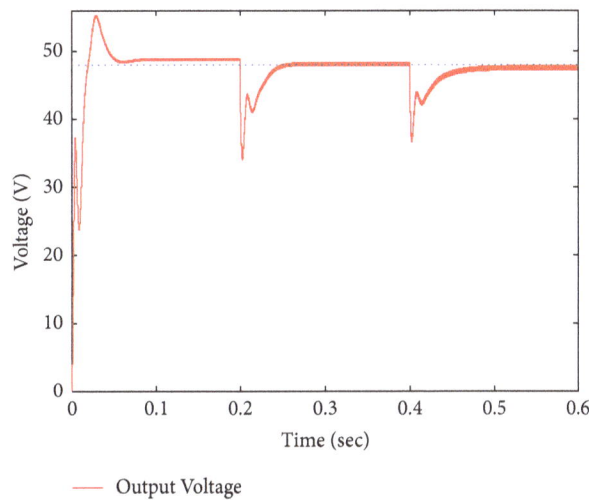

FIGURE 19: Output voltage with reference voltage.

in the load regulation characteristic for both converters as shown in Figure 20.

As indicated in Figure 20, when the output voltage is adjusted by the proposed algorithm to 48.7 V, the contribution of current for converters I and II to the total load current is 1.38 A and 1.62 A, respectively. However, before 0.4 sec, the load current sharing for converter I is 2.78 A and for converter II is 3.28 A. This is at a regulated output voltage of 48 V. Moreover, when the load is changed from 8 Ω to 5.33 Ω, the new regulated output voltage is 47.4 V. The load current sharing between converter I and converter II is 4.01A and 4.84A, respectively.

This paper presents a proof-of-concept methodology through simulations. In a subsequent paper, experimental results of the performance of two parallel-connected boost converters, which incorporate the proposed modified droop method based on the master current control, will be presented.

6. Conclusion

The simulation cases of the proposed method with its algorithm are presented. The modified droop method based on master current control is verified using Matlab/Simulink simulation. The proposed algorithm adjusts the output voltage according to the load regulation characteristic of the parallel-connected converters. The proposed method ensures that the output voltage level of the two parallel converters is identical, thus avoiding circulating current at the DC bus. Three cases involving no mismatch, 10% mismatch, and 20 % mismatch of the power stage are considered in order to examine the proposed method and its algorithm. The results demonstrate

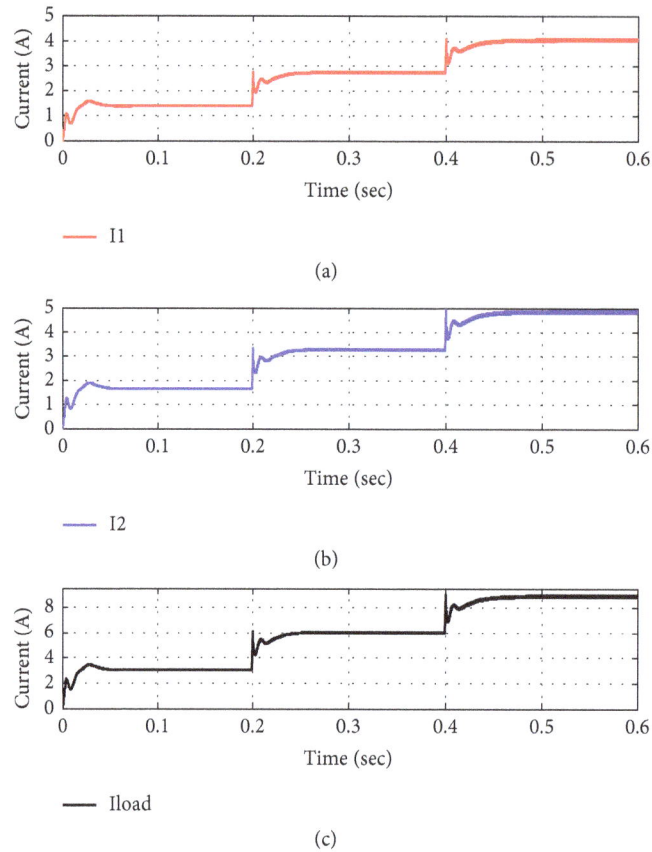

FIGURE 20: The output current from the simulation results for optimized control case. (a) Converter I output current; (b) converter II output current; (c) load current.

the effectiveness of the proposed method and its algorithm.

Conflicts of Interest

The authors declare that there are no conflicts of interest regarding the publication of this paper.

Acknowledgments

The authors would like to thank the Libyan Government for funding this research.

References

[1] M. Jovanovic, D. Crow, and . Lieu Fang-Yi, "A novel, low-cost implementation of "democratic" load-current sharing of paralleled converter modules," *IEEE Transactions on Power Electronics*, vol. 11, no. 4, pp. 604–611.

[2] J. Kim, H. Choi, and B. H. Cho, "A novel droop method for converter parallel operation," *IEEE Transactions on Power Electronics*, vol. 17, no. 1, pp. 25–32, 2002.

[3] J. B. Wang, "Parallel DC/DC converters system with a novel primary droop current sharing control," *IET Power Electronics*, vol. 5, no. 8, pp. 1569–1580, 2012.

[4] S. Anand and B. G. Fernandes, "Modified droop controller for paralleling of dc-dc converters in standalone dc system," *IET Power Electronics*, vol. 5, no. 6, pp. 782–789, 2012.

[5] S. K. Mazumder, A. H. Nayfeh, and D. Borojević, "Robust control of parallel dc-dc buck converters by combining integral-variable-structure and multiple-sliding-surface control schemes," *IEEE Transactions on Power Electronics*, vol. 17, no. 3, pp. 428–437, 2002.

[6] S. Hiti, "Robust Nonlinear Control for Boost Converter," *IEEE Transactions on Power Electronics*, vol. 10, no. 6, pp. 651–658, 1995.

[7] S. K. Mazumder and S. L. Kamisetty, "Design and experimental validation of a multiphase VRM controller," *IEE Journal on Electric Power Applications*, vol. 152, no. 5, pp. 1076–1084, 2005.

[8] G. Escobar, R. Ortega, H. Sira-Ramirez, J.-P. Vilain, and I. Zein, "An Experimental Comparison of Several Nonlinear Controllers for Power Converters," *IEEE Control Systems Magazine*, vol. 19, no. 1, pp. 66–82, 1999.

[9] C. H. Cheng, P. J. Cheng, and M. J. Xie, "Current sharing of paralleled DCDC converters using GA-based PID controllers," *Expert Systems with Applications*, vol. 37, pp. 733–740, 2010.

[10] J. Rajagopalan, K. Xing, Y. Guo, F. C. Lee, and B. Manners, "Modeling and dynamic analysis of paralleled dc/dc converters with master-slave current sharing control," in *Proceedings of the*

1996 IEEE 11th Annual Applied Power Electronics Conference and Exposition, (APEC'96), vol. 2, pp. 678–684, 1996.

[11] S. K. Mazumder, M. Tahir, and K. Acharya, "Master-slave current-sharing control of a parallel DC-DC converter system over an RF communication interface," *IEEE Transactions on Industrial Electronics*, vol. 55, no. 1, pp. 59–66, 2008.

[12] Y. M. Lai, S.-C. Tan, and Y. M. Tsang, "Wireless control of load current sharing information for parallel-connected DC/DC power converters," *IET Power Electronics*, vol. 2, no. 1, pp. 14–21, 2009.

[13] J.-J. Shieh, "Peak-current-mode based single-wire current-share multimodule paralleling DC power supplies," *IEEE Transactions on Circuits and Systems I: Fundamental Theory and Applications*, vol. 50, no. 12, pp. 1564–1568, 2003.

[14] L. Meng, T. Dragicevic, J. M. Guerrero, and J. C. Vasquez, "Optimization with system damping restoration for droop controlled DC-DC converters," in *Proceedings of the 5th Annual IEEE Energy Conversion Congress and Exhibition, ECCE 2013*, pp. 65–72, Denver, Colo, USA, September 2013.

[15] J.-W. Kim and P. Jang, "Improved droop method for converter parallel operation in large-screen LCD TV applications," *Journal of Power Electronics*, vol. 14, no. 1, pp. 22–29, 2014.

[16] M. M. Shebani, T. Iqbal, and J. E. Quaicoe, "Synchronous switching for parallel-connected DC-DC boost converters," in *Proceedings of the 2017 IEEE Electrical Power and Energy Conference (EPEC)*, pp. 1–6, Saskatoon, Canada, October 2017.

[17] B. M. Hasaneen and A. A. E. Mohammed, "Design and simulation of DC/DC boost converter," in *Proceedings of the 2008 12th International Middle East Power System Conference, MEPCON 2008*, pp. 335–340, Aswan, Egypt, March 2008.

[18] H. Abdel-Gawad and V. K. Sood, "Small-signal analysis of boost converter, including parasitics, operating in CCM," in *Proceedings of the 6th IEEE Power India International Conference, PIICON 2014*, India, December 2014.

[19] Bhowate. Apekshit and Deogade. Shraddha, "Comparison of PID Tuning Techniques for Closed Loop Controller of DC-DC Boost Converter," *International Journal of Advances in Engineering & Technology*, vol. 8, no. 1, pp. 2064–2073, 2015.

[20] M. Saoudi, A. El-Sayed, and H. Metwally, "Design and implementation of closed-loop control system for buck converter using different techniques," *IEEE Aerospace and Electronic Systems Magazine*, vol. 32, no. 3, pp. 30–39, 2017.

[21] L. Guo, J. Y. Hung, and R. M. Nelms, "Digital Controller Design for Buck and Boost Converters Using Root Locus Techniques," in *Proceedings of the The 29th Annual Conference of the IEEE Industrial Electronics Society*, vol. 2, pp. 1864–1869, November 2003.

[22] K. Ogata, *Discrete-Time Control Systems*, Prentice-Hall, Inc., Upper Saddle River, NJ, USA, 1987.

[23] I. Batarseh, K. Siri, and H. Lee, "Investigation of the output droop characteristics of parallel-connnected DC-DC converters," in *Proceedings of the 1994 Power Electronics Specialist Conference - PESC'94*, pp. 1342–1351, Taipei, Taiwan.

[24] B. T. Irving and M. M. Jovanovic, "Analysis, design, and performance evaluation of droop current-sharing method," in *Proceedings of the APEC 2000 - Applied Power Electronics Conference*, pp. 235–241, New Orleans, La, USA.

Permissions

List of Contributors

Jian-wei Yang, Man-feng Dou and Zhi-yong Dai
School of Automation, Northwestern Polytechnical University, Xi'an 710072, China

Damiano Patron, Yuqiao Liu and Kapil R. Dandekar
Department of Electrical and Computer Engineering, Drexel University, 3141 Chestnut Street, Philadelphia, PA 19104, USA

Jinpeng Qiu, Tong Liu, Xubin Chen, Yongheng Shang, Jiongjiong Mo, ZhiyuWang, Hua Chen, Jiarui Liu, Jingjing Lv and Faxin Yu
School of Aeronautics and Astronautics, Zhejiang University, No. 38 Zheda Road, Hangzhou 310027, China

Chong Wang and Qun Sun
School of Mechanical and Automotive Engineering, Liaocheng University, Liaocheng, China

Limin Xu
School of International Education, Liaocheng University, Liaocheng, China

Yu-Jun Mao and Chi-Seng Lam
State Key Laboratory of Analog and Mixed-Signal VLSI, University of Macau, Macau 999078, China

Sai-Weng Sin and Man-Chung Wong
State Key Laboratory of Analog and Mixed-Signal VLSI, University of Macau, Macau 999078, China
Department of Electrical and Computer Engineering, Faculty of Science and Technology, University of Macau, Macau 999078, China

Rui Paulo Martins
State Key Laboratory of Analog and Mixed-Signal VLSI, University of Macau, Macau 999078, China
Department of Electrical and Computer Engineering, Faculty of Science and Technology, University of Macau, Macau 999078, China
On Leave from Instituto Superior Técnico, Universidade de Lisboa, Lisbon 1649-004, Portugal

Takashi Kawamoto
Hitachi Central Research Laboratory, 1-280 Higashi-Koigakubo, Kokubunji-shi, Tokyo 185-8601, Japan

Masato Suzuki and Takayuki Noto
Renesas Electronics Corporation, Tokyo 185-8601, Japan

Pratibha Bajpai, Neeta Pandey and Jeebananda Panda
Department of Electronics and Communication Engineering, Delhi Technological University, Delhi, India

Kirti Gupta and Shrey Bagga
Department of Electronics and Communication Engineering, Bharati Vidyapeeth's College of Engineering, Delhi, India

Suying Zhou, Hui Lin and Bingqiang Li
School of Automation, Northwest Polytechnical University, Xi'an, China

Myeong-Eun Hwang
Memory Division, Samsung Electronics Inc., Hwaseong, Gyeonggi 18448, Republic of Korea

Sungoh Kwon
School of Electrical Engineering, University of Ulsan, Nam-gu, Ulsan 44610, Republic of Korea

Binbin Zheng and Qiangsheng Yue
College of Engineering, South China Agricultural University, Guangzhou 510642, China

Shengping Lv
College of Engineering, South China Agricultural University, Guangzhou 510642, China
Department of Industrial, Manufacturing and Systems Engineering, Texas Tech University, Lubbock, TX 79409, USA

Hoyeol Kim
Department of Industrial, Manufacturing and Systems Engineering, Texas Tech University, Lubbock, TX 79409, USA

Zhi Jiang, Yiqi Zhuang, Cong Li, Ping Wang and Yuqi Liu
School of Microelectronics, Xidian University, Xi'an 710071, China

Guangya Peng, Fuhong Min and Enrong Wang
School of Electrical and Automatic Engineering, Nanjing Normal University, Nanjing 210042, China

Hesheng Cheng and Qunli Zhao
College of Computer Science, Hefei Normal University, Hefei, China

Jin Zhang
School of Electronic and Information Engineering, Jinling Institute of Technology, Nanjing, China

Hexia Cheng
College of Computer & Information, Anqing Normal University, Anqing, China

Mona Moradi
Department of Computer Engineering, Islamic Azad University, Science and Research Branch, Tehran 1477893855, Iran

Reza Faghih Mirzaee
Department of Computer Engineering, Islamic Azad University, Shahr-e-Qods Branch, Tehran 37541-374, Iran

Keivan Navi
Faculty of Electrical and Computer Engineering, Shahid Beheshti University, G.C., Tehran 1983963113, Iran

Tian Ban
School of Electronic and Optical Engineering, Nanjing University of Science and Technology, Nanjing 210094, China
Department of Communications and Electronics, Institut Mines-Télécom, Télécom Paris Tech, 75013 Paris, France

Gutemberg G. S. Junior
Department of Communications and Electronics, Institut Mines-T´el´ecom, T´el´ecom ParisTech, 75013 Paris, France

Chengying Chen and Liming Chen
School of Opto-Electronic and Communication Engineering, Xiamen University of Technology, Fujian, Xiamen 361024, China

Jun Yang
School of Automation, Foshan University, Foshan 528000, China

Yun Huang
Asset Management Department, Chongqing University of Posts and Telecommunications, Chongqing 400065, China

Muamer M. Shebani, Tariq Iqbal and John E. Quaicoe
Department of Electrical and Computer Engineering, Faculty of Engineering and Applied Science, Memorial University of Newfoundland, St. John's, NL, Canada

Index